Ecological Studies

Analysis and Synthesis

Edited by

W. D. Billings, Durham (USA) F. Golley, Athens (USA)

O. L. Lange, Würzburg (FRG) J. S. Olson, Oak Ridge (USA)

Volume 26

Grassland Simulation Model

Edited by
George S. Innis

With 87 Figures

Springer-Verlag   NewYork   Heidelberg   Berlin

Library of Congress Cataloging in Publication Data
Main entry under title:
Grassland simulation model.
  (Ecological studies; v. 26)
  Includes bibliographies and index.
  1. Grassland ecology. 2. Grassland ecology—
Mathematical models. 3. Grassland ecology—Data
processing. I. Innis, George S., 1937-
II. International Biological Programme. III. Series.
QH541.5.P7G72   574.5'264   77-23016

©1978 by Springer-Verlag New York Inc.

Printed in the United States of America.

9 8 7 6 5 4 3 2 1

ISBN 0-387-90269-4   Springer-Verlag New York
ISBN 3-540-90269-4   Springer-Verlag Berlin Heidelberg

Foreword
Perspectives on the ELM Model and Modeling Efforts

This volume is the major open-literature description of a comprehensive, pioneering ecological modeling effort. The ELM model is one of the major outputs of the United States Grassland Biome study, a contribution to the International Biological Program (IBP). Writing this introduction provides welcome personal opportunity to (i) review briefly the state of the art at the beginning of the ELM modeling effort in 1971, (ii) to discuss some aspects of the ELM model's role in relation to other models and other phases of the Grassland Biome study, and (iii) to summarize the evolution of ELM or its components since 1973.

Pre-Program Historical Perspective

My first major contacts with ecological simulation modeling were in 1960 when I was studying intraseasonal herbage dynamics and nutrient production on foothill grasslands in southcentral Montana, making year-round measurements of the aboveground live vegetation, the standing dead, and the litter. Limitations in funding and the rockiness of the foothill soils prevented measuring the dynamics of the root biomass, both live and dead. Herbage biomass originates in live shoots from which it could be translocated into live roots or the live shoots could transfer to standing dead or to litter. Standing dead vegetation must end up in the litter and the live roots eventually transfer to dead roots. Obviously, the litter and the dead roots must decay away. This led to a conceptual model of a source, five compartments, and a sink for the carbon or biomass flow. I turned to compartmental models to develop equation systems with the hopes of utilizing field data on live shoots, standing dead, and litter to predict the dynamics of the root compartments. This led to an examination of the ecological modeling literature which I found was still in a "conceptual and linear" phase. Ecologists had available to them a long history of systems views of ecological system stretching as far back as Lotka (1925), strengthened by the work of Lindemann (1943), and further stimulated by work in the late 1950s and early 1960s by using radioisotopes as tracers for ecological field studies in nutrient cycling and food-chain determinations. The paper of Patten (1959) on an introduction to the cybernetics of ecosystems further stimulated thought and work on trophic-dynamic aspects.

Olson's (1963a) paper on the balance of producers and decomposers in ecological systems led directly to systems concepts of element cycling (Olson 1963b, 1964, 1965, Dahlman et al. 1969). But not many operational, numerically-implemented models had been developed. Surveys of the literature (Van Dyne 1966, Davidson and Clymer 1966) showed that publications on systems analysis in ecology, ecosystem modeling, and ecosimulation were relatively scarce. Watt (1966) discussed the meanings of systems analysis for ecologists and provided some practical guidelines for persons intending to develop simulation models. Holling (1964, 1966) conducted experiments on predator-prey relationships and outlined several essential elements which a biological system model should contain. Later Watt (1968) reviewed and discussed the applications of systems techniques to resource management problems.

A decade ago most of the attempts at simulating ecological systems utilized difference or differential equations. These equations usually were of first order and generally first degree (perhaps with cross-product terms) and were integrated using numerical procedures on digital computers (Olson 1965, Bledsoe and Olson 1970, Bellman et al. 1966, Garfinkle 1962) or on analog computers (Neel and Olson 1962, Olson 1963b, 1964, Davidson and Clymer 1966). Watt (1961, 1968) also had developed a key for selecting, for data from a biological problem, a differential equation of first order.

The emphasis in most biological systems modeling had been on first-order, constant-coefficient, linear differential equation systems using compartmental models. This was a gross simplification of biological reality to fit a model structure that could be handled easily by classical analytical methods and techniques. This was a necessary learning step, but Neel and Olson (1962), Bledsoe and Olson (1970), and Sollins (1971) provided for variable coefficients and nonlinear systems. Such compartmental model approaches were used relatively widely in ecological studies and provided useful information, especially in metabolic studies or in microcosmic experiments. These approaches were used by Neel and Olson (1962) and Olson (1963b, 1965) for describing element or radionuclide flow, by Patten and Witkamp (1967) for studying the redistribution of radiocesium in a microcosm, and by Smith (1968) in a study of interactive populations with various food-web structures. But few grazingland systems had been studied mathematically prior to the last decade. At that time no one or no group had developed a model representing all trophic levels for grazinglands. This was due to inadequate quantity and quality of data and due to lack of skills in the art of modeling biological systems.

Prior to the last decade none of the ecological models included spatial aspects (e.g., Goodall 1967) and very few have since then. Most early ecological models have been deterministic rather than stochastic and this still is the situation. Lucas (1960, 1964) and Olson and Uppuluri (1966) contrasted deterministic and stochastic models and discussed the sources of randomness in data. Garfinkle and Sack (1964) provided early simplified ecological simulations and Bergner (1965) reviewed compartmental analysis methods. And then the literature began to explode. Early leaders in grazingland model development were in the United States and Australia, but scientists from at least nine nations have contributed to the development of the first 100 models. Van Dyne and Abramsky (1975) review simulation and optimization models as applied to agricultural systems, including grazinglands, and their chapter and that of Wiegert (1975) summarize many of the aspects of techniques that have been developed and have proven useful in both simulation and optimization modeling.

The extensive development of grazingland system models awaited the advent of large, fast electronic computers. This is because of the complexity of the systems being analyzed and the nature of the models used to simulate systems. Depicting the dynamics of a grazingland system requires simulating biological and physiological processes. The characteristics of biological processes alone are such that one must consider great diversity in time and space, the existence of threshold processes, limiting values, and discontinuities when representing the system mathematically. Social and economic considerations even further complicate the problem. Thus, this level of complexity limits the use of simple linearized approximations to nonlinear models, adapted from the classical tools of mathematical physics, to the simplest situations.

For example, Kelly et al. (1969) implemented simulations of two humid temperate grasslands dominated by single species of cool and warm season growth habit, with extensive data on underground parts which had been missing in the native grassland studies; invertebrate food chains for the same area were treated by Van Hook et al. (1970). Gore and Olson (1967) treated the plant-peat subsystem with cotton grass *(Eriophorum)* and heather *(Calluna)* which dominates certain British moorlands; further modeling of this and similar systems has been treated within the Tundra Biome part of the IBP.

ELM's Role in IBP Grassland Biome Study

The Grassland Biome program begin in June 1968 but planning phases had gone on for at least two years. The origin of the role of mathematical models in this study is of special interest. The general role of models and systems analysis in such a large-scale study were introduced in a lecture given in June 1965 (Van Dyne 1966). The International Biological Program proposed in 1961–1963 was in several respects a call for systems ecology research, both experimental and theoretical. The Program could provide an incentive and means for ecologists from universities, experiment stations, and national laboratories to work cooperatively and share funds, research areas, and talent. As an example, the idea of studies on grasslands was put forward by considering the seminatural grassland ecosystems in the Great Plains. A more complete understanding of the structure and function of these ecosystems would become increasingly essential as these lands are called upon to provide food, water, and recreation for tomorrow's growing populations. Several state and federal agricultural experiment stations hold sizeable areas with representative variations of grasslands across the vast tracts of the Great Plains. But at none of these stations was there a team equipped with suitable manpower or funds for intensive total-ecosystem research.

Scientists at these experiment stations and nearby universities had accumulated considerable data and experience on and about these grassland ecosystems. But there was no national laboratory in this vast area and no university in the area could marshall all of the unique facilities and personnel that were necessary for systems ecology research. Yet the U.S. National Committee for the IBP felt that scientists in the universities in the area could provide much insight into these ecosystems and graduate students could help conduct ecosystem research. Eventually funds and planning became available which allowed us to combine the special skills and resources of scientists in universities, experiment stations, and national laboratories and bring them to bear on the problem of grassland systems.

So when the Grassland Biome study started in 1968, the broad conceptual groundwork of applying models and systems analysis had already been laid. One of the first efforts in the Grassland Biome study was a synthesis project in which the concepts of models were frequently introduced. The synthesis was undertaken in a "total ecosystem framework" in order to show the interconnectedness of the parts of the ecosystem. Some 32 review papers were developed, presented, and discussed in five workshops then reviewed, revised, and eventually published (Dix and Biedleman 1969). To facilitate these workshops a "picture model" was defined as a framework of reference. This picture model was in effect a crudely structured box-and-arrow diagram with some 13 compartments. At that time, however, the clear distinction of driving variables, state variables, and rate processes was not identified. In the first of our workshops a paper was presented to review examples of modeling grasslands and to discuss cautions, considerations, advantages, and needs of modeling (Van Dyne 1969a). The final paper in the last workshop also focused on mathematical models, in part incorporating into a quantitative framework many segments of information discussed in earlier papers (Bledsoe and Jameson 1969). Since that time several models cited below were developed in the Grassland Biome study. Additionally, several other important aspects of the overall modeling effort have been published including review papers, discussion of concepts of modeling, and discussion of conceptual structure for selected models.

Several of the simulation models preceding, including, and following ELM 1973 are outlined in Figure 1. The first operational model produced with partial support from the Grassland Biome study was PRONG, a system-level model for the shortgrass prairie developed as a training exercise by a team including several members from the Biome study (Van Dyne 1969b). This simulation model contained abiotic, autotrophic, and heterotrophic components but it was greatly simplified in detail both in number of state variables and mechanism in the process functions. Innis (1975 and this volume) summarized two major studies which resulted in models called PAWNEE 1 and LINEAR 1 in Figure 1. A team of scientists with special analytical and computer capabilities was brought together in Fall 1969 and they developed a total-system model (PAWNEE 1) based on a differential equation system and focusing on intraseasonal dynamics with a considerable degree of mechanism especially in the plant and soil components for the most intensively studied site at Colorado's Pawnee National

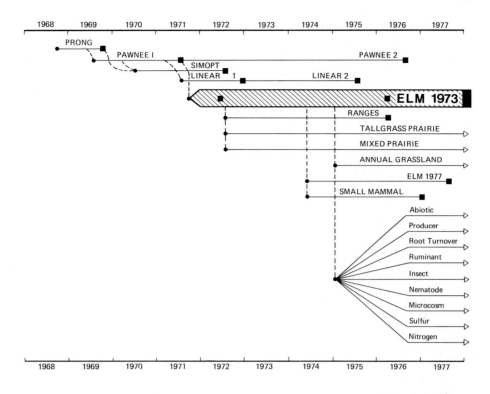

Figure 1. The ELM 1973 model and modeling effort in relation to antecedent and subsequent simulation models arising in or developing from US IBP Grassland Biome studies. The black boxes represent major reports on these models. The open triangles indicate the models are still changing rapidly or the work is still unpublished.

Grassland (Bledsoe et al. 1971). This model was later revised as PAWNEE 2 (Bledsoe 1976). A spin-off effort from PAWNEE 1 was the first version of a linearized model appropriately termed LINEAR 1 in Figure 1. This model was described by Patten (1972) and was revised into LINEAR 2 (Cale 1975). LINEAR has less detail in the aboveground description than PAWNEE but has an expanded series of state variables for the belowground components. Another model intiated in 1970 was SIMOPT which was a recursively coupled combination of difference equations and linear programming (Swartzman and Van Dyne 1972).

In 1971 another modeling team was mobilized and began the development of what we call ELM 1973 reported in this volume (see Figure 1). In 1972 the first major report on the model became available (Anway et al. 1972) and after a long gestation period the documentation was born (Cole 1976). The first technical report on ELM 1973 was, in effect, a base document around which comment could converge and planning develop. This led to subsequent models as discussed below, but it is important now to relate the ELM modeling effort to the overall program experimental design.

There was considerable incentive at the initiation of the IBP Biome studies for simultaneous development and interaction of field, laboratory, and modeling work. Preliminary field work began in the Grassland Biome study in 1968, primarily to test our sampling methods. Work on the Pawnee Site, the main field research location, greatly accelerated in 1969 and plans were further developed for other validation sites. The main block of field work in the Grassland Biome study was form 1970 through 1972. Most of the field studies were concerned with a collection of what we now call "validation data," i.e., time series on driving variables and state variables in the system for various experiment treatment conditions. Starting in 1969 and continuing separate "rate process studies" were conducted in the field and the laboratory. Information from these studies on each link in the network was utilized in part in building models and in part validating models. The important point here is that model development and completion did not precede field and laboratory studies. However, neither did field and labora-tory studies precede completely the various modeling studies. There was a reciprocal feedback through time between modeling, field, and laboratory studies.

The ELM 1973 model was developed primarily from 1971 through 1974 by a large, interdisciplinary team of scientists, the authors of the chapters in this volume. In 1972 after the first major report on the model, several other modeling efforts were accelerated (Fig. 1). A tallgrass prairie and a mixed prairie version of ELM were begun in 1972. Other variations were initiated for the desert grassland and the shrub-steppe grassland. Separate funding was obtained in 1975 by a group of California scientists for development of a version of ELM for the annual grassland. These various versions of the ELM model are to be utilized in the grassland type-oriented synthesis reports still in preparation, which will depend on the present volume for their logical basis.

The general level of resolution of ELM did not arise by chance. Early in 1970 in a series of program workshops we were experimenting with methods of evaluating priorities and the relative usefulness of different components of the overall research program. We had conducted some initial PERT analyses but found that to be an extremely difficult decision-making process because of the many subjective elements and because of the number of interactions among different phases of the program. We were asking questions about the systems analysis phase of our work such as "given the experimental results, which of the modeling efforts would be the most beneficial?" Alternately we asked, "given the kinds of models envisaged, which of the experimental projects are most useful?" We were also dealing with the problems of financial limitations. We eventually made a partially quantitative evaluation method useful in interrelating individual research projects and modeling efforts (see Van Dyne 1972).

At that time we had in mind three levels of resolution of a site-specific model, two general grassland models, and three levels of models each for abiotic, producer, consumer, and decomposer segments. We evaluated these 17 model structures and 14 areas of experimental work and concluded that a total-system, site-specific, intermediate-level model was worthy of a great deal of financial input. We felt that an ecosystem level model would help provide some guidance to integrated field and laboratory research and that it would be useful in evaluating ideas regarding grazingland management. Therefore, in 1971 we organized another modeling team including most of the scientists involved as authors of the present volume's chapters. The objectives that evolved for the ELM 1973 model are outlined herein.

Throughout the development of ELM 1973 liaison between the modeling staff and the field investigators was vigorously pursued. Regular meetings with field investigators were important in shaping the objectives and structure of the model. The significance of this liaison is partially illustrated by the fact that about two-thirds of the research projects proposed in the 1973 Grassland Biome Study Continuation Proposal were initiated by the modeling group. These initiatives were welcomed by the field investigators because they had come to appreciate the need for suggestions from the modelers to accomplish program objectives.

Several of the models outlined in Figure 1 developed as a result of changes in direction and structure of the overall program. In 1971 we expanded our proposed Biome studies to include a "utilization phase." However, this portion of our program was extracted from the Biome study, and eventually funded by the RANN program in NSF and was placed in the Range Science Department under the direction of Don Jameson who was formerly the Pawnee Site director in the Biome study. Within that application program, in part through subcontract to the main Biome study, another simulation model, RANGES, was developed and reported by Gilbert (1975) (Fig. 1). This was a low-resolution model of a grassland system.

In the process of work on the various models outlined in Figure 1 several Grassland Biome scientists made contributions to the literature on modeling.

Several review papers were published regarding grazingland models. The first review of grassland simulation models arose from the initial workshops discussed above. Subsequently, reviews were published on optimization models in the natural resource sciences with several grassland examples (Van Dyne et al. 1970), on analytical and operational approaches to grassland and grazingland models (Van Dyne et al. 1971) and on large-scale, macrosimulation models including many social-political-economic components (Van Dyne and Innis 1972). Innis and Van Dyne (1973) reviewed Grassland Biome modeling efforts in an international working meeting. Innis (1975) compared major aspects of the ELM, LINEAR, and PAWNEE models in a summary of some Biome modeling work. Van Dyne and Abramsky (1975) reviewed the world literature on agricultural system models and modeling and included numerous grazingland models. Yet to be published is another chapter for a book (Innis 1978) which reviews grassland and grazingland models.

Several Grassland Biome scientists published papers on concepts of the modeling process. In the initial workshops noted above, Bledsoe and Jameson (1969) presented a conceptual structure and approach for models of a shortgrass prairie ecological system. A series of working meetings are summarized by Swartzman (1970) in a technical report on concepts of a simulation. Further aspects of modeling with respect to grassland studies were summarized by Van Dyne (1970) in an international meeting of grassland workers. Jameson (1970a) reviewed basic concepts in mathematical modeling of grassland ecosystems at a symposium concerning systems analysis in range science. Jameson and Bledsoe (1970) discussed the relationship of models to selection of experiments. In another national workshop effort involving several Grassland Biome scientists, Clymer and Bledsoe (1970) provided a lucid guide to mathematical modeling of the ecosystem. In 1972 Innis published on modeling concepts including simulation of ill-defined systems, simulation of biological problems, and a discussion of the second derivative in population modeling (Innis 1972a,b,c). Subsequently he published an article on the defense of dynamic models in soft sciences (Innis 1974). In three overseas meetings Van Dyne presented papers which included discussions of the relationship of modeling to research management (Van Dyne 1972, 1975, and Van Dyne and Abramsky 1975). The first two of these papers discussed the steps of the modeling procedure as it evolved in our learning process. Two main reports on concepts of validation have been provided by Steinhorst (1973) and Garrett (1975).

Several important models of specific topics have been conceptualized by Grassland Biome scientists, but not implemented on the computer. Early conceptual efforts by Jameson (1970b) visualized stochastic elements for secondary succession models, Harris and Swartzman (1971) developed a preliminary model for consumer predation, and Harris and Francis (1972) developed a dynamic simulation model structure for an interactive herbivore community (subsequently implemented by Harris and Fowler 1975). In 1972 in a series of workshops Grassland Biome scientists reviewed and analyzed process functions necessary as input to simulation models. These are summarized in an internal technical report (Jameson and Dyer 1973) as well as in an overview (Dyer 1973). Several model structures have evolved out of these process workshops including the paper of Ellis et al. (1974) concerning a conceptual model of diet selection as an ecosystem process.

Some conceptual structures related to optimization techniques were published by Innis and Harris (1972) regarding brush control problems and Peden (1972) regarding optimization of combinations of large herbivores.

Several other major operational models, not strictly simulation or optimization techniques, have been produced by Grassland Biome scientists. Particularly, see the work of French and Sauer (1974) regarding phenology studies in grasslands and Grant and French (1975) regarding small mammal population dynamics and energetics.

Post-ELM 1973

The major thrust of the Grassland Biome study through early 1974 centered on a "total-system focus." However, in 1974, a change in program management led to considerable change in the modeling effort. A small segment of Biome work continued developing the ELM series eventually to what is referred to as ELM 1977 (Fig. 1). Many program participants, who had been working in an integrated total-system model structure previously, began modifying existing submodels of

ELM or developing new submodels in 1975. Thus separate subsystem models were developed on small mammals, producers, root turnover, ruminants, and so forth (Fig. 1). These two efforts are described below.

ELM 1977 refers to a version of the ELM family which has been used for a large number of hypothesis-testing experiments (Van Dyne et al. 1977a,b). ELM 1977 was derived from ELM 1973. Between ELM 1973 and ELM 1977 there were more than 70 program code changes and more than 400 parameter changes, but the overall structure is still the same. Additionally, there were major changes in the way the output was handled and separate sets of programs now were used to develop system-level output variables from the state variable and flow dynamics. In the operation of ELM 1977 some 1285 items are saved at each time step for post-run analyses and to provide inputs for a series of supplementary programs (Van Dyne et al. 1977b). Thus far, the model has been used for some 70 model runs including several 6-year run sequences.

ELM 1973 and early phases of ELM 1977 were funded by the Grassland Biome program. Much of the work on many phases of the "in progress" models, noted in the lower right portion of Figure 1, also was funded by Grassland Biome monies but now is funded also under separate projects supported by such agencies as NSF, Environmental Protection Agency (EPA), and Fish and Wildlife Service (FWS). Separate funding has reduced the centralized planning and management of modeling and let to quite diverse objectives. These models are discussed briefly as follows.

The abiotic model has been evolved over time by Parton with more mechanism being included. A version was also converted to a new stage of the biome-developed simulation compiler, SIMCOMP 4.0 (Gustafson 1977). The producer submodel has been restructured by Detling*, Parton, and Hunt and is more mechanistic than the producer section of ELM 1973. One major change has been the separate definition of young vs. fully expanded age categories of leaves. One portion of the producer submodel, including photosynthesis and respiration, may run in less than 1-day time steps. The root turnover model has been described in a technical report (Parton and Singh 1976). The ruminant model developed by Swift* is more detailed than the consumer model in ELM 1973 with consideration of the rumen as an organ, rumen function, nonprotein nitrogen and amino acid pools in the body, etc. The generalized insect model of Scott* is of lower resolution than the grasshopper model in ELM 1973. This insect model does not predict population dynamics, instead insects are put in as driving variables. In effect, this model is a detailed energy calculator. A nematode model had been developed by Coughenour* based in part on an early nematode model developed by Andrews*. One or more microcosm models have been developed by Hunt and others for studying bacteria, nematode, and protozoa relationships. These models are based on differential rather than difference equation systems. Coughenour has developed a sulphur-cycling submodel for inclusion in the basic ELM structure. A stand-alone nitrogen model has been developed by McGill**, Hunt,

---

*Current or recent Grassland Biome staff member.
**Consultant to the Grassland Biome study.

Woodmansee, and Reuss which is more mechanistic and more detailed than the nitrogen model described in ELM 1973. This model is highly theoretical and there are many parameters in the model that are not field measurable.

The philosophical thrust has been to develop separate subsystem models on various studies specifically for that project. The projects in general are not total-system projects and thus it is logical that the models developed in the way that they did. The hope is that some of these subsystem models can be coupled together on demand to produce a more wholistic model for a particular purpose. None of the models of the previous paragraph are published, and thus cannot be examined in detail, but it would appear there would be considerable difficulty in coupling them together. There is no evident philosophical superstructure into which all of these submodels fit. In general, these subsystem models are more mechanistic and have more compartments, more nonlinearities, require more parameters, driving variables, and shorter time steps than do the ELM series submodels. It is difficult at this time to determine whether they do any better job of predicting field-observed responses.

It is interesting to note the amount of time that has been used in the developing of various models shown in Figure 1. The ELM 1973 model had about 20 professional person years invested in it (Van Dyne and Anway 1976). Probably at least another 5 professional person years were expended through the completion of this volume. The small mammal model of Figure 1 has about a half a year in its development (Abramsky 1976). ELM 1977 has about 2.5 professional person years in its development and analysis (Van Dyne et al. 1977b). No detailed records are available as documentation is not yet complete, but rough estimates are about 15 professional person years have been spent to date in the development of the subsystem models listed in Figure 1 (based on personal communication with Woodmansee, Parton, Swift, Detling, and Gustafson). This estimate seems conservative when considering the following three items: over a dozen scientists have been heavily involved in the model development; other experimental scientists probably have contributed appreciably to the model development; and some of these programs were initiated as early as 1974. There are probably at least 20 professional person years spent to date on these submodels and by the time the documentation is completed and published there will be at least another 5 years of effort. Perhaps in the order of 20 professional person years have been expended collectively in models PRONG, SIMOPT, PAWNEE 1 and 2, LINEAR 1 and 2, RANGES, and ANNUAL GRASSLAND. This brings the total expenditure to over 70 professional person years and this does not include the optimization modeling work developed in the Grassland Biome studies or those studies derived directly from it.

We know that the ELM series has some serious limitations. Many aspects of the model and its performance were changed between ELM 1973 and ELM 1977, but some anomalous behavior still exists. For example, there is a rapid transfer of standing dead vegetation into the litter compartment in the spring of the year. There appears to be a cycling of net primary production from year to year under heavy use perhaps due to soil water buildup in alternate years or due to translocation impacts. Years of highest forage intake by consumers are not

necessarily the years of the highest forage production. There is a great temperature sensitivity in the grasshopper submodel. The amount of CO_2 evolved and the
decomposer production rates do not fit intuitive ideas about the experimental
treatments. There is a problem of low digestibility of forage late in the season.
There is greater phosphorus live shoot-to-crown flow under extra-heavy grazing
than under heavy grazing or light grazing. There are anomalous live root-to-shoot
flows of nitrogen as compared to phosphorus. Active decomposer respiration at
deep soil levels decreases under extra-heavy grazing with competing small
herbivore removal rather strangely! There is an unexplained variation in grasshopper production under sheep-grazing as compared to cattle-grazing situations.
Water dynamics in the lower soil depths are not entirely logical. Small mammals
go to extinction with the present parameterization of ELM after many years. The
model shows about 1% of the net primary production aboveground is used by
small mammals and insects, but is this realistic? Cattle gains under yearlong as
compared to seasonal grazing differ in unexplained ways.

A major lesson derived from development and analysis of the models in Figure
1 is the recognition of the need for clear definition and usage of a sound
philosophical structure for modeling. I believe the state variable approach which
clearly differentiates among driving variables, state variables, output variables,
and intermediate variables should be followed. We need a better graphic display
of model structures. The display should include a coupling matrix showing all the
flows and a flow: effects matrix showing what variables affect the calculation of
each flow. It is important also in the operation and development of these models
to state the flow functions with parameters rather than constants. There is a need
to develop yet better software systems to assist in the modeling process. The
SIMCOMP 4.0 compiler is only a step in this direction. Relatively few detailed
and quantitative sensitivity analyses have been reported for the models summarized herein. Better methods need to be developed and the work of Steinhorst
and others in this volume could provide some good leads. There is a strong need
for developing an approach that will accept replaceable modules of differing
levels of resolution similar to that initiated by Goodall in the Desert Biome
program. There is also a need then for a library of flow functions. I feel that the
principle of inclusiveness needs to be adhered to more closely, at least in
structuring a total-system model. We also need some automatic way to reduce
the voluminous output we obtain when we study model response under various
experimental conditions. The models developed thus far are not highly transport-
able nor decomposable. It has been very difficult to make transitions or changes.
Yet other modelers have been able to evolve models over time. For example
Goodall (1967, 1969, 1970a, 1970b, and 1972) evolved a grazingland model with
continual changes and documentation. Similarly Donnelly and Armstrong (1968),
Freer et al. (1970), Armstrong (1971), and Christian et al. (1972, 1974) evolved a
sheep simulation model for summer pasture on Australian grazinglands. The
longest stretch of evolution outlined in the models of Figure 1 is for the ELM
series and basically only ELM 1973 and ELM 1977 have undergone extensive
tests.

The reader should not be discouraged by the list of anomalies noted above for the ELM model for these are due to only a small part of the several hundred process functions in the model and identify topics for field and laboratory research projects. Likewise, the needs listed regarding the art and process of modeling should not detract from the tremendous gains made in the ecological modeling art since the inception of the Biome studies but delimits area of study. It has been an exciting and informative phase in the history of ecological research. We have moved many aspects of ecology from qualitative to quantitative phases. We are beginning to collect our knowledge into testable and semistructured theory. Considering the rate of modeling output on a world-wide scale in ecological studies in the past 10 years one almost shudders to look forward to the next 10. The ELM 1973 model, however, will go down as a landmark in the history of grassland ecology. This volume, and its supporting documents, will be of great interest for students of grassland ecology over the next decade.

George M. Van Dyne

Director US IBP Grassland Biome Study 1967–1974
September 1977
College of Forestry and Natural Resources
Colorado State University

References

Anway, J. C., Brittain, E. G., Hunt, H. W., Innis, G. S., Parton, W. J., Rodell, C. F., Sauer, R. H.: ELM: Version 1.0. US/IBP Grassland Biome Tech. Rep. No. 156. Fort Collins: Colorado State Univ., 1972, 285 pp.

Abramsky, Z. Z.: Small mammal studies in natural and manipulated shortgrass prairie. Ph.D. thesis, Colorado State Univ., Fort Collins, 1976, 197 pp.

Armstrong, J. S.: Modeling a grazing system. In: Proc. Ecol. Soc. Aust., Vol. 6., Canberra, Australia, 1971, pp. 194–202.

Bellman, R., Kagiwada, H., and Kalaba, R.: Inverse problems in ecology. J. Theor. Biol. **11**, 164–176 (1966).

Bergner, T. E.: 1965. Tracer theory: a review. Isotopes and Radiation Tech. **3**, 245–262 (1965).

Bledsoe, L. J.: Simulation of a grassland ecosystem. Ph.D. thesis, Colorado State Univ., Fort Collins, 1976, 346 pp.

Bledsoe, L. J., Francis, R. C., Swartzman, G. L., Gustafson, J. D.: PAWNEE: A grassland ecosystem model. US/IBP Grassland Biome Tech. Rep. No. 64. Fort Collins: Colorado State Univ., 1971, 179 pp.

Bledsoe, L. J., Jameson, D. A.: Model structure for a grassland ecosystem. The Grassland Ecosystem: A Preliminary Synthesis, R. L. Dix, R. G. Beidleman (eds.). Fort Collins: Range Science Dept. Science Series No. 3. Colorado State Univ., 1969, pp. 410–437.

Bledsoe, L. J., and Olson, J. S.: COMSYS 1: A stepwise compartmental simulation program, Oak Ridge National Lab. Tech. Memo. 2413, 1970.

Cale, W. G.: Simulation and systems analysis of a shortgrass prairie ecosystem. Ph.D. thesis, Univ. Georgia, Athens, 1975, 253 pp.

Christian, K. R., Armstrong, J. S., Davidson, J. L., Donnelly, J. R., and Freer, M.: A model for decision-making in grazing management system. Proc. Aus. Soc. Anim. Prod. **9**, 124–129 (1974).

Christian, K. R., Armstrong, J. S., Donnelly, J. R., Davidson, J. L., and Freer, M.: Optimization of a grazing management system. Proc. Aust. Soc. Anim. Prod. **9**, 124–129 (1972).

Clymer, A. B., and Bledsoe, L. J.: A guide to mathematical modeling of an ecosystem. In: Modeling and Systems Analysis in Range Science, R. G. Wright and G. M. Van Dyne (ed.). Fort Collins: Range Science Dept. Science Series No. 5. Colorado State Univ., 1970, pp. 1–75.

Cole, G. W. (ed.): ELM Version 2.0, Fort Collins Range Science Dept. Science Series No. 20. Colorado State Univ., 1976, 663 pp.

Dahlman, R. C., Olson, J. S., and Doxtader, K.: The nitrogen economy of grassland and dune soils. In: Biology and Ecology of Nitrogen Proceedings of a Conference, Washington D.C., National Academy of Sciences, 1969, pp. 64–82.

Davidson, R. S., and Clymer, A. B.: The desirability and applicability of simulation ecosystems. New York Acad. Sci. **128**, 790–794 (1966).

Dix, R. L., and Beidleman, R. G., (eds.): The grassland ecosystem: a preliminary synthesis. Range Science Dept. Science Series No. 3. Colorado State Univ., Fort Collins. 437 p. (1969).

Donnelly, J. R., and Armstrong, J. S.: Summer grazing. Proc. Second Conf. Appl. Simul., New York, December 24, pp. 329–332 (1968).

Dyer, M. I.: Process studies related to grassland ecosystem research, p. 669–683. *IN* Proc. of the 1973 Summer Computer Simulation Conference. Vol. II. La Jolla: Simulation Councils (1973).

Ellis, J. E., Wiens, J. A. Rodell, C. F. and Anway, J. C. A conceptual model of diet selection as an ecosystem process. J. Theoretical Biol. **60**, 93–108 (1974)

Freer, M., Davidson, J. L. Armstrong, J. S. and Donnelly, J. R.: Simulation of grazing system, p. 913–917. *IN* Proc. XI Inter. Grassland Congr., Surfer's Paradise, Queensland, Australia (1970).

French, N. R., and Sauer, R. H.: 4.3 Phenological studies and modelling in grasslands, p. 227–235. *In:* H. Lieth (ed.) Phenology and seasonality modeling. Vol. 8., New York: Springer-Verlag (1974).

Garfinkel, D.: Digital computer simulation of ecological systems. Nature **194**, 846–857 (1962).

Garfinkel, D., Sack, R.: Digital computer simulation of an ecological system, based on a modified mass action law. Ecology, **54/3** (1964), p. 502–507.

Garrett, M.: Statistical techniques for validating computer simulation models. US/IBP Grassland Biome Tech. Rep. No. 286. Colorado State Univ., Fort Collins. 68 p. (1975).

Gilbert, B. J.: RANGES: Grassland simulation model. Range Science Dept. Science Series No. 17. Colorado State Univ., Fort Collins. 119 p. (1975).

Goodall, D. W.: Computer simulation of changes in vegetation subject to grazing. Journal of the Indian Botanical Society **46**, 356–362 (1967).

Goodall, D. W.: Simulating the grazing situation, p. 211–236. *IN* F. Heinments (ed.) Concepts and models of biomathematics: Simulation techniques and methods, Vol. 1. New York: Marcel Dekker (1969).

Goodall, D. W.: Simulating of grazing systems, p. 51–74. *IN* D. A. Jameson (ed.) Modelling and systems analysis range science. Range Science Dept. Science Series No. 5. Colorado State Univ., Fort Collins (1970a)

Goodall, D. W.: Studying the effects of environmental factors of ecosystems, p. 19–26. *IN* D. E. Reichie (ed.) Analysis of temperate forest ecosystems. Berlin: Springer-Verlag (1970b).

Goodall, D. W.: Extensive grazing systems, p. 173–187. *In:* J. B. Dent and J. R. Anderson (ed.) Systems analysis in agricultural management. Sidney: Wiley (1971).

Goodall, D. W.: Potential applications of biome modelling. Tierre et la vie **1**, 118–138. (1972).

Gore, A. J. P., and Olson, J. S.: Preliminary models for accumulation of organic matter in an *Eriophorum/Calluna* ecosystem. Aquilo, Ser. Bot. **6**, 297–313 (1967).

Grant, W. E., and French, N. R.: The functional role of small mammals in grassland ecosystems. US/IBP Grassland Biome Preprint No. 163. Colorado State Univ., Fort Collins. 94 p. (1975).

Gustafson, J. D.: SIMCOMP 4.0 reference manual. (Unpublished report). 41 p. (1977).

Harris, L. D., and Fowler, N. K.: Ecosystem analysis and simulation of the Mkomazi Reserve, Tanzania. East African Wildlife J. **13**, 325–346 (1975).

Harris, L. D., and Francis, R. C.,: AFCONS: A dynamic simulation model of an interactive herbivore-community. US/IBP Grassland Biome Tech. Rep. No. 158. Colorado State Univ., Fort Collins. 88 p. (1972).

Harris, L. D., and Swartzman, G. L.: A preliminary model for consumer predation. US/IBP Grassland Biome Tech. Rep. No. 158. Colorado State Univ., Fort Collins. 88 p. (1971).

Holling, C. S.: The analysis of complex population processes. Can. Entomol. **96**, 335–347 (1964).

Holling, C. S.: The strategy of building models of complex ecological systems. *In*. K. E. F. Watt (ed.) Systems Analysis in ecology. New York: Academic Press (1966).

Innis, G. S.: Simulation of ill-defined systems: Some problems and progress. Simulation **19(6)**, 33–36 (1972a).

Innis, G. S.: Simulation of biological systems. Some problems and progresses, p. 1085–1089a. *In:* Proceedings of the 1972 Summer Computer Simulation Conference. Vol. II. La Jolla: Simulation Councils (1972b).

Innis, G. S.: The second derivative and population modelling: Another view. Ecology **53(4)**, 720–723. (1972c).

Innis, G. S.: 1974. Dynamic analysis in soft science studies: in defense of difference equations. p. 102–122 *IN* P. van den Driessche (ed.) Lecture notes in biomathematics. Vol. 2. New York: Springer-Verlag (1974).

Innis, G. S.: Role of total systems models in the Grassland Biome Study, p. 13–47. *IN* B. C. Patten (ed.) Systems analysis and simulation in ecology. Vol. III New York: Academic Press (1975).

Innis, G. S.: Simulation models of grasslands and grazinglands. (Submitted to "Ecology of grasslands and bamboolands in the World," edited by M. Numata. New York): Springer-Verlag (1978).

Innis, G. S., and Harris, L. D.: An economic model for brush control. US/IBP Grassland Biome Preprint No. 29. Colorado State Univ., Fort Collins. 18 p. (1972).

Innis, G. S., and Miskimins, R.: Ranges I grassland simulation model. Report No. 8. Regional Analysis of Grassland Environmental Systems. Colorado State University, Ft. Collins, CO., 86 p. (1973).

Innis, G. S., and Van Dyne, G. M.: Modelling in the US/IBP Grassland Biome Study. Ann. Univ. Abidjan, series (Ecologie, Tome VI, Fuse. **23**, 305–311). (1973).

Jameson, D. A.: Land management policy and development of ecological concepts. J. Range Manage. **23**, 316–322 (1970a).

Jameson, D. A.: Stochastic ability and secondary succession, p. 1-192–1-197. *In:* R. G. Wright and G. M. Van Dyne (ed.) Simulation and analysis of dynamics of a semi-desert grassland. An interdisciplinary workshop program toward evaluating the potential ecological impact of weather modification, Range Science Dept. Science Series No. 6. Colorado State Univ., Fort Collins (1970b).

Jameson, D. A., and Bledsoe, L. J.: Models and selection of experiments, p. 111–134 (Chap. 8). *In:* D. A. Jameson (ed.) Modelling and systems analysis in range science. Range Science Dept. Science Series No. 5. Colorado State Univ., Fort Collins (1970).

Jameson, D. A., and Dyer, M. I. (ed.): Process studies workshop report. US/IBP Grassland Biome Tech. Rep. No. 220. Colorado State Univ., Fort Collins. 444 p. (1973).

Kelly, J. M., Opstrup, P. A., Olson, J. S., Auerbach, S. I., and Van Dyne, G. M.: Models of seasonal primary productivity in eastern Tennessee, *Festuca* and *Andropogon* ecosystems (thesis). ORNL-4310, 1969, 305 pp.

Lindemann, R. L.: The trophic-dynamic aspect of ecology. Ecology **23**, 299–418 (1942).

Lotka, A. J.: Elements of physical biology. Baltimore: Williams and Wilkins (1925).

Lucas, H. L.: Theory and mathematics in grassland problems. Proc. Int. Grassland Congr. **8**, 732–736 (1960).

Lucas, H. L.: Stochastic elements in biological models; their sources and significance, p. 355–385. *In:* J. Gurland (ed.) Stochastic models in medicine and biology. Univ. Wisconsin, Madison (1964).

Neel, R. B., and Olson, J. S.: Use of analog computers for simulating the movement of isotopes in ecological system. Oak Ridge National Laboratory Report 3172. Oak Ridge, Tennessee (1962).

Olson, J. S.: Energy storage and the balance of producers and decomposers in ecological systems. Ecology **44**, 322–332 (1963a).

Olson, J. S.: Analog computer models for movement of nuclides through ecosystems, p. 121–125. *In:* V. Schultz and A. W. Klement, Jr. (ed.) Radioecology. New York: Reinhold (1963b).

Olson, J. S.: Gross and net production of terrestrial vegetation. J. Ecol. **52(Suppl)**, 99–118 (1964).

Olson, J. S.: Equations for cesium transfer in *Liriodendron* forest. Health Physics **11(12)**, 1965, pp. 1385–1392. (Reprinted in *Radiation and Terrestrial Ecosystems,* F. P. Hungate, (ed.). Oxford, Pergamon Press.

Olson, J. S.: Use of tracer techniques for the study of biogeochemical cycles. In: Functioning of

Terrestrial Ecosystems at the Primary Production Level, F. Eckardt (ed.). Paris: UNESCO, 1968, pp. 271–288. Paris: UNESCO

Olson, J. S.: Carbon cycles and temperate woodlands. In: Analysis of Temperate Forest Ecosystems, Analysis and Synthesis, vol. 1, D. E. Reichle (ed.). New York: Springer-Verlag, 1970, pp. 226–241. New York: Springer-Verlag

Olson, J. S., Uppuluri, V. R. R.: Ecosystem maintenance and transformation models as Markov processes with absorbing barriers. In Health Physics Division Annual Progress Report. ORNL-4007. Oak Ridge National Laboratory, Oak Ridge, 1966, pp. 104–105.

Parton, W. J. and Singh, J. S.: Simulation of plant biomass on a shortgrass and a tall grass prairie with emphasis on belowground processes. US/IBP Grassland Biome Tech. Rep. No. 300. Colorado State Univ., Fort Collins. 76 p. (1976).

Patten, B. C.: An introduction to the cybernetics of the ecosystem: the trophic-dynamic aspect. Ecology **40**, 221–231 (1959).

Patten, B. C.: A simulation of the shortgrass prairie ecosystem. Simulation **19**, 177–186 (1972).

Patten, B. C., and Witcamp, M.: Systems analysis of 134 cesium kinetics in terrestrial microcosms. Ecology **48**, 813–824 (1967).

Peden, D. G.: The trophic relations of *Bison bison* to the shortgrass prairie. Ph.D. Dissertation. Colorado State Univ., Fort Collins. 134 p. (1972).

Smith, F. E.: Effects of enrichment in mathematical models. Eutrophication, AAAS (1968).

Sollins, P.: CSS: A computer program for modeling ecological systems. ORNL/IBP-71/5, 1971, 96 pp.

Sollins, P., Reichle, D. E., and Olson, J. S.: Organic matter budget and model for a southern Appalachian *Liriodendron* forest. EDFB/IBP-73/2. 1973, 150 pp.

Steinhorst, R. K.: Validation tests for ecosystem simulation models. (manuscript submitted to Biometrics) (1973).

Swartsman, G. L. (ed.): Some concepts of modelling. US/IBP Grassland Biome Tech. Rep. No. 32. Colorado State Univ., Fort Collins. 142 p. (1970).

Swartsman, G. L., and Van Dyne, G. M.: An ecologically based simulation-optimization approach to natural resource planning. Annu. Rev. Ecol. System. **3**, 347–398 (1972).

Van Dyne, G. M.: Ecosystems, systems ecology, and systems ecologists. Oak Ridge National Laboratory Report ORNL 3957, Oak Ridge National Laboratory, Oak Ridge, Tennessee. 31 p. (1966).

Van Dyne, G. M.: Some mathematical models of grassland ecosystems. pp. 3–26. *IN* Dix, R. L. and R. G. Beidleman (eds.) The grassland ecosystem: a preliminary synthesis. Range Science Dept. Science Series No. 3. Colorado State Univ., Fort Collins. 437 p. (1969a).

Van Dyne, G. M.: Grasslands management, research, and training viewed in a systems context. Range Science Dept. Science Series No. 3. Colorado State Univ., Fort Collins. 50 p. (1969b).

Van Dyne, G. M.: Examples of trophic level and total ecosystem models, p. 191–198. *in:* R. T. Coupland, and G. M. Van Dyne (ed.) Grassland ecosystems: Review of research. Range Science Dept. Science Series No. 7. Colorado State Univ., Fort Collins (1970).

Van Dyne, G. M.: Organization and management of an integrated ecological research program—with special emphasis on systems analysis, universities, and scientific cooperation, p. 111–172. *In:* J. N. R. Jeffers (ed.) Mathematical models in ecology. Oxford: Blackwell Sci. Pub., (1972).

Van Dyne, G. M.: Some procedures, problems, and potentials of systems-oriented, ecosystem-level research programs, p. 4–58. *IN* Procedures and examples of integrated ecosystem research. Barrsklogsladskapets Ecologi, Swedish Coniferous Project Tech. Rep. 1. Uppsala, Sweden (1975).

Van Dyne, G. M., and Abramsky, Z.: Agricultural systems models and modelling. An overview, p. 23–106. *In:* G. E. Dalton (ed.) Study of agricultural systems. London: Applied Science Publishers (1975).

Van Dyne, G. M., and Anway, J. C.: A program for and the process of building and testing grassland ecosystem models. J. Range Manage. **29**, 114–122 (1976).

Van Dyne, G. M., Frayer, W. E., and Bledsoe, L. J.: Some optimization techniques and problems in the natural resurce sciences, p. 95–124. *IN* Society for Industrial and Applied Mathematics. Studies in Optimization I: Symposium on optimization. Philadelphia, Pennsylvania. 137 p. (1970).

Van Dyne, G. M., and Innis, G. S. (ed.): Macrolevel ecosystems models in relation to man: A

developing dynamic models of ecological systems, p. 9–26. *In:* M. Lilllywhite and C. Martin (ed.) Environmental awareness. Inst. Environ. Sci., Proc., Second Annu. Session, Colorado Chapter (1971).

Van Dyne, G. M., Joyce, L. A., and Williams, B. K.: Models and the formulation and testing of hypotheses in grazingland ecosystem management. *In:* Holdgate, M. and M. Woodman (eds.) Ecosystem restoration—principles and case studies. London: Plenum Press (1977a).

Van Dyne, G. M., Joyce, L. A., Williams, B. K., and Kautz, J. E.: Data, hypotheses, simulation and optimization models, and model experiments for shortgrass prairie grazinglands: a summary of recent research. Report to the Council on Environmental Quality (1977b).

Van Hook, R. I., Jr., Reichle, D. E., and Auerbach, S. I.: Energy and nutrient dynamics of predator and prey arthropod populations in a grassland ecosystem (thesis). ORNL-4509, 1970, 110 pp.

Watt, K. E. F: Mathematical models for use in insect pest control. Can. Entomol. **88 (Suppl. 19).** 62 p. (1961).

Watt, K. E. F. (ed.): Systems analysis in ecology. New York: Academic Press (1966).

Watt, K. E. F.: Ecology and resource management—A quantitative approach. New York: McGraw-Hill 450 p. (1968).

Wiegert, R. G.: Simulation models of ecosystems. Ann. Rev. Ecol. Syst. **6**, 311–338. (1975).

Preface

Managing the world's renewable natural resources is difficult because of the uncertainties of the environment and because of our current state of knowledge about ecological systems. These points were recognized by the founders of the International Biological Program (IBP) whose objective was "to examine the biological basis of productivity in human welfare."

National parts of the IBP took different directions, and within nations different studies had their particular emphases. The US Grassland Biome study made an early and steady commitment to the use of simulation models as one of the techniques for organizing research to increase our knowledge while striving to develop tools for predicting ecosystem response to environmental change—whether man-made or natural.

The model presented in this volume is the third in a series produced by that program. This model is built on its predecessors and has been elaborated and improved subsequently. The *model objectives,* described in Chapter 1 and mentioned often thereafter, were only partially achieved. These model objectives address the desire to predict the response of the system to change. The *objectives of the modeling effort,* of which construction of this model was only a part, were those of program management designed to make effective use of support resources while advancing our knowledge of the grassland systems we studied.

This modeling effort objective forced a close integration between modeler and field investigator. Achievement of this integration was difficult because of the backgrounds of the people needed to do the diverse jobs and the size of the task. We were fortunate to be able to attract a staff of postdoctoral fellows (Anway, Hunt, Parton, Rodell, Sauer, and Woodmansee) who could bridge the gap between model building and field and laboratory research; to have scientists of the calibre of Cole and Reuss, who were anxious to try the tools of simulation in their fields; to have a systems programmer of Gustafson's ilk who was interested in providing the programming language capabilities that markedly increased the effectiveness of the modelers; to have contact with excellent scientists from all over the world, such as Stewart and Haydock, as well as the staff and associates of the Grassland Biome study. These people freely gave of themselves in a team effort to address issues of scientific and social import.

The fact that the program was a team effort cannot be overemphasized. Part of the team activity was modeling, and because of this, we have in this volume one of the most thorough integrations of ecosystem science available anywhere. Yes, there are errors for which we, the authors and editor, assume full responsibility; but there is also a lot of grassland ecological science summarized here.

A successful team must be lead and financed. Dr. George M. Van Dyne provided excellent leadership for these modeling studies. Much of the operation's structure and execution is the result of his insight and direction. The effort was financed by the National Science Foundation on grants GB31862X, GR31862X2, GB41233X, and BMS73-02027 A02.

G. S. INNIS

Contents

Contributors

ANWAY, J. C. US/IBP Grassland Biome, Natural Resource Ecology Laboratory, Colorado State University, Fort Collins, Colorado 80523, USA Present address: School of Applied Science, Canberra College of Advanced Education, P. O. Box 381, Canberra City A.C.T. 2601, Australia

COLE, C. V. Phosphorus Laboratory, Agricultural Research Service, U.S. Department of Agriculture, Fort Collins, Colorado 80521, USA

GUSTAFSON, J. D. US/IBP Grassland Biome, Natural Resource Ecology Laboratory, Colorado State University, Fort Collins, Colorado 80523, USA

HAYDOCK, K. P. US/IBP Grassland Biome, Natural Resource Ecology Laboratory, Colorado State University, Fort Collins, Colorado 80523, USA Present address: CSIRO, The Cunningham Laboratory, Division of Mathematics and Statistics, Mill Road, St. Lucia, QLD. 4067, Australia.

HUNT, H. W. Natural Resource Ecology Laboratory, Colorado State University, Fort Collins, Colorado 80523, USA

INNIS, G. S. US/IBP Grassland Biome, Natural Resource Ecology Laboratory, Colorado State University, Fort Collins, Colorado 80523, USA Present address: Department of Wildlife Sciences, Utah State University, Logan, Utah 84322, USA

PARTON, W. J. US/IBP Grassland Biome, Natural Resource Ecology Laboratory, Colorado State University, Fort Collins, Colorado 80523, USA

REUSS, J. O. Agronomy Department, Colorado State University, Fort Collins, Colorado 80523, USA and Natural Resource Ecology Laboratory, Colorado State University, Fort Collins, Colorado 80523, USA

RODELL, C. F. US/IBP Grassland Biome, Natural Resource Ecology Laboratory, Colorado State University, Fort Collins, Colorado 80523, USA Present address: Department of Biology, Vanderbilt University, Nashville, Tennessee 37235, USA

SAUER, R. H. US/IBP Grassland Biome, Natural Resource Ecology Laboratory, Colorado State University, Fort Collins, Colorado 80523, USA Present address: Terrestrial Ecology, Battelle-Northwest Laboratory, Richland, Washington 99352, USA

STEINHORST, R. K. US/IBP Grassland Biome, Natural Resource Ecology Laboratory, Colorado State University, Fort Collins, Colorado 80523, USA Present address: Institute of Statistics Texas A & M University, College Station, Texas 77843, USA

STEWART, J. W. B. Department of Soil Science, University of Saskatchewan, Saskatoon, Saskatchewan, Canada S7N OWO

VAN DYNE, G. M. US/IBP Grassland Biome, Natural Resource Ecology Laboratory, Colorado State University, Fort Collins, Colorado 80523, USA Present address: College of Forestry and Natural Resources, Colorado State University, Fort Collins, Colorado 80523, USA

WOODMANSEE, R. G. US/IBP Grassland Biome, Natural Resource Ecology Laboratory, Colorado State University, Fort Collins, Colorado 80523, USA

1. Objectives and Structure for a Grassland Simulation Model

GEORGE S. INNIS

With Appendix by JON D. GUSTAFSON

Abstract

The objectives of the modeling efforts in the US/IBP Grassland Biome were to stimulate the precise formulation of ideas about the behavior of the system, help direct the research effort through the identification of critical areas for study, and produce a tool for investigating the consequences of management alternatives. The objectives of the model were to simulate biomass dynamics in a variety of grassland types and the response of the system to irrigation, fertilization, and cattle grazing. The model was constructed over a period of several years by a team of investigators, each of whom assumed responsibility for one of the component submodels: i.e., abiotic, producers, mammals, grasshoppers, decomposers, nitrogen, and phosphorus. The primary information source used in developing this model was the US/IBP Grassland Biome intensive study site, Pawnee.

1.1 Introduction

The US/IBP Grassland Biome study began modeling, laboratory, and field studies in 1968. Models were incorporated early to provide research guidance and a synthesis framework. Research direction evolved from the need to fill information gaps identified by the models, and synthesis involved their use for hypothesis testing.

Three total systems models had been built by 1974: (a) PWNEE (Bledsoe et al., 1971), (b) a model built at the University of Georgia (Patten, 1972), and (c) ELM, which is reported here (see also Anway et al., 1972; Innis, 1975). More of the history and relationship among these models is reported by Innis (1975).

1.2 Objective

The objective of this modeling activity was to develop a total-system model of the biomass dynamics for a grassland that, via parameter change, could be representative of the sites in the US/IBP Grassland Biome network and with which there could be relatively easy interaction.

There are several key points in this objective that deserve elaboration. First, the term *total-system model* refers to the inclusion of abiotic, producer, consumer, decomposer, and nutrient subsystems. This requirement was imposed to assure that the modeling effort played the integrative role delegated to it. Program directors feared that a collection of process or subsystem models would not provide the basis for programmatic decisions as would an overall model.

Second, *biomass dynamics* identifies our principal concern with carbon or energy flow through the system. Focus on biomass facilitated the comparison of model and data but turned out to be unfortunate because it is not conserved. The model, therefore, tracks carbon and converts it to biomass (and vice versa) in a number of places. We are concerned with dynamics as part of the general objective of the International Biological Program (IBP).

Third, *representative* expresses our desire to have the model apply, with minimal effort, to sites in the US/IBP Grassland Biome study. Changes of parameters are certainly necessary as these describe site characteristics (among other things). The representation was to depict "normal" dynamics as well as the response of the system to a variety of perturbations.

Finally, *relatively easy interaction* was a desideratum because of the role the effort was to play in program direction. We anticipated working sessions with field investigators wherein the investigators would review a portion of the model and suggest changes. We wanted to be able to incorporate these changes and evaluate their effects quickly (in a matter of hours). Achievement of this goal depended in part on the construction of the simulation (pre)compiler SIMCOMP described in Appendix 1.A.

This objective provides only the broadest guidelines to the modelers as to their respective functions. The purpose of the objective is to found the decisionmaking processes that accompany model building. This involves clarification as to how many producers and consumers should be included, the amount of detail required in a representation of a producer, and whether a phosphorus, calcium, or lead model is required. Refinement of this objective was accomplished with the help of program scientists. It was difficult, however, to sharpen this objective without causing some investigators to doubt the utility of their efforts in the program.

On the other hand, the most ardent scientist realizes that to undertake such a task, goals must be narrowed and specific questions subsumed under the objective addressed. In 1970 it was agreed that this objective would stand, with the first model addressing four specific questions:

1. What is the effect on net or gross primary production as the result of the following perturbations: (a) variations in the level and type of herbivory, (b) variations in temperature and precipitation or applied water, and (c) the addition of nitrogen or phosphorus?

2. How is the carrying capacity of a grassland affected by these perturbations?

3. Are the results of an appropriately driven model run consistent with field data taken in the Grassland Biome Program, and if not, why?

4. What are the changes in the composition of the producers as a result of these perturbations?

These questions were further specified with definitions of terms such as "variations," "level," and "type"; acceptance criteria were chosen. Woodmansee reconsiders this objective and collection of questions in Chapter 10.

1.3 Model Construction

Work on ELM (not an acronym) began late in 1971, taking the models of Bledsoe et al. (1971) and Patten (1972) as a starting point. A team of postdoctoral fellows joined the staff in 1971 (J. Anway, H. W. Hunt, W. Parton, C. Rodell, and R. Sauer) and 1972 (R. G. Woodmansee). Each accepted the responsibility for the development of one portion of the model and for the adaptation of ELM to one of the biome study sites. From Fort Collins, both C. V. Cole, Agricultural Research Service, and J. O. Reuss, Agronomy Department, Colorado State University, cooperated by developing phosphorus and nitrogen models, respectively. Simultaneously, J. D. Gustafson proceeded with the development of SIMCOMP, a simulation language (Appendix 1.A). One version for use at the central program was quite elaborate. The model runs on this system. A simpler version, more easily adapted to other computers and used by the postdoctoral fellows at the study sites, was also developed.

Each of the fellows spent about 5 mo in the central program learning the modeling philosophy, SIMCOMP, and becoming acquainted with one another and the central program staff. Following this period each was sent to a field site for 1 yr (see Fig. 1.1). During this year the team met at monthly intervals to assure the development of compatible components and to maintain the *esprit de corps* that had developed during the first months. Following this year at field sites the fellows returned to Fort Collins for the final 7 mo of their appointments. This portion of time was designed to allow the team to bring together not only the models but also the knowledge and experience that had been gained.

The philosophy of Jay W. Forrester as expressed in his book Industrial Dynamics (1961) was chosen. The model is written in difference equations (about 120) using a time step of 1–7 days for solution.

1.4 Pawnee Site and Data

Although the ELM model is designed for easy adaptation to sites in the Grassland Biome network (see Fig. 1.1), actual implementation at a site has been a large task. In the following papers attention is focused on the Pawnee Site, described by Jameson (1969), which is the field research facility of the Natural Resource Ecology Laboratory of Colorado State University, located on the USDA Agricultural Research Service Central Plains Experimental Range in northeastern Colorado. Figures 1.2 and 1.3 are photographs of study areas

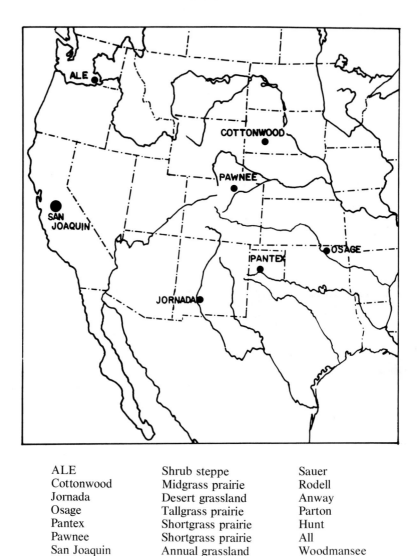

ALE	Shrub steppe	Sauer
Cottonwood	Midgrass prairie	Rodell
Jornada	Desert grassland	Anway
Osage	Tallgrass prairie	Parton
Pantex	Shortgrass prairie	Hunt
Pawnee	Shortgrass prairie	All
San Joaquin	Annual grassland	Woodmansee

Fig. 1.1.   Sites in the US/IBP Grassland Biome study, grassland type, and the postdoctoral fellow who worked at one of the sites

discussed below and shown on the map in Figure 1.4. Table 1.1 gives long-term
monthly precipitation and temperature. The vegetation of this shortgrass prairie,
described by Klipple and Costello (1960), is dominated by *Bouteloua gracilis*
(blue grama) and *Buchloe dactyloides* (buffalo grass). Table 1.2 identifies the
locales numbered in Figure 1.4.

The ecosystem stress area (Fig. 1.5 and ESA in Fig. 1.4) was the site of four
pairs of treatments: (a) control, (b) nitrogen (fertilizer) added, (c) water added,

Fig. 1.2.   General view of the Pawnee Site; the Pawnee Buttes, in background, are about
11 km to the north

Table 1.1.   Long-term average monthly temperature and precipitation for Pawnee

Month	Precipitation (mm)	Air temperature (°C)
January	8	−4
February	6	−1
March	15	1
April	28	7
May	53	12
June	56	18
July	47	21
August	42	20
September	23	15
October	20	10
November	5	2
December	2	−2
Annual	305 (total)	8 (average)

Table 1.2.   Identification[a] of points in Figure 1.4

	Map location	Treatment	Aspect	Slope (%)	Soil series
ESA	9	Irrigated	—	0	Ascalon
	9	Fertilized	—	0	Ascalon
	9	Irrigated and fertilized	—	0	Ascalon
	9	Control	—	0	Ascalon
	2	Ungrazed replicate 1 (1970–1971)	SE	2	Ascalon
	8	Ungrazed replicate 2 (1970–1971)	E	3	Ascalon
	10	Ungrazed replicate 1 (1972)	W	1–3	Manter
	10	Ungrazed replicate 2 (1972)	S	2–5	Manter
	4	Lightly grazed replicate 1	NE	5	Ascalon
	5	Lightly grazed replicate 2	NE	5	Ascalon
	1	Heavily grazed replicate 1	N	3	Ascalon
	3	Heavily grazed replicate 2	S	3	Ascalon
	6	Litter-bag experiments	E	4	Ascalon
	7	Record daily maximum and minimum temperature	—	—	Ascalon

[a]   From Jameson (1969) and Smith and Striffler (1969).

Fig. 1.3.   Close up of vegetation and microtypography on the Pawnee site

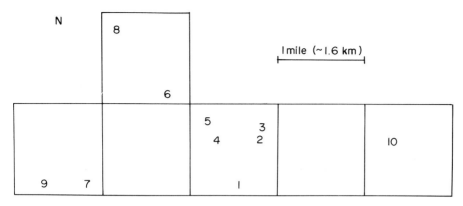

Fig. 1.4. Map of experimental units of the Pawnee Site (see Table 1.2 for identification of the units)`

Fig. 1.5. Aerial view of the eight 1-ha plots used in the irrigation (I) and fertilization (F) experiments. The eight areas are (beginning in the top row and proceeding from left to right): control, I, F+I, F; (Second row) F, F+I, control, I

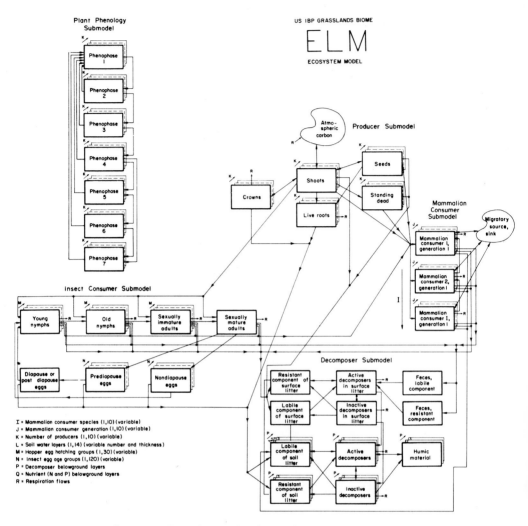

Fig. 1.6.   Flow diagram for the total ELM ecosystem model

and (d) nitrogen and water added (Lauenroth and Sims, 1973). These plots and both the lightly and heavily grazed treatments were used for the validation runs. Rainfall and temperature data were collected at points 2 and 7 in Figure 1.4, respectively.

The soil for all plots except the one in section 12 (Fig. 1.4) was classified as Ascalon (neé sandy loam); however, recent work by Woodmansee (in press) indicates four soil series to be represented on the ESA site; thus variability within the soils had observable effects on the herbage. An example of problems caused by soil variability is illustrated by noting the differences between the "ungrazed" treatment located on the Ascalon series in 1970–1971 and the

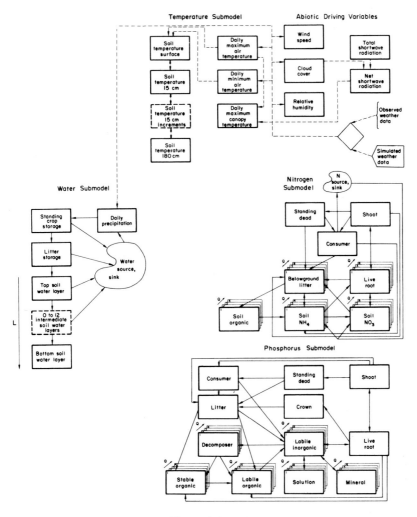

Fig. 1.6. (cont.)

Manter in 1972 (see Table 1.2 and Fig. 1.4). The problems of heterogeneity and simultaneity as they relate to interpreting model and experimental results are discussed later.

1.5 The Model

A diagram of ELM is presented in Figure 1.6. Much detail is omitted from the figure to assure legibility. One can gain, however, a sense of its size and complexity. The subsystems discussed in the subsequent chapters are easily identified.

In building ELM, deviations from the models described later were dictated by considerations of compatibility, data requirements, and ease of implementation. The only major change was in the nitrogen model where the block diagram shown as Figure 7.1 was replaced by the block diagram for nitrogen shown in Figure 1.6. Differences include the relabeling of the "dead-root-N" compartment as "below-ground-litter," elimination of a "litter-N" compartment, and the insertion of a "consumer" compartment.

In other instances, linking submodels often involved replacing "forcing functions" by dynamic compartments that reacted to events in the driven subsystem. For example, the water-flow model was developed using plant-growth data and simple production submodels. These simple submodels provided limited response to soil-water dynamics. When the soil-water model was joined with the production model, wherein soil water and production interacted, parameters had to be adjusted.

A detailed description of the program that was used to derive the results presented in this book is found in Cole (1976). All of the data, listings, definitions of parameters, and driving variables needed to run the model are found in the appendices to that work. Conversion from SIMCOMP 3.0 to some other language would be a substantial effort.

1.6 The Structure of ELM

Figure 1.6 provides an overview of the entire model. The various submodels are easily identified in the figure and are described in detail in Chapters 2–8. A brief description of the submodels is presented here to provide the background needed to discuss a series of model experiments. The abiotic model contains two main submodels: a heat model and a water-flow model. Both simulate conditions in the canopy, litter, and soil profile. Input to this model includes site-characterization parameters (soil information, latitude, etc.) and data on air temperature, rainfall, and other variables (see Chap. 2).

The producer model simulates plant growth, reproduction, senescence, and phenology. Input to this model includes site-characterization parameters (plant types, growth characteristics, etc.) and such things as soil water and grazing loss as determined in other submodels (see Chap. 3).

The mammalian consumer submodel simulates the demography, bioenergetics, and diet selection of up to 10 classes of mammals. Input parameters determine the animal classes present, and simulated food available and temperature are among the variables that influence simulated mammalian dynamics (see Chap. 4).

The grasshopper model (Chap. 5) simulates demography and food intake by grasshoppers. Like the mammalian model, available food and temperature influence the dynamics that the model simulates.

The decomposer model subdivides material lost from other models, such as dead plant material and feces, into readily and slowly decomposed fractions. These fractions are decomposed under the control of simulated factors such as soil water and temperature (see Chap. 6).

In Chapters 7 and 8 the nutrient models for nitrogen and phosphorus are described. They track the cycling of these nutrients as functions of soil type, soil water, temperature, and other factors.

These models were developed using their own driving submodels. To construct ELM we had to replace the driving submodels by the appropriate part of the total model. This process involved changing parameters and some modification to submodel structure.

After these interfacing problems' were solved, runs were made (Table 1.3), which included comparisons with field experiments (experiments 1–6) and investigation of the model's properties (experiments 7–13). These results are discussed and analyzed in detail in Chapter 10. Here a general description of model output serves as an introduction to the more careful presentation of the model and its analysis.

In Table 1.3 each experiment is described under treatment description. A bit of elaboration is in order, however. Treatment D ("control") refers to the replicated ESA control treatments during 1972. In treatment E ("water added") water was sprayed on two ESA treatments at night sufficient to maintain a soil-water potential in excess of −0.8 bars. In treatment F ("nitrogen added") nitrogen fertilizer was added to maintain a difference of at least 50 kg/ha of mineral nitrogen between the nitrogen and control treatments. In treatment G ("water and N added") water and nitrogen fertilizer were added as described in treatments E and F. Initial conditions (January 1, 1972) for these experiments were estimated from conditions at the end of 1971 and on the initial sampling date of 1972. These conditions were the same for treatments D, E, and F but differed for treatment G (see Lauenroth and Sims, 1976).

Treatment 2 ("light grazing," experiment 5) was in a pasture located in the area indicated in Figure 1.4. Very little effect on the pasture is visible at this grazing intensity. Treatment 4 ("heavy grazing"; see Fig. 1.4) had a clear effect, although the "heavy" grazing term is relative. The pasture certainly did not appear to have been abused.

The next seven experiments in Table 1.3 have no field counterpart. In experiment 7, treatment D is simulated without any consumers at all—neither small mammals nor grasshoppers. In experiment 8, treatment D is simulated with 0.04 cows/ha for a year. In experiment 9, treatment D is simulated with 0.3 cows/ha for 140 days. Experiments 10 and 11 simulate changes in the environment as represented by 2°C increments and decrements, respectively, in the air temperature. In experiment 12, treatment D is simulated with 25% of the 1972 rainfall—each event reduced to 25% of the recorded value. In all of the experiments to this point the difference equations were solved with a time step of 2 (days). In experiment 13, treatment D is simulated with a time step of 1 (day). Experiments 1 and 13 should give similar results.

Table 1.3. Results of a series of model experiments with comparisons to data where available

Experiment number	Treatment heading	Treatment description	Water total (cm)	Net primary production (g dry wt/m²)	Gross primary production (g dry wt/m²)	Light[d] interception (cal/cm²)	Live[b,c] peak (g dry wt/m²)	Date	WSC[b,c] peak (g dry wt/m²)	Deviation (%)[a]	Date
1	Treatment D; control	1972 weather	27.1	707.0	1086.0	24,950.0	170.0 / 187.0	256	65.25 / 47.0	38% / 25%	258
2	Treatment E; water added	Water added to keep tension >−0.8 bars	88.0	3494.0	5151.0	50,643.0	620.0 / 388.0	286	382.0 / 175.7	62% / 45%	288
3	Treatment F; N added	N fertilizer applied in June	27.1	612.0	1009.0	23,978.0	161.0 / 290.0	258	43.0 / 44.8	27% / 15%	260
4	Treatment G; water and N added	E and F combined—different initial conditions	85.9	3190.0	4673.0	46,902.0	604.0 / 863.0	288	360.0 / 465.2	60% / 54%	288
5	Treatment 2; light grazing	1972 weather 0.082 cows/ha for 6 mo	27.1	715.0	1087.0	21,166.0	139.0 / 182.0	258	94.5 / 101.5	68% / 56%	258
6	Treatment 4; heavy grazing	0.2 cows/ha for 6 mo	27.1	823.0	1294.0	25,440.0	169.0 / 170.0	258	117.2 / 73.3	69% / 43%	258
7	Treatment D + no consumers	All consumers removed 1972 weather	27.1	711.0	1092.0	25,147.0	172.0	256	66.0	38%	258
8	Treatment D + light grazing all year	0.04 cows/ha for 365 days	27.1	689.0	1071.0	24,302.0	165.0	256	62.8	38%	258
9	Treatment D + short heavy grazing	0.3 cows/ha for 140 days	27.1	552.2	901.0	19,000.0	119.0	188	37.0	31%	188
10	Treatment D + heat	Temperature raised 2°C	27.1	649.0	1060.0	24,920.0	163.0	258	62.0	38%	260
11	Treatment D + cool	Temperature lowered 2°C	27.1	706.0	1031.0	24,452.0	170.0	254	59.5	35%	256
12	Treatment D + drought	Rainfall reduced to 25% of 1972 level	6.86	49.0	134.0	7,507.0	63.2	174	12.2	19%	174
13	Treatment D DT = 1	1972 weather	27.1	713.0	1090.0	25,084.0	170.0	256	66.5	39%	258

Table 1.3. (continued)

Experiment number	CSG[b,c] peak (g dry wt/m²)	Deviation (%)[a]	Date	Forb[b,c] peak (g dry wt/m²)	Deviation (%)[a]	Date	Shrub[b,c] peak (g dry wt/m²)	Deviation (%)[a]	Date	Cactus[b,c] peak (g dry wt/m²)	Deviation (%)[a]	Date	Date	CO₂[e] evolved (g)	Mammal producer	Grasshopper producer	Secondary producer
1	4.3	3%	254	12.9	8%	256	61.5	36%	254	26.8	16%	254	188	898.9	0.0245	0.0199	0.0444
	14.5	8%		20.0	11%		70.3[f]	38%		35.4[f]	19%						
2	20.7	3%	286	42.0	7%	288	145.8	24%	280	31.9	5%	280	272	1,591.0	0.025	0.029	0.054
	21.9	6%		41.3	11%		101.9[f]	26%		41.9[f]	11%						
3	10.3	6%	254	33.0	20%	256	38.2	24%	256	37.2	23%	256	262	1,034.0	0.025	0.021	0.046
	44.5	15%		69.0	24%		60.6[f]	21%		70.9[f]	24%						
4	20.4	3%	288	26.2	4%	288	150.2	25%	280	50.6	8%	280	181	2,130.0	0.025	0.025	0.050
	52.6	6%		68.0	8%		246.6[f]	29%		30.6[f]	4%						
5	2.5	2%	254	6.43	5%	256	23.6	17%	254	12.6	9%	254	150	953.0	0.093	0.019	0.112
	10.4	6%		14.7	8%		38.6[f]	21%		16.7[f]	9%						
6	4.8	3%	174	9.6	6%	256	2.1	1%	256	35.8	21%	256	198	902.0	0.201	0.018	0.219
	16.6	10%		13.6	8%		1.02[f]	0%		65.4[f]	38%						
7	4.9	3%	254	13.2	8%	256	61.2	36%	254	26.8	16%	254	188	891.0	0.000	0.000	0.000
8	3.5	2%	252	11.6	7%	188	61.5	37%	254	26.8	16%	254	188	906.0	-0.0337	0.0197	-0.014
9	3.2	3%	174	9.0	8%	144	58.8	49%	252	26.8	22%	252	188	953.0	0.260	0.020	0.280
10	7.8	2%	188	9.0	6%	146	67.0	41%	258	26.8	16%	258	188	1,016.0	0.025	0.000	0.025
11	6.4	4%	252	15.6	9%	254	62.0	36%	252	27.0	16%	252	262	774.0	0.0212	0.0010	0.022
12	1.41	2%	172	6.4	10%	174	16.8	27%	174	26.5	42%	174	134	543.0	0.0254	0.0256	0.051
13	4.2	2%	253	12.4	7%	187	60.2	35%	256	26.8	16%	256	146	862.0	0.020	0.029	0.049

[a] Percentage of live peak (see text).
[b] Where multiple entries occur, model values are given above and field-determined means are given below.
[c] Model computes g C (grams of carbon), whereas g dry weight are quoted; g dry wt = g C · 2.5.
[d] Total available for all runs is 112,552.0 cal/cm².
[e] Model computes gC, whereas g CO₂ are quoted: g CO₂ = g C · 3.67.
[f] Averages of samples—through the growing season.

1.7 Questions Reviewed

The first question addresses variations in the level and type of herbivory. These effects are best seen in experiments 1 and 5–9 of Table 1.3. Of these, experiments 1, 5, and 6 corresponded to field experimental situations, and observed values are tabulated with model output. There are two principal differences among these three experiments: (a) In experiment 1 no domestic herbivores were included, but in 5 and 6 cattle were present and (b) initital conditions varied with each site. Net and gross primary production were essentially the same; in 1 and 5, however, the grazed systems made better use of the light that they received. Secondary production was higher under the grazed system, and the slight shift from grasshoppers to mammals is expected.

Experiments 7–9 simulated the absence of mammalian consumers, perennial grazing, and short-season, intensive grazing, respectively. Leaving cattle on all year long would depress mammalian consumer production because of inadequate food supplies in the winter. The improved (over experiment 6) secondary production of a short intensive grazing season in experiment 9 had been hypothesized by local range managers. The depression of net primary production (experiments 8 vs. 7) for light grazing is slight. This is explained by light grazing early in the growing season reducing the photosynthetic machinery, an effect that is compounded over the year. Evidence for this is provided by the light-intercepted column of Table 1.3. Likewise, evidence for growth stimulation by grazing in native grasslands is difficult to muster (Jameson, 1963).

Variations in precipitation and applied moisture are studied in experiments 1, 2, and 12. Experiments 1 and 2 correspond to ESA plots, but experiment 12 does not. While the direction of the response of the system is correct, the simulated response of warm season grass to large water additions is excessive. A possible explanation of this problem, centered on representation of phenology, is discussed in Chapters 3 and 10. The simulated drought is severe (25% of 1972 precipitation). The direction of the response of each variable is appropriate, but the magnitudes are questioned.

Variations in temperature are studied in experiments 1, 10, and 11. There were no field data to compare with experiments 10 and 11. Net and gross primary productivity are reduced relative to control (experiment 1) in both treatments. Many of the modeled mechanisms are temperature-dependent, and temperature-dependence is one of the most thoroughly investigated variables. Slightly lower temperatures produced little response, but higher temperatures invoked sharper reactions. The system is water limited, and higher temperatures resulted in greater system stress. On the other hand, slightly cooler temperatures did not significantly improve the water economy. Decomposers are more active at the higher temperatures. Mammalian consumers fared slightly better at the higher temperatures, and the grasshopper results indicate high sensitivity of that submodel to temperature.

Variations in applied nitrogen are treated in experiments 1 and 3 and 2 and 4. Field data are available in each case, and it is clear that the system's response to nitrogen additions is inappropriate. Nitrogen and phosphorus effects in the producer system are poorly known.

The second question "How is the carrying capacity of the grassland affected by these perturbations?" is answered in part by the discussion of the first question. Table 1.3 indicates proper direction of response of the system to all but the fertilization experiments. From these studies certain changes in domestic cattle-carrying capacity of the system under limited management alternatives are inferred. No simulations were conducted of grazing fertilized systems because the individual effects of fertilization were inadequately represented.

Question (3) is answered by the observation that the direction and magnitude of model response is consistent with field observation save for the fertilization treatments. The model is not as fully integrated as we would like; insufficient interfacing activity interations were made. The wide confidence bands seen in the subsequent chapters apply also to initial conditions. Destructive sampling combined with the biological heterogeneity means that each plot from each sample date is from a slightly (or greatly) different system. Logical and coding errors, as well as biological misconceptions, are known to remain in the system. As a focal point for ecosystem studies, such a model is without peer. It leaves much to be desired as a managerial tool and is not the final product of the US/IBP Grassland Biome.

The fourth question is addressed in Table 1.3, wherein computed values are compared to the live peak value. The comparison is not entirely appropriate because: (a) the peak of a given producer category did not always occur at the time of the peak of total live material and (b) shrub and cactus observed values are averages of samples taken throughout the sampling season.

Drought (experiment 12) shifts dominance from warm season grass to cactus. Nitrogen amendment (experiment 3) equalizes the contributions among producers. Water and grazing sharpens the dominance of the warm season grasses, except in experiment 9, where shrubs are favored over the warm season grasses. Heating and cooling (experiments 10 and 11) show little effect on the dominance relations of the producers. It is noteworthy that in experiments 2 and 4 the relative contributions to the peak are fairly well represented.

Experiments 1 and 13 are the same, except that DT, the time step for the solution of the difference equations, was set to 1 rather than 2 (days). The effect was nominal save for the consumers where 20% to 50% changes were recorded.

In experiment 13 the concepts of verification, validation, and sensitivity analysis meet. Changing DT is a legitimate step for verification (it must be determined whether the mechanisms were truly represented with DT a legitimate parameter), validation (confidence is increased by the similarity of the results of experiments 1 and 13), and sensitivity analysis (a 50% parameter reduction produces a certain amount of system response). It was a desideratum of ELM that DT should be able to assume the values 1, 2, . . . , 7. For the range of DT, 1 ≤ DT ≤ 4, the model meets this design goal except for the consumers.

1.8 Additional Trials

The study of the questions provides an impression of the model's characteristics. Other experiments were used to probe further.

1.8.1 Tex's Hypothesis

J. K. (Tex) Lewis (personal communication) of South Dakota State University proposed that certain prairie ecosystems cannot be seriously damaged by heavy domestic cattle grazing during the second year of two successive good years. To test this hypothesis, ELM was used for the Pawnee Site by running the 1972 weather pattern for two consecutive years. That year was a good one, with 27.1-cm rainfall.

Two simulations were run: (a) the control with no grazing by domestic herbivores and (b) the treatment with light grazing the first year (0.088 animals/ha for the 6-mo grazing season) and heavy grazing the second year (initially 0.316 animals/ha for the 6-mo grazing season). The 6-mo season runs from late April to late October and overlaps the growing season.

Animal weights increased from the initial value of 25 kg of carbon in late April to a final weight of 34 kg in late October. During the second year the animals gained only 2–3 kg in the first 70 days (as compared to 6 kg/animal for the same period the first year). By this time the previous year's dead material was exhausted, and a significant weight loss occurred. During the period from early June to late August there was a continual loss of cattle. By late August the cattle were reduced to the point that remaining animals survived and weight gains were observed. While this grazing strategy might be unreasonable, it tests the hypothesis. The ranch manager would remove a fraction of the stock before letting them starve. Supplemental feeding could maintain herd condition.

The hypothesis, however, addresses the impact of cattle on the plants. First, the impact on the total live shoot biomass is shown in Figure 1.7. Comparison of the control to the treatment during the first year indicates little influence. In the second year growth was delayed almost 3 mo, during which time the cattle numbers were reduced. The flush of growth in the late summer and fall was due to a greater soil-water store that was used early in the control. Both treatment and control returned to 40 g/m² by the end of the second year.

Standing dead started the heavy grazing period at a lower level in the treatment because of grazing effects from the previous year. Grazing during the second year delayed the accumulation of standing dead material, but by the end of the trial there was little difference in these levels.

Crown growth was delayed but returned to a level comparable with the control by the end of the trial. Root dynamics were phased differently in the treatment, but the final value was comparable with that of the control.

In conclusion, if the above experiments are accepted as a test of the hypothesis, we cannot reject Tex's proposition using this model.

1.8.2 Decomposers

Table 1.3 shows that the decomposers responded (CO_2 evolved) to the nitrogen amendment. Nitrogen allowed an 11% increase in CO_2 evolution (experiments 1 and 3) and in the presence of adequate water, a 31% increase (experiments 2 and 4).

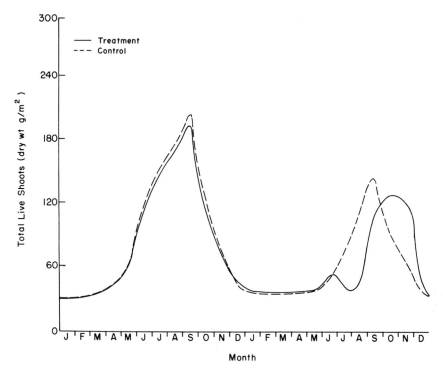

Fig. 1.7.   Simulated test of Tex's hypothesis. Total live shoots for the two years are shown for both the control and treatment (see text)

In Figure 1.8a litter dynamics are presented for the four litter layers (soil surface, 0–4 cm, 4–15 cm, and 15–60 cm) for the control (experiment 1). In Figure 1.8(b) the same variables are shown for the nitrogen-amended plot (experiment 3). The dynamics of the surface and deep layers are similar (note the difference in surface litter initial condition). The litter values for top and middle soil layers are affected by the nitrogen. The surface layer is not affected directly because decomposition of surface litter is assumed independent of soil nitrogen and the fertilizer was applied in the soil. The bottom layer is unaffected because nitrogen is added only to the top three soil nutrient layers, and movement between layers is not incorporated in the model. Each graph shows obvious limits to the decomposition rate in the top two soil layers from late June to late July. Soil-water graphs indicate that water is limiting at these depths and times. The shallower layer (0–4 cm) benefited from light rains. Thus by the year end, comparing Fig. 1.8a to 1.8b, litter of 0–4 cm was reduced by about 17%, whereas litter of 4–15 cm was reduced by 33%.

Similar situations obtain when water is added, as shown in Figure 1.8(c,d). Figure 1.8(c) depicts the effect of decomposition and growth where water limits neither. In Figure 1.8(d) the influence of the nitrogen amendment on decomposition is seen. Initial conditions were different for the two treatments, but the

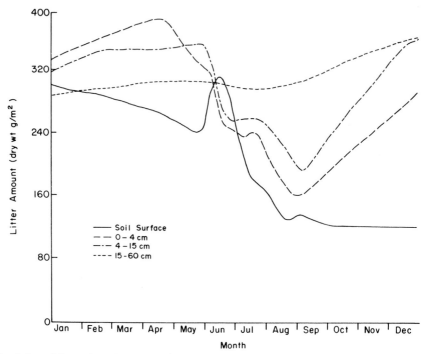

Fig. 1.8a.   Litter dynamics in the control (treatment D). Soil surface, 0–4 cm, 5–15 cm, and 15–60 cm litter densities are illustrated

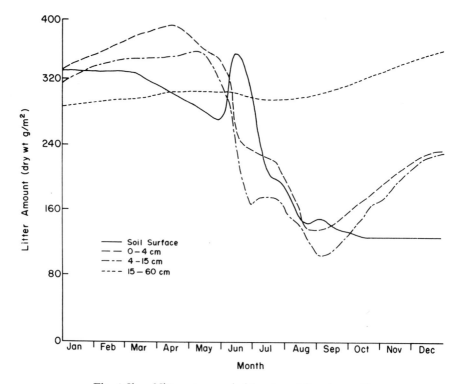

Fig. 1.8b.   Nitrogen-amended treatment (treatment F)

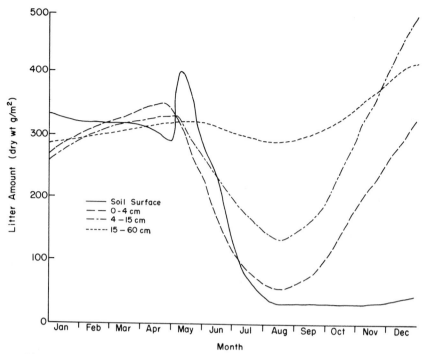

Fig. 1.8c.   Water-amended treatment (treatment E)

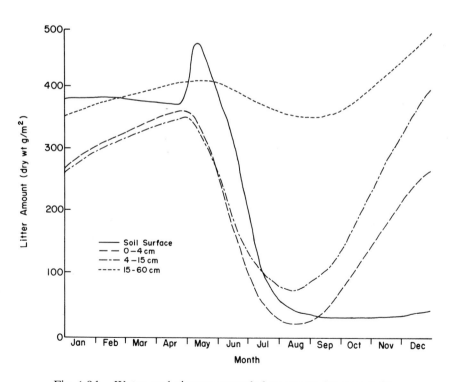

Fig. 1.8d.   Water- and nitrogen-amended treatment (treatment G)

effects of nitrogen on the 0–4 and 4–15-cm levels are clear. The lows to which the litter in the upper (0–4 cm) and middle (4–15 cm) soil-layer drops are much deeper in Figure 1.8(d).

1.9 Conclusion

In Chapters 2–8 detailed descriptions of the submodels are provided by the investigators who developed them. In Chapter 9 a sensitivity analysis is described, and in Chapter 10 the entire effort is critiqued by one of the participants.

Acknowledgments. Thanks are due Drs. M. I. Dyer, N. R. French, and H. W. Hunt for providing some of the material in the figures and tables.

This chapter reports on work supported in part by National Science Foundation Grants GB-41233X and BMS73-02027 A02 to the Grassland Biome, U.S. International Biological Program, for "Analysis of Structure, Function, and Utilization of Grassland Ecosystems."

References

Anway, J. C., Brittain, E. G., Hunt, H. W., Innis, G. S., Parton, W. J., Rodell, C. F., Sauer, R. H.: ELM: Version 1.0. US/IBP Grassland Biome Tech. Rep. No. 156. Fort Collins: Colorado State Univ., 1972, 285 pp.

Bledsoe, L. J., Francis, R. C., Swartzman, G. L., Gustafson, J. D.: PWNEE: A grassland ecosystem model. US/IBP Grassland Biome Tech. Rep. No. 64. Fort Collins: Colorado State Univ., 1971, 179 pp.

Cole, George W. (ed.). ELM: Version 2.0. Range Science Department Science Series No. 20, Colorado State University, Ft. Collins, Colorado, April 1976. 663 pp.

Innis, G. S.: The role of total systems models in the Grassland Biome study. In: Systems Analysis and Simulation in Ecology, B. C. Patten (ed.). New York: Academic Press, 1975, Vol. III. pp 13–47.

Jameson, D. A.: Responses of individual plants to harvesting. Bot. Rev. **29**, 532–594, 1963.

Jameson, D. A. (coordinator): General description of the Pawnee Site. US/IBP Grassland Biome Tech. Rep. No. 1. Fort Collins: Colorado State Univ., 1969, 32 pp.

Klipple, G. E., Costello, D. F.: Vegetation and cattle responses to different intensities of grazing on short-grass ranges of the Central Great Plains. USDA Tech. Bull. 1216. 1960, 82 pp.

Lauenroth, W. K., Sims, P. L.: Effects of water and nitrogen stresses on a shortgrass prairie ecosystem. US/IBP Grassland Biome Tech. Rep. No. 232. Fort Collins: Colorado State Univ., 1973, 117 pp.

Lauenroth, W. K., Sims, P.L.: Evapotranspiration from a shortgrass prairie subjected to water and nitrogen treatments. Wat. Resources Res. **12,** 437–442, 1976.

Patten, B. C.: A simulation of the shortgrass prairie ecosystem. Simulation **19,** 177–186, 1972.

Smith, F. M., Striffler, W. D.: Pawnee Site microwatersheds: Selection, description, and instrumentation. US/IBP Grassland Biome Tech. Rep. No. 5. Fort Collins: Colorado State Univ., 1969, 29 pp.

Woodmansee, R. G: Soils of North American grasslands. In: Grassland Ecosystems of North America. J. E. Ellis, (ed.). Stroudsburg, Pennsylvania: Dowden, Hutchinson, and Ross, In press, 1977.

Appendix 1.A.　SIMCOMP 3.0

Detailed discussions of versions of SIMCOMP are available (Gustafson and Innis, 1972, 1973a, 1973b; Stevens and Gustafson, 1973). This appendix provides a brief description of SIMCOMP 3.0 and illustrates its use.

The Forrester (1961) symbolism and philosophy provided a starting point for development of SIMCOMP. The flow orientation (of carbon, nitrogen, phosphorus, water, etc. or heat, energy, etc.) was compatible with the intuitive view of hydrologists, botanists, zoologists, agronomists, and others regarding ecologic systems. The system is event oriented with events scheduled every time step for "continuous" simulation and at arbitrary intervals for "discrete" simulations.

1.A.1 Minimum Specification of a Dynamic System

Consider a set of first-order ordinary forward difference equations as

$$\mathbf{x}_{t + \delta t} = \mathbf{x}_t + \delta t \cdot \mathbf{f}(\mathbf{X}_t, \mathbf{Z}_t, \mathbf{P}, t, \delta t) \tag{1.A.1}$$

where $\mathbf{X}$, $\mathbf{Z}$, and $\mathbf{P}$ are vectors of state variables, driving variables, and parameters, respectively; t is time, δt is time increment, and $\mathbf{f}$ is a vector function of rates. Then a minimal set of user-provided information consists of: (a) initial conditions (values of the state variables at time t_0, the time of the beginning of the simulation), (b) flows ($\mathbf{f}$), (c) driving variables ($\mathbf{Z}$), (d) parameters ($\mathbf{P}$), (e) the time increment (δt), and (f) the times of the beginning and end of the simulation (t_0 and t_1, respectively). Equation (1.A.1) is derivable from the flows and hence redundant.

Consider the Lotka–Volterra system (Lotka, 1925) as the example. In difference form, equations for this system are:

$$H_{t + \delta t} = H_t + \delta t \cdot (a_1 - b_1 \cdot P_t)$$
$$P_{t + \delta t} = P_t + \delta t \cdot (-a_2 + b_2 \cdot H_t)$$

where H is the host population numbers; P is the parasite population numbers; the a and b are constants.

A Forrester diagram of this system is shown in Figure 1.A.1. Unique numbers from 1 to 999 are assigned to each state variable, source, and sink. These indexed compartments are the components of the $\mathbf{X}$ vector in Eq. (1.A.1).

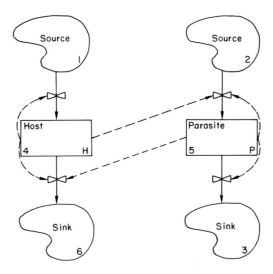

Fig. 1.A.1   A Forrester diagram of a simple Lotka–Volterra system (the difference equations govern these dynamics)

Let t_0 be 0, $H_0 = X_0(4) = 25$, and $P_0 = X_0(5) = 10$. Initial conditions for sources and sinks must be specified, but are arbitrary. Thus $X_0(1) = 100$, $X_0(2) = 100$, $X_0(3) = 100$, and $X_0(6) = 100$. Input values of X are initial conditions.

Flows in SIMCOMP 3.0 are specified in the directive defined by placing the "from" and "to" indices in the Forrester diagram within parentheses separated by a dash and followed by a period. Thus the directive for host-population gain is (1-4) (see below).

Following the directive, any block of FORTRAN IV code, including logical operations, subroutine calls, and so on, may be used subject to the following restrictions:

1. The variable FLOW is set somewhere within this block of code. If FLOW is set several times, the final one determines the rate for that flow for the next time step.

2. The FORTRAN statement RETURN should not be used since all flows are combined into a single subroutine called by the system.

For the Lotka–Volterra example the host-population gains are written:

$$(1\text{-}4).\ \text{FLOW} = A1 * X(4)$$

Host-population losses are written:

$$(4\text{-}6).\ \text{FLOW} = B; * X(4) * X(5)$$

Parasite-population gains are written:

$$(2\text{-}5).\ \text{FLOW} = B2 * X(4) * X(5)$$

Parasite-population losses may, for illustrative purposes, be written:

$$(5\text{-}3).\ \text{TEMP} = \text{ATAN}\ [X(5) * X(4) + 1]$$
$$\text{IF (TEMP. GT. FLOW) 2, 3}$$
$$2\ \text{TEMP} = 0$$
$$\text{GO TO 4}$$
$$3\ \text{TEMP} = 1$$
$$4\ \text{FLOW} = \text{A2} * X(5)$$

The system scans the directives used to describe the flows and constructs a scalar difference equation [one component of Eq. (1.A.1)] for each counting number found in a flow description. For this example the system would establish six difference equations—one for each numbered compartment in Figure 1.A.1. The resulting equations are as follows:

$$X(1) = X(1) - \text{A1} * X(4) * \text{DT}$$
$$X(2) = X(2) - \text{B2} * X(4) * X(5) * \text{DT}$$
$$X(3) = X(3) + \text{A2} * X(5) * \text{DT}$$
$$X(4) = X(4) + \text{A1} * X(4) * \text{DT} - \text{B1} * X(4) * X(5) * \text{DT} \qquad (1.A.2)$$
$$X(5) = X(5) + \text{B2} * X(4) * X(5) * \text{DT} - \text{A2} * X(5) * \text{DT}$$
$$X(6) = X(6) + \text{B1} * X(4) * X(5) * \text{DT}$$

The conservative nature of the system is evident from Eqs. (1.A.2), where each flow term appears in two equations but with opposite signs.

There are no driving variables in this example. The chapters that follow describe several ways in which driving variables may be used in the system.

User-defined parameters (as opposed to temporary variables) must be identified to SIMCOMP. For this purpose the directive STORAGE is provided. Any user-defined parameter or variable must be declared in storage if it is to be read in the data section, printed, plotted, or automatically incorporated into each subroutine (see other features below). Thus cards of the form

$$\text{STORAGE. var}_1, \text{var}_2, \ldots, \text{var}_n$$

with A1, B1, A2, and B2 as the variables (var_i) would need to appear in the source deck to allow the data card below to be read and interpreted in SIMCOMP.

Data are entered, beginning in any column with items separated by dollar signs:

$$\text{A1} = 1.0\ \$\ \text{B1} = 0.1\ \$\ \text{A2} = 0.5\ \$\ \text{B2} = 0.02\ \$$$

The time of the run is determined by two items: (a) defined user-set variables TSTRT (t_0) and (b) TEND (t_1). The time step DT (δt) is also system-defined and user-set. These values are entered as data, but need not be put in storage as they are defined by the system.

This is all that is required to set up a run. All of this (save STORAGE) would be required to describe the dynamics of the system. Generally, much more is both desired and needed.

1.A.2 Other Features

SIMCOMP 3.0 has a large number of features, default capabilities, and debug facilities. Only those useful in reading the chapters that follow are described.

Subroutines and Events. In Figure 1.A.2 the system-called, user-provided subroutines START, CYCL1, CYCL2, and FINIS are shown in crosshatch. As indicated in Figure 1.A.2, both START and FINIS are called only once in each simulation, whereas CYCL1 and CYCL2 are called once for each time step.

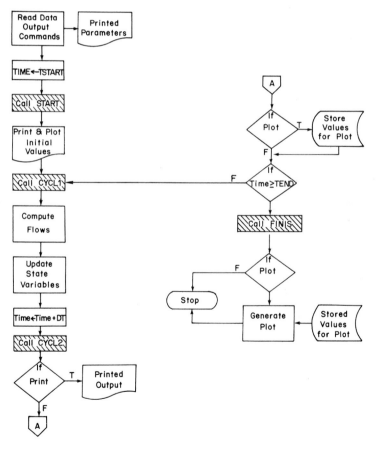

Fig. 1.A.2 Processing sequence for SIMCOMP 3.0 showing the user-provided subroutines in crosshatch; default routines are provided by the system if needed

User-defined subroutines may be coded, placed in the source deck, and called as needed. This is identical to FORTRAN in logic, organization, and execution.

The SIMCOMP system automatically places all STORAGE variables as well as the state variables and time variables in COMMON blocks, which are inserted at the beginning of all subroutines, whether system-defined or user-defined. This feature was incorporated because of the difficulty of developing and maintaining a large number of supposedly identical common blocks in a collection of programs and subroutines.

Indexed Flows. Similar structures that vary in parameter values are common. The Lotka–Volterra system in which many pairs of host/parasite systems were operating simultaneously would be an example. It becomes laborious to write each of the flows needed to describe the system. Thus the directive for specifying the flows has been expanded (after the pattern of the FORTRAN-implied DO loop) to include variable indices. The forms for this directive are:

$$(n - I = n_1, n_2) \text{ or } (I = n_1, n_2 - n)$$
$$(n - I = n_1, n_2, n_3) \text{ or } (I = n_1, n_2, n_3 - n)$$
$$(I = n_1, n_2, - J = m_1 * I \pm m_2) \text{ or}$$
$$(I = n_1, n_2, n_3 - J = m_1 * I \pm m_2) \text{ or } (I = n_1, n_2 - J = m_1, m_2)$$

In the first form the flow is from compartment n to compartment I for interger values of I between n_1 and n_2, inclusive. The second form merely provides for flows in the opposite direction to that of the first form. In the third form, I assumes values $n_1, n_1 + n_3, n_1 + 2 * n_3, \ldots, n_1 + h * n_3$, where $n_1 + h * n_3 \leqslant n_2$ and $n_1 + (h + 1) * n_3 > n_2$. The fourth form provides for the reverse of the flows in the third. The fifth form allows both indices to be computed albeit in a simple manner. As I takes the values $n_1, n_1 + 1, \ldots, n_2$, J assumes the values $m_1 * n_1 \pm m_2, m_1 * (n_1 + 1) \pm m_2, \ldots, m_1 * n_2 \pm m_2$, respectively. In the sixth form the variable I is incremented by n_3 as it changes from n_1. Finally, in the seventh form I and J vary from n_1 to n_2 and m_1 and m_2, respectively, independently of one another. The third variable (for increment) may be used in specifying either or both indices.

As examples the first form might be used to describe the losses suffered by grass $[X(n)]$ as a result of consumption by herbivores $[X(I), I = n_1, n_2, 1]$ whereas the second form might be used to describe the diet selection of herbivore $[X(n)]$ from available feeds $(X(I), I = n_1, n_2, 1)$.

Output. Printouts, printer-plots, and microfilm plots are available from SIM-COMP 3.0. Any system variable or STORAGE. variable may be displayed with the SIMCOMP directive PRINT. $vbl_1, vbl_2, \ldots, vbl_n$ where vbl is the variable of interest. All state variables may be printed by issuing the directive PRINT.. The frequency with which such information is printed is controlled by the user specified variable DTPR.

The plotting of results is invoked by issuing the directive

PLOT. $(vb1_1),(vb1_2),(vb1_2)$ or PLOT. $(vb1_1, vb1_2, vb1_3)$

(where vb1 is the variable of interest) or combinations of these. Each pair of parentheses identifies a different scale for the variable contained therein. Thus, in the first case three variables would be plotted against the independent variable (time), and each would be plotted to its own scale (chosen to provide an essentially full-scale display of each variable). In the second case each variable is plotted against time, but only one scale is used.

Another plot feature allows for the plotting of one dependent variable against another. The SIMCOMP directive reads

PLOT. $(vb1_1)/vb1_2$

and results in variable 1 being plotted as the ordinate and variable 2 serving as the abscissa.

In any case, the frequency of plotted points is under the control of the variable DTPL, which if not provided by the user is set by the system to fill the graph (if possible) without using extra storage.

1.A.3 The Lotka–Volterra Example: SIMCOMP 3.0 Deck and Model Run

In this section the Lotka–Volterra example is presented. Model output in printed and plotted forms are shown.

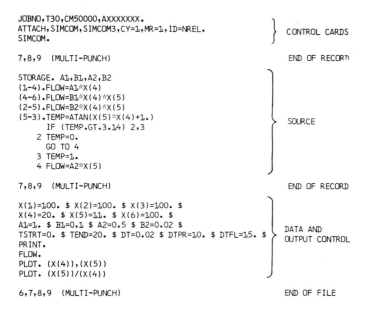

```
JOBNO,T30,CM50000,AXXXXXXX.
ATTACH,SIMCOM,SIMCOM3,CY=1,MR=1,ID=NREL.      }  CONTROL CARDS
SIMCOM.

7,8,9  (MULTI-PUNCH)                             END OF RECORD

STORAGE. A1,B1,A2,B2
(1-4).FLOW=A1*X(4)
(4-6).FLOW=B1*X(4)*X(5)
(2-5).FLOW=B2*X(4)*X(5)
(5-3).TEMP=ATAN(X(5)*X(4)+1.)
      IF (TEMP.GT.3.14) 2,3                       SOURCE
    2 TEMP=0.
      GO TO 4
    3 TEMP=1.
    4 FLOW=A2*X(5)

7,8,9  (MULTI-PUNCH)                             END OF RECORD

X(1)=100. $ X(2)=100. $ X(3)=100. $
X(4)=20. $ X(5)=11. $ X(6)=100. $
A1=1. $ B1=0.1 $ A2=0.5 $ B2=0.02 $              DATA AND
TSTRT=0. $ TEND=20. $ DT=0.02 $ DTPR=10. $ DTFL=15. $   OUTPUT CONTROL
PRINT.
FLOW.
PLOT. (X(4)),(X(5))
PLOT. (X(5))/(X(4))

6,7,8,9  (MULTI-PUNCH)                           END OF FILE
```

Listing of SIMCOMP deck for Lotka–Volterra example

```
SIMCOMP VERSION 3.0          PARAMETER VALUES

                              - SIMULATION CONTROL PARAMETERS -

                             TIME  = NOT INITIALIZED
                             TSTRT =          0
                             TEND  =  20.0000000
                             DT    =  .200000000E-01
                             DTPR  =  10.0000000
                             DTPL  = NOT INITIALIZED
                             DTFL  =  15.0000000

                               - STATE VARIABLES -

       X( 1- 3) =   100.000000      X( 4) =   20.0000000     X( 5) =  11.0000000      X( 6) =  100.000000

                        - PRIMARY USER DEFINED VARIABLES -

         A1 =  1.00000000         A2 =  .500000000         B1 =  .100000000      B2 =  .200000000E-01

  *****DTPL .LE.  0  OR UNDEFINED AND WILL BE GIVEN THE VALUE, DTPL = .200000000



SIMULATION RESULTS

TIME = 0.
       X(1) =   100.000000        X(4) =   20.0000000        X(6) =   100.000000        X(2) =   100.000000
       X(5) =   11.0000000        X(3) =   100.000000

VALUES OF FLOWS. TIME = 0.              TO  .200000000E-01
       FLOW( 1,  4) =  20.0000000   FLOW( 4,  6) =  22.0000000    FLOW( 2,  5) =  4.40000000   FLOW( 5,  3) =  5.50000000
TIME =  10.0000000
       X(1) =  -144.289405        X(4) =   18.9547173         X(6) =   345.334687        X(2) =   50.9330625
       X(5) =   9.70633553        X(3) =   150.360602

VALUES OF FLOWS. TIME =  15.0000000       TO  15.0200000
       FLOW( 1,  4) =  31.6954180   FLOW( 4,  6) =  34.4536474   FLOW( 2,  5) =  6.89072949   FLOW( 5,  3) =  5.43511486
TIME =  20.0000000
       X(1) =  -388.258057        X(4) =   20.5757570         X(6) =   587.682300        X(2) =   2.45343996
       X(5) =   8.59499526        X(3) =   199.941465
```

SIMCOMP Version 3.0 output for Lotka–Volterra example showing printed output format

PLOT NO. 1

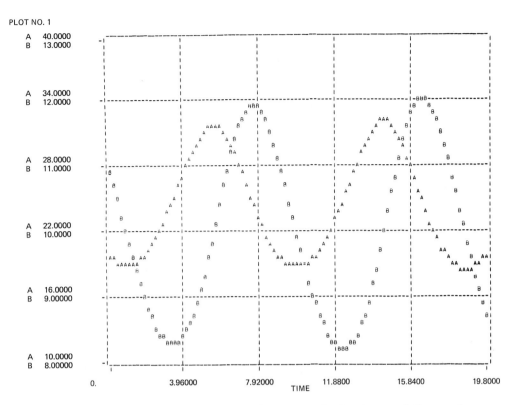

SIMCOMP Version 3.0 output for Lotka–Volterra example showing population numbers
plotted against time

PLOT NO. 2

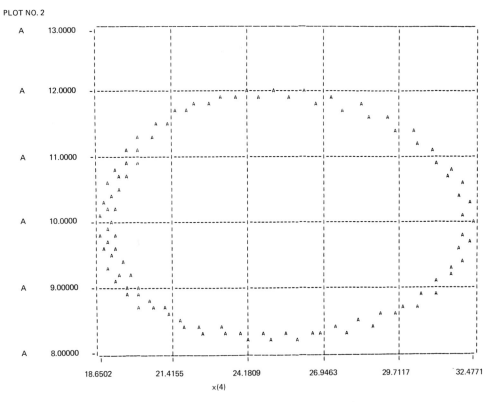

SIMCOMP Version 3.0 output for the Lotka–Volterra example: parasite population A is plotted against host population

References

Forrester, J. W.: Industrial Dynamics, Cambridge, Mass.: MIT Press, 1961, 464 pp.

Gustafson, J. D., Innis, G. S.: SIMCOMP Version 2.0 user's manual. US/IBP Grassland Biome Tech. Rep. No. 138. Fort Collins: Colorado State Univ., 1972, 62 pp.

Gustafson, J. D., Innis, G. S.: SIMCOMP Version 2.1 user's manual and maintenance document. US/IBP Grassland Biome Tech. Rep. No. 217. Fort Collins: Colorado State Univ., 1973a, 96 pp.

Gustafson, J. D., Innis, G. S.: SIMCOMP Version 3.0 user's manual. US/IBP Grassland Biome Tech. Rep. No. 218. Fort Collins: Colorado State Univ., 1973b, 149 pp.

Lotka, A. J.: Elements of Physical Biology. Baltimore: Williams and Wilkins, 1925, 460 pp. (Reprinted 1956 as: Elements of Mathematical Biology. Dover Publications)

Stevens, K. L., Gustafson, J. D.: SIMCOMP Version 3.0 maintenance document. US/IBP Grassland Biome Tech. Rep. No. 219, Fort Collins: Colorado State Univ., 1973

2. Abiotic Section of ELM

WILLIAM J. PARTON

Abstract

A simulation model for abiotic variables influencing grassland ecosystems including water flow and temperature profile submodels is presented. The water-flow submodel treats flow in the plant canopy and soil, while the temperature submodel includes solar radiation, canopy air temperature, and soil temperature. The atmospheric driving variables are either daily weather observations or stochastic weather-simulator results. A preliminary validation of the model has been performed.

2.1 Introduction

This model is structured to simulate soil and canopy abiotic variables needed to achieve the ELM objectives and is presented in two parts: (a) a water-flow submodel and (b) a temperature-profile submodel. The water-flow submodel simulates the flow of water through the plant canopy and soil layers. Allocation of rainfall and evapotranspiration of water are the important processes considered. The structure of this model is similar to that of other hydrologic models (Cartmill, 1970; Kohler and Richards, 1962; Palmer, 1965; Crawford and Linsley, 1966; Nimah and Hanks, 1973; Goldstein and Mankin, 1972; Cowan, 1965) but handles an arbitrary number of soil layers where depth and soil type are specified in each. The water-flow model is structured to provide feedback mechanisms between biotic and abiotic state variables.

The temperature-profile submodel simulates daily solar radiation, maximum canopy air temperature, and soil temperature at 13 points in the soil profile. Soil temperature is computed from a modified finite difference solution for the one-dimensional Fourier heat equation (Fluker, 1958; Penrod et al., 1960; Langbein, 1949) in which the temperatures at the upper (0 cm) and lower (180 cm) boundaries are specified. The temperature at the upper boundary is calculated as a function of air temperature at 2 m, potential evapotranspiration rate, and standing-crop biomass, whereas monthly average temperatures at 180 cm are used for the lower boundary. Solar radiation is simulated as a function of cloud cover and time of year, using equations presented by Sellers (1965) and Haurwitz (1941), whereas maximum canopy air temperature is a function of solar radiation and maximum air temperature at 2 m. This model differs from canopy heat-flow

Table 2.1.   List of symbols used in this chapter (*, input parameter required by model; ND, nondimensional)

Symbol	Definition
A_e	Fraction of evapotranspiration water loss due to bare soil evaporation (ND)
A_i	Soil water content of ith layer at field capacity (cm of water)·
A_{ij}	Probability for the occurrence or nonoccurrence of rainfall ($j = 1, j = 2$, respectively) given the occurrence or nonoccurrence of rainfall ($i = 1, i = 2$, respectively) at time $t - \Delta t$.
A_1	Leaf-area index of live aboveground biomass (ND)
A_t	Fraction of evapotranspiration water loss due to transpiration (ND)
B_i	Evaporative soil water loss from ith layer (cm/day)
B_1	Live aboveground biomass (g/m²)
C_a	Average daily fractional cloud cover (ND)
C_d	Average daytime canopy air temperature (°C)
*C_i	Volumetric soil-water content at field capacity for ith layer without any dead-root biomass present (%)
C_1	Live aboveground cactus biomass (g/m²)
C_m	Maximum canopy air temperature (°C)
C_s	Soil thermal conductivity (cal · cm⁻¹ · s⁻¹ · °C⁻¹)
D	Constant used in Eq. (2.8) (0.033 cm m²/g)
D_a	Air density (g/cm³)
D_d	Depth of air column used for dew formation (cm)
*D_i	Depth of ith soil-water layer (cm)
E_p	Potential evapotranspiration rate (cm/day)
E_s	Potential bare-soil evaporation rate (cm/day)
E_S^T	Total bare-soil evaporation water-loss rate (cm/day)
E_t	Potential transpiration rate (cm/day)
E_t^T	Total transpiration water-loss rate (cm/day)
E_z	Actual evapotranspiration water loss (cm/day)
F_i	Volumetric soil-water content at field capacity for ith layer (%)
*G_i	Bare-soil water-absorption coefficient for ith layer (ND)
G_1	Live aboveground grass biomass (g/m²)
*H_i	Transpiration water-absorption coefficient for ith layer (ND)
I_1	Litter interception (cm/day)
I_s	Standing-crop interception (cm/day)
K_i	Transpiration from ith soil-water layer (cm/day)
L	Litter biomass (g/m²)
L_a	Leaf-area index of standing-crop biomass (ND)
M_i	Soil-water content of ith layer (cm of water)
M_i^1	Soil-water content of ith layer at time $t - 24$ h (cm of water)
*N	Total number of soil-water layers (ND)
*N_a	Number of soil water layers above 4 cm (ND)
N_p	Number of points where observed and simulated data are compared (ND)
NR	No rain (ND)
P	Probability that the statistic has a more extreme value given the H_0 (NOD)
P_a	Calculated parameter [Eq. (2.3); ND]
P_b	Calculated parameter [Eq. (2.4); ND]
P_c	Fraction of ground covered by vegetation (ND)
P_d	Calculated parameter [Eq. (2.5); ND]

Table 2.1 [*cont.*]

Symbol	Definition
P_e	Ratio of difference between observed soil water and soil water at wilting point to difference between soil water at field capacity and soil water at wilting point (ND)
P_h	Height of plant canopy (cm)
P_i	Soil water tension of the ith layer (-bars)
P_r	Atmospheric pressure (mb)
$*Q_i$	Slow drainage coefficient for the ith layer (0.02 cm/day)
R	Rain (ND)
R_a	Daily rainfall (cm/day)
R_c	Shortwave solar radiation above the plant canopy (langleys/day)
R_e	Root-mean-square error (ND)
R_f	Reflectivity of plant canopy (0.18; ND)
R_i	Dead-root biomass in ith layer (g/m²)
R_n	Net solar radiation at ground level (langleys/day)
R_t	Daily solar radiation on a clear day (langleys/day)
S_a^V	Average water-vapor pressure (mb)
S_b	Green shrub biomass (g/m²)
S_c	Standing crop grass biomass (g/m²)
$*S_d$	Bulk density of the soil (1.40 g/cm³)
S_h	Specific heat capacity of soil (cal · g⁻¹ · °C⁻¹)
S_i	Slow drainage from ith layer (cm/day)
S_m	Volumetric soil-water content (fraction; ND)
S_m^V	Saturation vapor pressure at minimum daily air temperature (mb)
S_4	Average soil-water potential in top 4 cm (-bars)
S_{75}	Average soil-water potential in top 75 cm (-bars)
T_i	Average daily soil temperature at ith point (°C)
T_i^1	Average daily soil temperature at ith point at time $t - 24$ h (°C)
T_m	Air temperature at the mth hour after sunrise (°C)
$*T_n$	Minimum air temperature at 2 m (°C)
T_p	Average daily air temperature at 2 m (°C)
$*T_x$	Maximum air temperature at 2 m (°C)
T_1	Average daily air temperature at top of soil (°C)
$*U_{\Delta t}$	Transition matrix (ND)
V_i	Volumetric soil-water content for ith layer (%)
W_d	Dew formed on ground and plant canopy (cm/day)
X_i^0	Observed values at ith point (ND)
X_i^S	Simulated values at ith point (ND)
Y	Number of daylight hours (h)
Z_r	Sum of weight factors for bare soil-water loss (ND)
Z_t	Sum of weight factors for transpiration (ND)
$*\Delta t$	Time step (24 h)
$*\Delta X$	Distance between points where soil temperature is calculated ($\Delta X = 15$ cm)

models (Goudriaan and Waggoner, 1972; Waggoner and Reifsnyder, 1968; Murphy and Knoerr, 1970; Shawcroft et al., 1974) as it does not consider the heat budget within the plant canopy. The level of resolution of ELM and the modeling objectives did not require such detail.

Temperature profile and water-flow submodels use daily rainfall, cloud cover, wind speed (2 m), maximum and minimum air temperatures (2 m), and relative humidity (2 m) as variables. These variables were taken from either daily weather observations or a stochastic weather simulator. The weather simulator randomly determines daily values for the driving variables using Monte Carlo techniques in which first-order Markov chains and conditional probabilities are utilized. Monte Carlo simulation techniques have been used by a variety of authors to simulate time series of weather events. Gringorten (1966) used first-order Markov chains to simulate air temperatures while Wiser (1966), Caskey (1963), Gabriel and Newman (1962), and Lowry and Guthrie (1968) simulated rainfall events. Although these authors found Monte Carlo techniques useful in studying the frequency and duration of weather events, stochastically simulated weather variables for driving ecosystem models is relatively new (Randell and Gyllenberg, 1972; Parton, 1972).

Validation procedures were performed and the results are presented. A summary of definition of the symbols used in the paper is presented in Table 2.1, where an asterisk beside a symbol indicates an input parameter.

2.2 Model Development

2.2.1 Water-flow Submodel

This submodel simulates flow of water through the plant canopy and soil layers. Rainfall interception, infiltration of water into soil, rapid and slow soil-water drainage, and evaporation of water from the plant canopy and the soil layers are included (Fig. 2.1). The submodel can treat an arbitrary number of soil layers in each of which depth and soil type are specified. The description of the submodel is presented in two parts: (a) allocation of precipitation and (b) evapotranspiration water loss.

Allocation of Precipitation. Rainfall may be intercepted by the standing crop or the litter, infiltrated into the soil, or transported from the area as runoff. Several authors (Corbet and Crouse, 1968; Clark, 1940; Zinke, 1967) have demonstrated that interception by plant canopies is primarily a function of rainfall amount and plant-cover density. Corbet and Crouse (1968) developed equations that predict standing crop and litter interception for an annual grassland. A generalized form of these equations [Eqs. (2.1)–(2.5)] are used to calculate standing-crop interception (I_s in cm/day) as a function of the rainfall amount (R_a in cm/day), height of the plant canopy (P_h in cm), the fraction of ground area covered by vegetation (P_c), and litter interception (I_l in cm/day) as a function of R_a and the litter biomass (L in g/m^2).

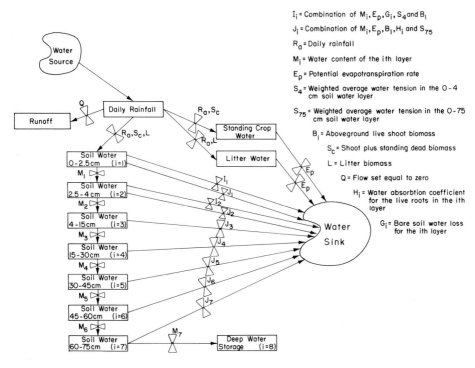

Fig. 2.1. Flow diagram of waterflow submodel

$$I_S = 0.026\ P_a R_a + 0.094\ P_b \left.\right\}\ \text{if } R_a > 0,\ I_S = I_1 = 0 \text{ if } R_a = 0 \quad (2.1)$$
$$I_1 = (0.015\ R_a + 0.0635)\ P_d \qquad\qquad\qquad\qquad (2.2)$$

$$P_a \begin{cases} = 0.9 + 0.016\ P_c P_h & \text{if } P_c P_h \leqslant 21.6 \\[2mm] = 1.24 + (P_c P_h - 21.6)\ 0.138 & \text{if } P_c P_h > 21.6 \end{cases} \quad (2.3)$$

$$P_b \begin{cases} = P_c P_h\ 0.13 & \text{if } P_c P_h \leqslant 7.6 \\[2mm] = 1.0 + (P_c P_h - 7.6)\ 0.072 & \text{if } P_c P_h > 7.6 \end{cases} \quad (2.4)$$

$$P_d = \exp[-1.0 + 0.45 \log_{10}(L + 1.0)]\ \ln(10.0) \qquad (2.5)$$

P_c is estimated as a function of leaf-area index (L_a) as follows:

$$P_c \begin{cases} = L_a/3 & \text{if } L_a/3 \leqslant 1.0 \\ = 1 & \text{if } L_a/3 > 1.0 \end{cases} \qquad (2.6)$$
$$L_a = S_c/100 \qquad\qquad\qquad (2.7)$$

and P_h is determined as a function of the standing-crop biomass (S_c in g/m²) using the relationship shown in Figure 2.2 (Conant, 1972). The standing-crop biomass

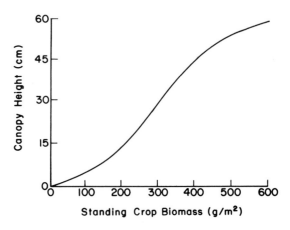

Fig. 2.2.   Relationships of canopy height (P_h) to the standing-crop biomass (S_c) for grasslands based on data taken in northeastern Oklahoma (Conant, 1972)

is the total live and dead grass and shrub biomass plus one percent of the cactus biomass. Cactus is given a low weight factor because of its morphological characteristics. The relationship of leaf-area index to standing-crop biomass is based on data presented by Květ and Marshall (1971), Conant (1972), and Conant and Risser (1974). A comparison of data presented by Květ and Marshall (1971), Conant (1972), Conant and Risser (1974), Knight (1973), and Brown (1974) shows the leaf-area-index conversion factor [(Eq. 2.7)] to vary as a function of plant species; however, this model uses a single expression for all grass species.

Net runoff from a particular area is a function of the rainfall rate, slope, soil types, infiltration rate, and many other system characteristics. A variety of models have been developed to simulate runoff (Cartmill, 1970; Kohler and Richards, 1962; Chow, 1964; Palmer, 1965; Smith, 1971; Parton and Marshall, 1973). At Pawnee it was observed that there was little runoff, and thus the Pawnee model ignores runoff.

Rainfall not intercepted by vegetation infiltrates into the top soil layer. If the soil water in this layer is greater than field capacity, the excess water flows into the second layer (rapid drainage). This process continues until there is insufficient water to bring the next lower layer to field capacity. The volumetric soil-water content at field capacity for the ith soil layer (F_i in %) depends on soil type and dead-root biomass as follows:

$$F_i = C_i + DR_i/D_i \tag{2.8}$$

where C_i is the field capacity of the ith soil layer ignoring dead-root biomass in the ith soil layer (g/m²); and D_i, the depth of the ith soil layer (cm). Equation (2.8) is based on the assumption that the field capacity of a soil layer is directly proportional to the dead-root biomass in the layer, whereas the value of D is estimated from data at the Pawnee Site. It is assumed that wilting point is not a function of root biomass and that the soil-water potential at field capacity is −0.3 bars. The relationship of soil-water potential (P_i in bars) to volumetric water content (V_i in %) at the Pawnee Site is shown in Figure 2.3.

Slow drainage from the ith layer (S_i in cm/day) occurs when the water content is less than field capacity. It is a function of the soil-water content and is represented by a modified version of the equation according to Black et al. (1969):

$$S_i = Q_i \exp[(M_i - A_i) \, 15/D_i] \tag{2.9}$$

where Q_i is the drainage coefficient for the ith layer (0.02 cm/day) and M_i and A_i, respectively, are soil-water content (cm of water) and soil-water content at field capacity (cm) of the ith soil layer. Slow and rapid drainage into the bottom layer is ground-water recharge.

Evapotranspiration Water Loss. The evaporation of intercepted water, evaporation from bare ground, and transpiration are the mechanisms used to estimate total water loss by evapotranspiration. Bare-soil evaporation and transpiration are determined separately because data indicate that water loss from

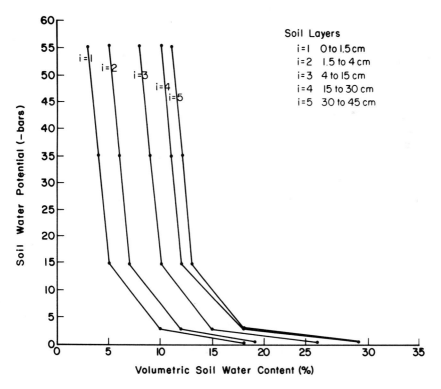

Fig. 2.3. Relationship of soil-water potential (P_i) to the volumetric soil-water content (V_i) for two soil layers at Pawnee site (Van Haveren and Galbraith, 1971)

bare soil decreases rapidly as the top layer (2–15 cm) dries (Lemon, 1956; Philip, 1957), whereas transpiration water loss occurs from all layers that have living roots. The division of water loss due to bare-soil evaporation (A_e) and transpiration ($A_t = 1 - A_e$) is calculated as a function of the leaf-area index of live biomass (A_1) using relationships shown in Figure 2.4 (Mihara, 1961 as cited by Chang, 1968; Ritchie and Burnett 1971). Using the conversion factor in Eq. (2.7), A_1 is calculated from live biomass (B_1 in g/m²). This factor was determined for grass biomass (G_1); however, the influence of green shrub (S_b) and live cactus biomass (C_1) was incorporated as follows:

$$B_1 = S_b + G_1 + 0.01 \, C_1 \tag{2.10}$$

Water intercepted by the standing crop and litter (I_s, I_1) evaporates at the potential evapotranspiration rate (E_p in cm/day) and is removed first from I_s and then from I_1. If the intercepted water supply is less than the daily water loss estimated from the potential evapotranspiration rate, then water is removed from the soil. The potential evapotranspiration rate is determined from average daily air temperature, relative humidity, cloud cover, and wind speed using Penman's (1948) equation.

Information presented by Lemon (1956) shows that bare-soil evaporation decreases rapidly as top soil layers dry. The depth to which bare-soil evaporation is limited depends on soil type; this depth is proportional to the clay content of the soil (sandy soil, 0–5 cm; clay soil, 5–15 cm). Bare-soil evaporation is limited to the top soil layers (0–4 cm at the Pawnee site) and is computed from the potential evapotranspiration rate (E_p), standing-crop biomass (S_c), and a weighted average soil-water potential in the top 4 cm (S_4). The effect of E_p and

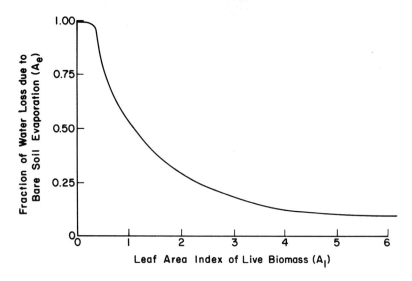

Fig. 2.4.   Influence of leaf-area index of live biomass on fraction of water loss due to bare-soil evaporation (Mihara, 1961, as cited by Chang, 1968)

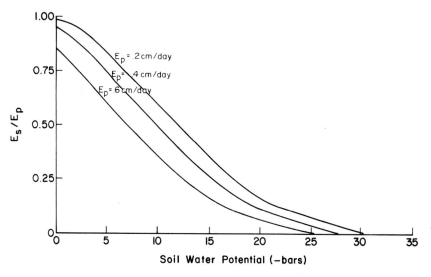

Fig. 2.5a.   The functional relationship of E_p and soil-water potential (S_4) on $E_s:E_p$ ratio

S_4 on the potential bare-soil evaporation rate (E_s) is based on information from Lemon (1956), and the effect of standing-crop biomass (S_c) on E_s is derived from soil-water data (Risser and Kennedy, 1972) that shows evaporation to be inversely proportional with S_c. The functional relationship of E_p and S_4 on E_s presented in Figure 2.5a reveals $E_s: E_p$ to depend on the inverse of the soil-water potential (-bars) and E_p. The fraction of the bare-soil evaporation coming from the different layers is proportional to the bare-soil water-absorption coefficient

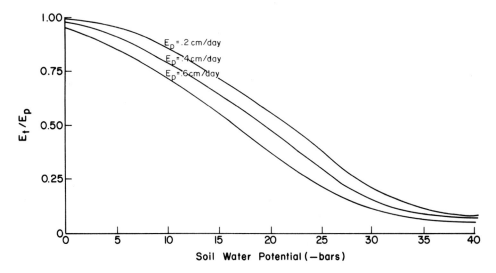

Fig. 2.5b.   Functional relationship of E_p and soil-water potential (S_{75}) on $E_t:E_p$ ratio

for the layer $G_i = 0.75, 0.25$ (ND) and inversely proportional to the water tension in the layer (P_i). Equations (2.12) and (2.13) show the effect of G_i and P_i on water loss from the ith layer, and Eq. (2.11) calculates the total bare-soil evaporation water loss (E_s^T in cm/day):

$$E_s^T = E_s A_e \, (1 - S_c/999) \tag{2.11}$$

$$Z_r = \sum_{i=1}^{N_a} G_i/P_i \tag{2.12}$$

$$B_i = E_s^T[(G_i/P_i)/Z_r] \qquad \text{for } i = 1, 2 \tag{2.13}$$

$$M_i = M_i^1 - B_i \qquad \text{for } i = 1, 2 \tag{2.14}$$

where M_i and M_i^1 are soil water in the ith layer (cm) at times t and $t - 1$ respectively; B_i is the evaporative water loss from the ith soil-water layer (cm/day); Z_r is the sum of the weight factors for bare soil-water loss; and N_a ($N_a = 2$) is the number of layers above 4 cm. The water-absorption coefficients (G_i) are the weighting factors used to calculate S_4.

The potential transpiration rate (E_5 in cm/day) is calculated as a function of the weighted average of the soil-water potential in the top 75 cm of soil (S_{75}) and potential evapotranspiration rate (E_p) by using a relationship based on data presented by Denmead and Shaw (1962) and Ritchie (1973). The ratio of E_t to E_p is inversely related to the soil-water potential (-bars) and E_p as seen in Figure 2.5b. The fraction of the transpiration water loss coming from the ith soil layer is assumed to be proportional to the transpiration water-absorption coefficient of the layer [$H_i = 0.13, 0.14, 0.37, 0.18, 0.17, 0.06, 0.05$ (ND)] and inversely proportional to the soil-water tension (P_i) in the layer. The values of H_i are estimated as the fraction of the total live roots in the ith layer, which decrease exponentially with depth in most grassland sites (Singh and Coleman, in press). The basis for estimating the fraction of transpiration and bare soil-water loss removed from each layer is a paper by Gardner (1964). Equations (2.15), (2.16), and (2.18) show the effect of H_i and P_i on the transpiration water loss from the ith layer, and Eq. (2.17) shows the effect of A_t on the total transpiration water loss (E_t^T in cm/day).

$$M_i = M_i^1 - K_i \qquad \text{for } i = 1, 2, \ldots, N \tag{2.15}$$

$$K_i = E_t^T(H_i/P_i)Z_t \qquad \text{for } i = 1, 2, \ldots, N \tag{2.16}$$

$$E_t^T = E_t A_t \tag{2.17}$$

$$Z_t = \sum_{i=1}^{N_1} H_i/P_i \tag{2.18}$$

where K_i is transpiration from the ith layer (cm/day); N, the total number of soil water layers; and Z_t, the sum of the weight factors for transpiration. The transpiration water absorption coefficients are used as the weighting factors in calculating S_{75}.

The amount of dew (Wd in cm/day) deposited on the grassland and plant canopy is calculated as follows:

$$
W_d = \begin{cases} D_a(S_a^V - S_m^V)0.622\ D_d/P_r & \text{if } S_a^V > S_m^V \\ \\ 0.0 & \text{if } S_a^V \leqslant S_m^V \end{cases} \tag{2.19}
$$

where D_a is the density of the air (g/cm³); P_r is the atmospheric pressure (mb); D_d is the depth of the air column (366 cm) considered for dew formation; S_a^V is the average water-vapor pressure (mb) during the day; and S_m^V is the saturation-vapor pressure (mb) at the minimum daily air temperature. The density of the air is calculated using the ideal gas law, and S_m^V and S_a^V are calculated using the Clausius–Clapeyron equation (Hess, 1959). Equation (2.19) assumes that dew will be formed on days when the average daily mixing ratio is greater than the saturation mixing ratio at the minimum daily air temperature.

2.2.2 Temperature-profile Submodel

The temperature-profile submodel simulates temperatures at 13 points in the soil, average maximum air temperature in the canopy, average daytime air temperature in the canopy, average daytime air temperature at 2 m, and total daily solar radiation. The average daily air temperature at the top of the soil (T_1 in °C) is predicted from the average daily air temperature (T_p in °C), the potential evapotranspiration rate (E_p), the standing-crop biomass (S_c), and the ratio of the actual evapotranspiration rate (E_z) to E_p:

$$
T_1 = \begin{cases} T_p + 15\ E_p(1 - E_z/E_p(1 - S_c/300) & \text{if } S_c \leqslant 300 \text{ g/m}^2 \\ \\ T_p + -4(S_c - 300)/600 & \text{if } S_c > 300 \text{ g/m}^2 \end{cases} \tag{2.20}
$$

The influence of E_p on T_1 is based on an analysis of observed data at the Pawnee Site, whereas the influence of vegetation biomass is derived from a comparison of soil temperatures at 2.5 cm and the standing-crop biomass at different sites. The effect of E_z/E_p on T_1 uses the concept that sensible heat flow decreases with increasing latent heat flow.

Soil Temperature. The soil-temperature model computes temperatures at the upper (0 cm) and lower (180 cm) boundaries and solves for soil temperatures at intervening points (15 cm, 30 cm, . . . , 165 cm) using the one-dimensional Fourier heat equation (Munn, 1966). The soil temperature at the upper boundary is T_1, and the temperature at the lower boundary is the mean daily soil temperature at 180 cm. The finite-difference scheme used to solve the Fourier heat equation is:

$$
\frac{T_i - T_i^1}{\Delta t} = \frac{C_s}{S_h S_d} \frac{T_{i-1} - 2T_i^1 + T_i^1 + T_{i+1}^1}{\Delta X^2} \qquad \text{for } i = 2,\ 12 \tag{2.21}
$$

where T_i and T_i^1 are average daily soil temperatures at point i, i = 2-(15 cm), 3-(30 cm), . . . , 12-(165 cm) at times t and (t-1) respectively; C_s is the soil thermal conductivity (cal cm^{-1} s^{-1} °C^{-1}); S_h is the specific heat capacity (cal g^{-1} °C^{-1}); S_d is the bulk density of the soil (1.40 g/cm^3); and ΔX is the distance between profile points ($\Delta X = 15$ cm). The values of C_s and S_h for each soil layer as a function of the soil-water content are determined by Eqs. (2.22) and (2.23), and then C_s and S_h are averaged for the soil-water layers that enter into calculating T_i.

$$C_s = 0.00070 + P_e\, 0.00030 \qquad\qquad (2.22)$$
$$S_h = S_m + 0.18\, (1 - S_m) \qquad\qquad (2.23)$$

where P_e is the ratio of the difference between volumetric soil-water content, S_m, and soil-water content at the wilting point to the difference between soil water content at field capacity and soil-water content at wilting point. These equations are due to Munn (1966), and C_s and S_h are 0.00061 and 0.30, respectively, at points in the profile where soil water is not simulated.

Solar Radiation.   The daily solar radiation on a clear day (R_t in langleys/day) is calculated as a function of time of year, latitude, and the transmission coefficient using an equation due to Sellers (1965), and the influence of cloud cover on net solar radiation (R_n in langleys/days) is determined using Eq. (2.24) (Haurwitz, 1941)

$$R_n = R_t\, [1 - (R_f + 0.53\, C_a)] \qquad\qquad (2.24)$$

where R_f is the reflectivity of the plant canopy (0.18) and C_a is the average daily fractional cloud cover.

Air Temperature.   The average canopy air temperature during the daylight hours (C_d) is simulated by integrating a truncated sine wave in which the minimum air temperature (T_n) occurs at sunrise and the maximum canopy air temperature (C_m in °C) occurs at 2:00 P.M.:

$$T_m = (C_m - T_n) \sin \frac{2\pi m}{2(Y + 4.0)} + T_n \qquad\qquad (2.25)$$
$$C_d = \frac{(C_m - T_n)\,(Y + 4)}{Y} \frac{1}{\pi}\left[\cos\left(\frac{\pi Y}{Y + 4}\right) - 1\right] + T_n \qquad\qquad (2.26)$$

where T_m is the air temperature at the mth hour after sunrise (°C) and Y, the number of daylight hours. The minimum canopy air temperature is the same as the minimum air temperature at 2 m (T_n). The maximum canopy air temperature (C_m) is calculated from the maximum air temperature at 2 m (T_x) and the shortwave solar radiation above the plant canopy (R_c in langleys/day):

$$C_m = \begin{cases} T_x + (R_c - 300)0.035 & \text{if } R_c \geqslant 300 \\[2mm] T_x & \text{if } R_c < 300 \end{cases} \qquad\qquad (2.27)$$
$$R_c = R_t(1 - 0.53 C_a) \qquad\qquad (2.28)$$

These equations are based on the assumption that C_m is independent of plant-canopy height and primarily a function of T_x and R_c. These assumptions are supported by observations of soil and air temperatures at Biome study sites and canopy air temperatures presented by Old (1969).

2.2.3 Driving-weather Variables

The temperature and water flow submodels require daily values for atmospheric driving variables: (a) rainfall, (b) relative humidity, cloud cover, and wind speed, and (c) maximum and minimum air temperatures at 2 m. The variables were determined from records of daily weather observations or simulated weather determined by a simulator. At the Pawnee site daily average values of cloud cover, relative humidity, and wind speed were not available and were calculated assuming the daily average values of these parameters to be the long-term monthly average values observed at Cheyenne, Wyoming (30 km to the north).

The procedure used by the weather simulator is illustrated in Figure 2.6. The relative humidity is forecast randomly at the start of each time step. The rainfall

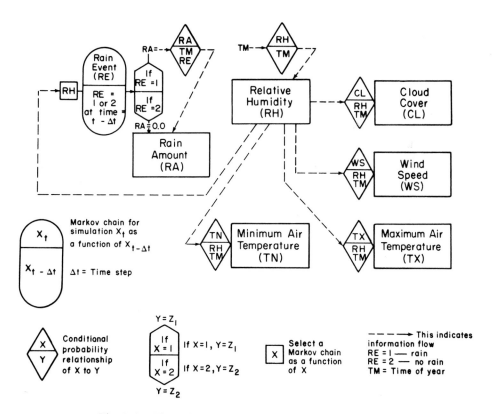

Fig. 2.6.   Flow diagram for stochastic weather simulator

is then simulated as a function of forecast relative humidity. Cloud cover, wind speed, and maximum and minimum air temperatures are subsequently forecast using conditional probabilities that relate these variables to relative humidity. The statistical weather relationships used were summarized for monthly intervals from 7 yr of daily weather observations at Cheyenne, Wyoming. The model is structured so that predicted relative humidity influences the values simulated for the other atmospheric variables in accordance with the Pawnee Site weather characteristics. The model of Randell and Gyllenberg (1972) differs from this model in its utilization of the correlation of the atmospheric variables to the air-mass type.

Relative humidity is forecast to be in one of 10 class intervals (0–10%, 10–20%, . . . , 90–100%) by randomly selecting from the frequency distribution for the month under consideration. A computer-generated "random" number (0–1 rectangular distribution) is the ordinate of the cumulative frequency distribution, and the class interval of relative humidity is taken from the abscissa. Any further reference in this chapter to the random selections from a frequency distribution implies a process similar to this one. The relative-humidity values used (nearest 1%) are randomly selected from the frequency distribution in which each relative humidity value in the class intervals has equal probability.

Rainfall is simulated as a first-order Markov chain stratified into five class intervals of relative humidity (0–20%, 20–40%, . . . , 80–100%). The Markov chain is specified by the transition matrix ($U_{\Delta t}$):

$$
U_{\Delta t} = \quad
\begin{array}{c c}
 & R \quad NR \\
\hline
R & A_{11}\ A_{12} \\
\\
NR & A_{21}\ A_{22}
\end{array}
$$

where R is rain; NR is no rain; and A_{ij} is the probability for the occurrence or nonoccurrence of rainfall ($j = 1, j = 2$) given the occurrence or nonoccurrence of rainfall ($i = 1, i = 2$) the previous day. Rainfall occurrence is simulated by randomly selecting from the Markov chain that corresponds to the forecast relative humidities. The class interval for rainfall amount is selected randomly from a frequency distribution in which there are 10 class intervals of rainfall (0–0.05, 0.05–0.10, 0.1–0.3, 0.3–0.5, 0.5–0.7, 0.7–0.9, 0.9–1.5, 1.5–2.5, 2.5–4.5, and 4.5–6.5 in.). The rainfall amount (nearest 0.01 in.) is then calculated by randomly selecting from a frequency distribution derived from linear interpolation between the probabilities for rainfall in the class intervals on either side of the predicted class interval.

The average daily cloud cover and wind speed and the maximum and minimum air temperatures are determined from the forecast relative humidity by using frequency distributions calculated for five class intervals of relative humidity. The class intervals of cloud cover (0–10%, 10–20%, . . . , 90–100%), wind speed (0–5, 5–10, . . . , ≥25 mph), maximum air temperature (≤0.0°F, 0–5°F, 5–

10°F, . . . , >100°F), and minimum air temperature (≤−30°, −30° to −25°C, 111, >70°F) are randomly selected from frequency distributions corresponding to the forecast relative humidity. The exact values of variables (nearest 1%, 1 mph, 1°F) are calculated as is the relative humidity. The values for maximum air temperature <0°F or >100°F and minimum air temperature <−30°F or >70°F are calculated by randomly selecting from a frequency distribution in which the probability for temperatures outisde the extreme values decreases linearly to zero at 10°F beyond these values.

2.3 Validation of the Model

Validation of the temperature and water-flow submodels involved comparing simulated time series of certain state variables with data observed at the Pawnee Site during the 1972 field season. Note that the 1972 data were reserved specifically for this validation and were not used in formulating the model. Simulated values were determined by using daily values of the maximum and minimum air temperature and rainfall observed at the Pawnee Site for 1972 (no domestic herbivores included). The soil-water-content profile was measured at 10 points (Smith and Stiffler, 1969) using a neutron probe, and the average values for these 10 profiles was compared with the simulated soil water. Comparison of the simulated soil water at 0–15, 15–30, 30–45, 45–60, and 0–75 cm with observed values (Fig. 2.7) yields a root-mean-square error (R_e) of 0.42, 0.36, 0.20, 0.20, and 0.94 cm, respectively.

The R_e value was used for comparison with the standard deviation of the observed (1972) soil-water content (0.29 − 0 to 15 cm, 0.25 − 15 to 30 cm, 0.33 − 30 to 45 cm, 0.40 − 45 to 60 cm, 0.38 − 60 to 75 cm). These results show R_e to be greater than the observed standard deviation for the top two layers but smaller than the standard deviation for the bottom three layers. A detailed study of Figure 2.7 indicates that the most significant deviation between the observed and simulated values is at the end of the year (October to December). This error is partially attributed to the fact that the producer section of the total systems model was simulating excessive live-shoot biomass during the latter part of the growing season, which caused too much water to be transpired from the soil water layers during the latter half of the growing season. Another contributing factor is the model does not consider the effect of plant phenological stage on transpiration water loss. In general, the transpiration from a plant will decrease as the phenological stage of the plant advances toward senescence. Other factors influencing the difference between the observed and simulated values include: (a) sampling error in observing soil-water content and (b) distance to points where the driving-air temperature (24 km) and rainfall (30.48 m) and observed soil-water contents were measured. The distance between the rainfall and the soil-water-content observations were particularly important because most precipitation during the summer occurs as convective showers that vary widely.

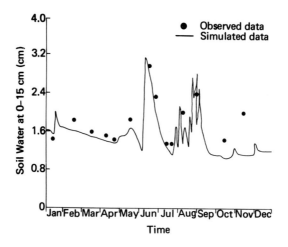

a

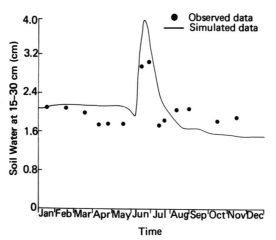

b

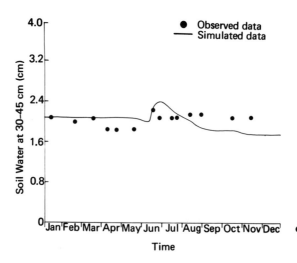

c

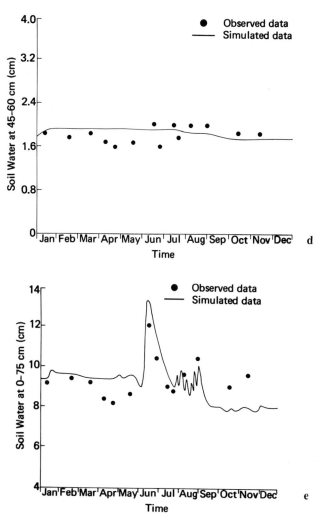

Fig. 2.7. Comparison of simulated and observed soil-water data for: (a) 0–15 cm, (b) 15–30 cm, (c) 30–45 cm, (d) 45–60 cm, and (e) 0–75 cm soil-water layers for 1972 at Pawnee site

Computed average daily soil temperatures at 0, 15, and 45 cm are compared with observed values at 2.54, 20.32, and 50.8 cm, respectively (Fig. 2.8). The R_e value for the soil temperature at 0, 15, and 45 cm, respectively, was 4.2°, 3.6°, and 3.2°C. Careful analysis of Figure 2.8 and comparison of R_e by month (see Table 2.2) reveals the greatest deviation between the observed and simulated soil temperatures to be during November through February. In fact, 55–72% of the yearly R_e occurs during these four months. During this period the ground is either freezing, thawing, or covered with snow. This observed error was

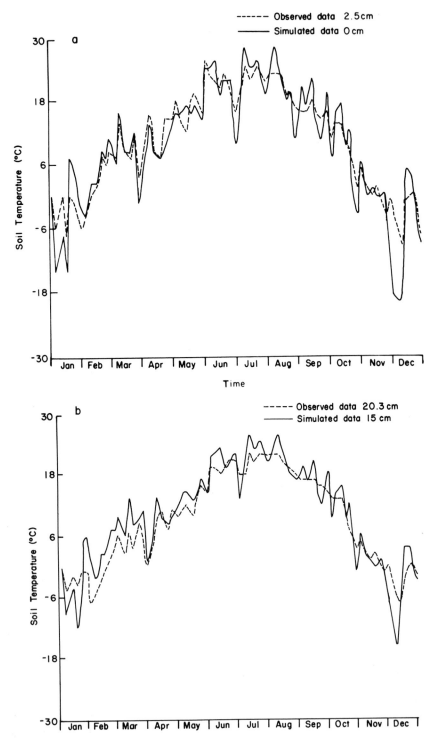

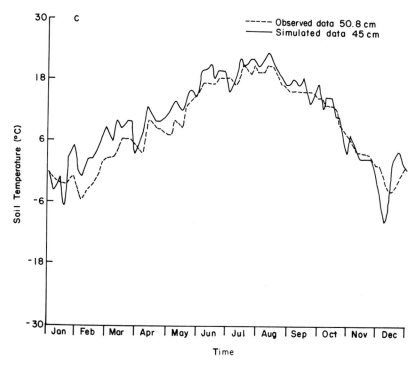

Fig. 2.8. Comparison of simulated soil temperatures at: (a) 0 cm, (b) 15 cm, and (c) 45 cm with the observed soil temperatures at 2.5, 20.3, and 50.8 cm, respectively, for Pawnee site during 1972

expected since the effect of snow-cover freezing and thawing were not considered.

A comparison of the monthly average observed and simulated soil temperatures and the solar radiation above the plant canopy (R_c) is shown in Table 2.3. Observed and simulated soil temperature values differ by less than 1.5°C, except for differences of 2–4.5°C at 15 and 45 cm during January through April. The reason for this discrepancy is unknown. A significant improvement in the predictive capability of the temperature submodel could be accomplished by a more mechanistic representation of the physical processes involved. The discrepancy between the observed and simulated solar radiation above the canopy was attributed to the fact that the long-term monthly average cloud cover at Cheyenne, Wyoming was used as the driving variable for cloud cover (daily average cloud cover was not available at the Pawnee Site).

In general, the results of the validation procedure demonstrated that the output from the temperature and water-flow models compare well with observations. The most significant differences between the observed and simulated values appear to be related to omission of effects of snow, freezing, and thawing and the influence of the plant phenological stage on transpiration. These improvements have been incorporated in later versions of the model.

Table 2.2. Monthly root-mean-square error (°C) for the soil temperatures at 0, 15, and 45 cm.

Soil depth	Month											
	Jan	Feb	Mar	Apr	May	Jun	Jul	Aug	Sep	Oct	Nov	Dec
0 cm	6.9	2.5	3.3	2.7	2.7	2.3	3.0	2.0	2.6	2.9	3.7	8.9
15 cm	5.7	4.5	3.9	3.5	2.3	2.1	2.5	1.8	1.9	2.8	2.7	6.3
45 cm	3.8	5.1	3.9	3.8	2.8	1.7	2.1	1.6	1.5	2.0	1.8	4.7

Table 2.3. Comparison of observed and simulated monthly average soil temperature (°C) and solar radiation above plant canopy (langleys/day)

Soil temperature and radiation	Depth (cm)	Month											
		Jan	Feb	Mar	Apr	May	Jun	Jul	Aug	Sep	Oct	Nov	Dec
Simulated temperature	0.0	-1.7	3.6	9.8	12.5	16.1	22.3	24.0	23.3	18.3	12.3	0.9	-4.1
Observed temperature	2.5	-2.5	3.1	10.3	13.8	17.1	22.9	23.0	22.2	18.1	11.7	2.0	-3.3
Simulated temperature	15.0	-0.6	3.5	9.7	11.9	15.3	21.2	23.0	22.6	18.3	12.7	2.1	-2.8
Observed temperature	20.3	-2.6	-0.4	6.8	9.6	13.9	20.7	21.8	21.4	17.4	12.1	2.8	-2.5
Simulated temperature	45.0	0.6	3.6	9.3	11.0	13.8	19.3	21.0	21.2	17.7	13.1	4.0	-0.9
Observed temperature	50.8	-1.6	-1.1	5.8	7.8	11.7	18.0	19.7	20.1	16.9	12.8	4.8	-0.9
Observed solar radiation		237.0	341.0	424.0	467.0	545.0	616.0	545.0	490.0	434.0	283.0	222.0	182.0
Simulated solar radiation		204.0	275.0	389.0	511.0	565.0	622.0	634.0	567.0	437.0	316.0	217.0	178.0

2.4 Verification of the Weather Simulator

The stochastic weather simulator was verified (Nolan, 1972) using statistical techniques to compare simulated values with the observed weather data used as input for the model. Daily values from a 20-yr run of the weather simulator were used to generate frequency distributions of cloud cover, relative humidity, wind speed, and maximum and minimum air temperatures that were compared with observed frequency distributions of these variables using the χ^2 goodness-of-fit test (Dixon and Massey, 1969). The χ^2 test showed no significant deviation from any of the observed frequency distributions ($p \geqslant 0.80$). Similarly, the simulated monthly average rainfall was compared to observed monthly rainfall using "student's t test" (Dixon and Massey, 1969) and showed no significant difference at the 90% confidence level for all months except October.

Acknowledgments. This chapter reports on work supported in part by National Science Foundation Grants GB-31862X, GB-31862X2, GB-41233X, and BMS73-02027 A02 to the Grassland Biome, U.S. International Biological Program, for "Analysis of Structure, Function, and Utilization of Grassland Ecosystems."

References

Black, T. A., Gardner, W. R., Thurtell, G. W.: The prediction of evaporation, drainage and soil water storage for a bare soil. Soil Sci. Soc. Am., Proc. **33**, 655–660 (1969)

Brown, L.: Photosynthesis of two important grasses of the short-grass prairie as affected by several ecological variables. Ph.D. thesis, Colorado State Univ., Fort Collins, 1974, 191 pp.

Cartmill, R.: Forecasting the volume of storm runoff using meteorological parameters. Ph.D. thesis, Univ. Oklahoma, Norman, 1970, 130 pp.

Caskey, J. E.: A Markov chain model for the probability of precipitation occurrence in intervals of various lengths. Mon. Weath. Rev. **91**, 299–301 (1963)

Chang, Jen-Hu: Climate and Agriculture. Chicago: Aldine Publ. Co., 1968, 304 pp.

Chow, V. T.: Handbook of Applied Hydrology. New York: McGraw-Hill, 1964, ca. 2000 pp.

Clark, O. R.: Interception of rainfall by prairie grasses, weeds, and certain crops plants. Ecol. Monogr. **10**, 243–277 (1940)

Conant, S.: Vegetation structure of a tallgrass prairie. M.S. thesis, Univ. Oklahoma, Norman, 1972, 47 pp.

Conant, S., Risser, P. G.: Canopy structure of a tallgrass prairie. J. Range Mgmt. **27**, 313–318 (1974)

Corbet, E. S., Crouse, R. P.: Rainfall interception by annual grass and chaparral. U.S. For. Serv. Res. Paper PSW-48 1968, 12 pp.

Cowan, I. R.: Transport of water in the soil–plant–atmosphere system. J. Appl. Econ. **2**, 221–239 (1965)

Crawford, N. H., Linsley, R. K.: Digital simulation in hydrology: Stanford watershed model IV, Tech. Rep. No. 39. Dept. Civil Engineering. Palo Alto: Stanford Univ., 1966, 186 pp.

Denmead, P. T., Shaw, R. H.: Availability of soil water to plants as affected by soil moisture content and meteorological conditions. Agron. J. **54,** 385–390 (1962)

Dixon, W. S., Massey, F. J.: Introduction to statistical analysis, 3rd ed. New York: McGraw-Hill, 1969, 638 pp.

Fluker, B. J.: Soil temperature. Soil Sci. **86,** 35–46 (1958)

Gabriel, K. R., Newman, J.: A Markov chain model for daily rainfall occurrence at Tel Aviv. Quart. J. Roy. Meteorol. Soc. **88,** 90–95 (1962)

Gardner, W. R.: Relationship of root distribution to water uptake and availability. Agron. J. **56,** 41–45 (1964)

Goldstein, R. A., Mankin, S. B.: Prosper: A model of atmosphere–soil–plant water flow. Proc. Summer Computer Sim. Conf., La Jolla, Calif.: Simulations Councils, Inc., 1972, pp. 176–181

Goudriaan, J., Waggoner, P. E.: Simulating both aerial microclimate and soil temperature from observations above the foliar canopy. Neth. J. Agr. Sci. **20,** 104–124 (1972)

Gringorten, I. I.: A stochastic model of the frequency and duration of weather events. J. Appl. Meteorol. **5,** 606–624 (1966)

Haurwitz, B.: Dynamic Meteorology. New York: McGraw-Hill, 1941, 365 pp.

Hess, S. L.: Introduction to Theoretical Meteorology. New York: Holt, 1959, 362 pp.

Knight, D. H.: Leaf area dynamics of a shortgrass prairie in Colorado. Ecology **54,** 891–896 (1973)

Kohler, M. A., Richards, M. M.: Multi-capacity basic accounting for predicting runoff from storm precipitation. J. Geophys. Res. 67, 5187–5197 (1962)

Květ, J., Marshall, J. K.: Assessment of leaf area and other assimilating plant surfaces. Sěstk, A., Čatský, J., Jarvis, P. G. (eds.). In: Plant Photosynthetic Production: Manual of Methods. The Hague, The Netherlands: Dr. W. Junk, 1971, pp. 517–555

Langbein, W. B.: Computing soil temperatures. Am. Geophys. Union, Trans. **30,** 543–547 (1949)

Lemon, E. R.: The potentialities for decreasing soil moisture evaporative loss. Soil Sci. Soc. Am., Proc. **20,** 120–125 (1956)

Lowry, W. P., Guthrie, D.: Markov chains of order greater than one. Mon. Weath. Rev. **96,** 798–801 (1968)

Mihara, R.: The microclimate of paddy rice culture and the artificial improvement of the temperature factor. Pres. 10th Pac. Sci. Congr., Honolulu, Hawaii, 1961

Munn, R. E. 1966. Descriptive micrometeorology. New York: Academic Press, 1966, 245 pp.

Murphy, C. E., Knoerr, K. R.: A general model for the energy exchange and microclimate of plant communities. Proc. 1970 Summer Computer Sim. Conf., La Jolla, Calif.: Simulation Councils, Inc., 1970, pp. 786–797

Nimah, M. N., Hanks, R. J.: Model for estimating soil water, plant, and atmospheric interrelations: II. Field test of model. Soil Sci. Soc. Am., Proc. **37,** 528–532 (1973)

Nolan, R. L. Verification/validation of computer simulation models. In: Proc. 1972 Summer Computer Sim. Conf., Vol. II. La Jolla, Calif.: Simulation Councils, Inc., 1972, pp. 1254–1265

Old, S. M.: Microclimate, fire and plant production in an Illinois prairie. Ecol. Monogr. **39,** 355–384 (1969)

Palmer, W. C.: Meteorological drought. Washington, D.C.: U.S. Dep. Com., Weath. Bur., Res. Paper No. 45, 1965, 58 pp.

Parton, W. J.: Development of an urban–rural ecology model. Ph.D. thesis, Univ. Oklahoma, Norman, Ok. 1972, 340 pp.

Parton, W. J., Marshall, J. K.: MODENV: A grassland ecosystem model. Proc. 1973 Summer Computer Sim. Conf., Montreal, Can., La Jolla, Calif.: Simulation Councils, Inc., 1973, pp. 769–776

Penman, H. L.: Natural evaporation from open water, bare soil and grass. Royal Soc. (Lond.), Proc. A. **193,** 120–145 (1948)

Penrod, E. B., Elliott, J. M., Brown, W. K.: Soil temperature variations. Soil Sci. **90,** 275–283 (1960)

Philip, J. R.: Evaporation and moisture and heat fields in the soil. J. Meteorol. **14,** 354–366 (1957)

Randell, R. L., Gyllenberg, G.: Weather simulation models, In: Matador Project (IBP), 5th Ann. Rep. (1971–1972), Saskatoon, Saskatchewan, 1972, pp. 111–140

Risser, P. G., Kennedy, R. K.: Herbage dynamics of a tallgrass prairie, Osage, 1971. US/IBP Grassland Biome Tech. Rep. No. 173. Fort Collins: Colorado State Univ., 1972, 75 pp.

Ritchie, J. T.: Influence of soil water status and meteorological conditions on evaporation from a corn canopy. Agron. J. **65,** 893–897 (1973)

Ritchie, J. T., Burnett, E.: Dry land evaporation flux in a subhumid climate. Agron. J. **63,** 56–62 (1971)

Sellers, W. D.: Physical Climatology. Chicago: Univ. Chicago Press, 1965, 272 pp.

Shawcroft, R. W., Lemon, E. R., Aller, L. H., Stewart, D. W., Jensen, S. E.: The soil-plant-atmospheric model and some of its predictions. Agr. Meteo. **14,** 287–307 (1974)

Singh, J. S., Coleman, D. C.: Evaluation of functional root biomass and translocation of photoassimilated carbon-14 in a shortgrass prairie ecosystem. In: The Belowground Ecosystem: A Synthesis of Plant-associated Processes. Marshall, J. K. (ed.). Range Science Series, Fort Collins: Colorado State Univ., in press, 1977

Smith, F. M.: Volumetric threshold infiltration model. Ph.D. thesis, Colorado State Univ., Fort Collins, 1971, 234 pp.

Smith, F. M., Striffler, W. D.: Pawnee Site microwatersheds: Selection, description, and instrumentation. US/IBP Grassland Biome Tech. Rep. No. 5. Fort Collins: Colorado State Univ., 1969, 29 pp.

Van Haveren, B. P., Galbraith, A. F.: Some hydrologic and physical properties of the major soil types on the Pawnee Intensive Site. US/IBP Grassland Biome Tech. Rep. No. 115. Fort Collins: Colorado State Univ., 1971, 46 pp.

Waggoner, P. E., Reifsnyder, W. E.: Simultion of the temperature humidity and evaporation profiles in a leaf canopy. J. Appl. Meteorol. **7,** 400–409 (1968)

Wiser, E. H.: Monte Carlo methods applied to precipitation-frequency analysis. Am. Soc. Agr. Eng., Trans. **9,** 538–540 (1966)

Zinke, P. J.: Forest interception studies in the United States. In: Forest Hydrology: Proceedings. Internat. Symp. For. Hydrol., Pennsylvania State Univ., 29 Aug.–10 Sept., 1965. Sopper, W. E., Lull, H. W. (eds.). Oxford: Pergamon Press, 1967, pp. 137–161

3. A Simulation Model for Grassland Primary Producer Phenology and Biomass Dynamics

RONALD H. SAUER

Abstract

The two primary producer submodels described in this chapter, plant phenology and carbon flow, are components of the grassland ecosystem model, ELM. The objectives of these submodels are to simulate biologically realistic phenology and growth responses to conditions varying from drought to irrigation and fertilization. These submodels simulate five primary producer groups and interact extensively with each other and the other ELM submodels.

The phenology submodel simulates phenological change in each producer group. The rate of phenologic change is determined from maximum air temperature, insolation, soil-water potential, soil temperature, day length, and shoot weight.

The carbon submodel simulates dynamics of live shoots, dead shoots, live roots, seeds, and crowns for each producer group. One litter and three dead-root state variables are represented for all producers. The processes modeled for each producer group are growth, respiration, and death of shoots, roots, and crowns, gross photosynthesis, seed production and germination, and the fall of standing dead shoots.

Comparison between observed data and simulations suggest the model to be most sensitive to the belowground system, the fall of standing dead, and the relationship between phenology and environment, and that these areas need additional refinement.

3.1 Introduction

This chapter describes the primary producer simulation submodels developed to integrate data and hypotheses on the biomass (carbon) and phenology dynamics of grassland ecosystems. These models are part of the grassland ecosystem model ELM. The objective of this study was to simulate biologically reasonable responses to drought, irrigation, and fertilization. Accordingly, the model addresses the effects of temperature, soil-water potential, nitrogen, phosphorus, and insolation on important biological processes. In the version described here, the 323 plant species on the Pawnee Site (Dickinson and Baker, 1972) are divided into five ecologic groups with representative species as follows: (a) warm-season grasses, *Bouteloua gracilis,* (b) cool-season grasses, *Agropyron smithii,* (c) forbs, *Sphaeralcea coccinea,* (d) shrubs, *Artemisia frigida,* and (e) cactus, *Opuntia polyacantha.*

The unique features of these submodels include the simulation of seven phases of plant phenology, the use of plant phenology in regulating biological processes, and the representation of carbon flow through a whole plant. The information bases (concepts and data) on which these submodels were developed varied considerably in detail. In some cases available detail on well-known processes (photosynthesis) was omitted, whereas in other instances untested hypotheses were used in poorly understood processes (fall of standing dead). Parameter values with units and the FORTRAN notations in the model are given in Table 3.1.

3.2 Literature Review

Phenology publications have usually focused on the relationships between an indicator of plant development, usually anthesis, and a combination of weather factors. A cumulative sum of degree-hours was correlated with the timing of anthesis for perennial species in Indiana (Lindsey and Newman, 1956; Jackson, 1966), and rainfall and temperature patterns were related to vegetative and productive development of native species in Montana and Idaho (Blaisdell, 1958; Mueggler, 1972). Lilac flowering date in Montana was correlated with the product of insolation and nonfreezing temperatures accumulated at the time of flowering (Caprio, 1971). Phenology has been modeled by moving all biomass from one state to another (e.g., dormancy to germination) in one time step when weather requirements were met (Bridges et al., 1972).

Many available producer models were too detailed to be applied to different sites and producer species or were inapplicable to herbaceous grassland species. Models have been constructed of the interactions of: (a) temperature and insolation on shrub and tree productivity (Cline, 1966; Botkin, 1969), (b) temperature and soil water on tree production (Zahner and Stage, 1966), (c) root zone and soil-water content on wheat yield (Bridge, 1976), and (d) light, temperature, soil water, and biologic competition on tree growth (Botkin et al., 1972). Photosynthesis regulated by light interception was modeled by de Wit (1965), Duncan et al. (1967), and Lemon, et al. (1971). A detailed model of corn photosynthesis and growth utilizing light, water, nitrogen, and leaf age (de Wit et al., 1970) lacks the necessary flexibility for adaptation to different species or species groups. Primary production models based on energy budgets (Murphy and Knoerr, 1972; Miller and Tieszen, 1972) and potential evapotranspiration (Rosenzweig, 1968; Rose et al., 1972) also seemed inappropriate for the objectives addressed.

3.3 Phenology Submodel

The objectives of the phenology submodel were to combine hypotheses relating changes in plant phenology to plant characteristics and changes in weather. This phenology model is thought to be the first attempt to simulate plant development through a complete growth cycle and use the results to regulate other ecosystem processes.

Growth and development of plants and their consumers differ from year to

Table 3.1.   Parameter values for Pawnee version of Elm

Parameter	FORTRAN Symbol	Warm-season grass	Cool-season grass	Forb	Shrub	Cactus
c_1	PRDC	0.09	0.09	0.07	0.07	0.20
c_3	PGA	70.0	90.0	70.0	90.0	200.0
c_4	PPC	50.0	28.0	40.0	50.0	300.0
c_5	PGRZM	0.09	0.16	0.16	0.16	0.16
c_6	PLRT	0.3	0.4	0.4	0.4	0.3
c_7	PURT	0.9	0.9	0.9	0.8	0.9
c_8	PTPRT	4.0	5.0	5.0	5.0	5.0
c_9	PTPGM	0.0	0.0	0.0	0.0	0.0
c_{10}	PUSTR	0.04	0.02	0.06	0.06	0.04
c_{11}	PSIGR	0.002	0.002	0.002	0.002	0.002
c_{12}	PBINC	0.1	0.3	0.3	0.3	0.3
c_{13}	PCRES	0.00005	0.000015	0.00005	0.00007	0.00005
c_{14}	PBDEC	0.00025	0.00005	0.00075	0.00015	0.000006
c_{15}	PSDTH	0.02	0.04	0.04	0.04	0.0005
c_{16}	PFALR	0.99	0.99	0.99	0.99	0.99
i_1	PSAT	450.0	400.0	380.0	400.0	450.0
p_1	PP1	5,000.0	4,000.0	4,000.0	4,000.0	4,000.0
p_2	PP2	7,000.0	7,000.0	6,000.0	7,000.0	7,000.0
p_3	PP3	9,000.0	8,000.0	7,000.0	9,000.0	9,000.0
p_4	PP4	9,500.0	8,500.0	8,000.0	11,000.0	8,500.0
p_5	PP5	10,000.0	9,000.0	8,500.0	12,000.0	9,000.0
p_6	PP6	12,000.0	12,000.0	11,000.0	11,000.0	12,000.0
p_7	PP7	6,000.0	6,000.0	6,000.0	6,000.0	6,000.0
p_8	PP8	2.0	6.0	5.0	3.0	10.0
p_9	PP9	0.008	0.008	0.008	0.008	0.008
Q	PQEFF	0.07	0.04	0.06	0.05	0.008
r_1	PCNG			all phenophases and producers—0.2		
r_2	PFRT	0.0006	0.0006	0.0006	0.0006	0.0006
r_3	PROT	0.3	0.3	0.3	0.3	0.3
r_4	PRTRS	0.000012	0.000007	0.000015	0.000015	0.0000004
r_5	PRTDM	0.0030	0.0004	0.0021	0.0015	0.00015
r_6	PSLOS	0.30	0.30	0.40	0.10	0.002
t_5	PMIN	7.0	3.0	7.0	8.0	7.0
t_6	POPT	32.0	26.0	26.0	35.0	25.0
t_7	PMAX	45.0	45.0	45.0	45.0	45.0
t_{10}	PFSG	10.0	7.0	7.0	6.0	7.0
t_{11}	PFDAM	−10.0	−10.0	−10.0	−10.0	−40.0
w_1	PWWP	−30.0	−30.0	−30.0	−30.0	−40.0
w_2	PETDY	125.0	125.0	125.0	125.0	125.0

year as a result of contrasting weather patterns. The onset of anthesis in *Agoseris glauca* varied 45 days in a 5-yr period, and the date of first bloom in *Agropyron spicatum* varied 31 days in a 23-yr period (Blaisdel, 1958; Mueggler, 1972), indicating that time of year is a poor index of plant phenological state. Forage quality and hence herbivore activities depend on the producer phenology. It is well known that young tissue is more palatable and nutritious than dry straw. Reproductive activity and success in birds (Snow, 1968) and antelope (Ellis, 1970) are related to the phenology of the food plants.

In this model a producer category can be distributed over more than one phenophase (Table 3.2) at any time. Flows between phenophases (Fig. 3.1) changes this distribution and thereby simulates phenological change. In nature, differences in genotype, age, and microclimate result in individuals within a species being phenologically asynchronous. In *Bouteloua gracilis*, less than one-

Table 3.2   Definitions of the Seven Phenophases

1. Winter quiescence or first visible growth
2. First leaves fully expanded
3. Middle leaves fully expanded
4. Late leaves fully expanded; first floral buds
5. Flowering (floral buds, open flowers, and ripening fruit)
6. Fruiting (buds, flowers, green and ripe fruit, and dispersing seeds)
7. Dispersing seeds and senescence

half of the tillers flowered in a year with average rainfall, but with more summer rain a higher proportion of tillers flowered (Sims et al., 1973; Dickinson and Dodd, 1976).

3.3.1 Phenophases

Plant development is divided into seven phenophases (Table 3.2) applicable to species with vegetative growth followed by flowering, the sequence most com-

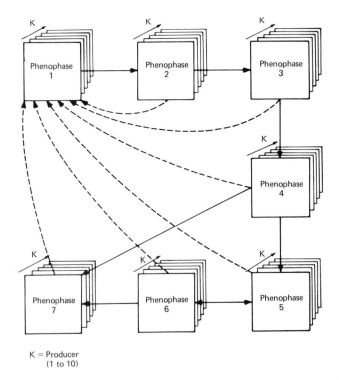

Fig. 3.1.   State variables and flows in phenology model (dashed lines indicate senescence-ending flows)

mon in grassland species. In the first phenophase, winter quiescence, growth, and development are limited by low temperatures. Phenophases 2, 3, and 4 represent progressive leaf expansion and maturation. Phenophase 4 also describes the transition between vegetative and reproductive growth with floral bud initiation. Phenophase 5 is a flowering phenophase with flowers more numerous than buds and ripening fruit. Phenophase 6 represents a fruiting plant that might have some buds, flowers, and ripe seeds. Phenophase 7 describes senescence including both death of annuals and shoot death of perennials.

3.3.2 Interphenophase Flows

Air Temperature and Insolation. Interphenophase flows depend on a 10-day moving average of the product of insolation and maximum air temperature, t_1 (cal $\cdot$ cm^{-2} $\cdot$ °C). Table 3.3 contains definitions of terms and functions and their units. This formulation is a compromise among estimates of the effect of temperature and insolation on plant development. The cumulative sum of the product of insolation and average air temperature was correlated with the flowering time of lilac (*Syringa vulgaris*) (Caprio, 1971). Maximum air temperature, rather than the mean of the daily maximum and minimum, and a moving average, rather than a cumulative sum, were correlated with flowering time in the annual forb *Clarkia tembloriensis* (Sauer, 1976). Plant development, therefore, depends on metabolic rate (temperature) and carbohydrate accumulation as related to insolation. Each simulated phenophase has a lower limit of t_1, below which no outflow occurs, and an upper limit, above which flow to the next phenophase is not limited by t_1. Between these limits, flow to the next phenophase increases linearly with t_1. The 10-day moving average is used to simulate delay in plant response to weather changes.

Soil-water Potential. The effect of soil-water potential on phenological change and carbon flow is a function of the water-absorbing capacity of a producer's roots in each soil layer and the soil-water potential. The soil-strata thicknesses, from the surface, are 1.5, 2.5, 11, 15, 15, 15, and 15 cm. Water-absorbing capacity is an estimate of the fraction of a producer's total water requirements obtainable from a soil layer. For example, grasses have near-surface roots, whereas dicotyledonous species are more deeply rooted (Weaver, 1958). The water-absorbing capacity of roots is assumed to decrease exponentially with increasing soil depth. The water-absorbing capacity is calculated first in Eq. (3.1) and then normalized so that the w values sum to 1.0.

$$w(j) = \exp(-c_1 a_{j-1}) - \exp(-c_1 a_j) \qquad (3.1)$$

In Eq. (3.1) a_j is the depth at the bottom of soil layer $j (a_0 = 0)$. In Eq. (3.1) and all subsequent equations, parameters and variables that describe plant characteristics are species (functional group)-specific. A subscript, k, indicating such dependence is omitted. See Table 3.1 for values of the parameters for each species. This function was determined emperically from data as in Cline et al. (1977). The

Table 3.3.   Definitions of terms and functions used in this chapter

Symbol	Definition
a	Temporary variable (Depends on use)
B	Live crown weight (g/m^2)
B_t	Total live crown (g/m^2)
c_1	Root-distribution coefficient (ND)
c_2	Shoot weight when basal area is 1 m^2 (g/m^2)
c_3	Shoot weight when leaf area (one side) is 1 m^2 (g/m^2)
c_4	Ratio of grazing removal and net photosynthesis that will cause a pristine pasture to become overgrazed in three years (ND)
c_5	Minimum fraction of photosynthate leaving shoots (ND)
c_6	Maximum fraction of photosynthate leaving shoots (ND)
c_7	Root:shoot ratio at equilibrium (ND)
c_8	Maximum ratio of live crown to live shoot, crown, and root for translocation from crown to shoot (ND)
c_9	Fraction of photosynthate moved from shoots going to crown (ND)
c_{10}	Relative effect of precipitation to effect water content in the fall of standing dead (ND)
c_{11}	Smallest storm size causing fall of standing dead (cm H$_2$O)
c_{12}	Largest storm size causing an increase in rate of fall of standing dead (cm H$_2$O)
D	Standing dead shoots (g/m^2)
e_a	Effect of phenology (ND)
e_g	Effect of grazing (ND)
e_N	Effect of nitrogen (ND)
e_P	Effect of phosphorus (ND)
e_t	Effect of temperature (ND)
e_w	Effect of soil-water potential and root distribution (ND)
g	Gross photosynthetic rate (g C/m^2/day)
$h(x_1, x_2, y_1, y_2, x)$	an s-shaped piecewise linear function

$$
h(x_1,x_2,y_1,y_2,x) = \begin{cases} y_1 & x \leqslant x_1 \\ (y_2 - y_1)/(x_2 - x_1)(x - x_1) + y_1, & x_1 < x < x_2 \\ y_2, & x \geqslant x_2 \end{cases}
$$

Symbol	Definition
i_1	Light saturation for photosynthesis (ly)
i_2	Fraction of intercepted light available (ND)
i_3	Fraction of light intercepted by canopy (ND)
k	Producer group (ND)
l	Length of the photoperiod (h)
L	Live shoots removed by grazing (g/m^2/day)
M	Mean phenophase (ND)
n	Net photosynthetic rate (g/m^2/day)
$P(1-7)$	Phenophase contents (fraction) (ND)
p_1-p_9	Values of t_1 for phenology change (ly °C)
Q	Quantum efficiency (ND)
R	Live roots (g/m^2)
r_1	Phenophase transfer coefficient (time^{-1})
r_2	Shoot-to-seed transfer coefficient (time^{-1})
r_3	Seed-to-shoot–root transfer coefficient (time^{-1})

Table 3.3. (continued)

Symbol	Definition
r_4	Crown to shoot transfer coefficient (time^{-1})
r_5	Crown respiration coefficient (time^{-1})
r_6	Crown death coefficient (time^{-1})
r_7	Shoot death coefficient (time^{-1})
r_8	Root respiration coefficient (time^{-1})
r_9	Root death coefficient (time^{-1})
r_{10}	Standing dead:litter coefficient (time^{-1})
S	Seeds (g/m^2)
T	Live shoots (g/m^2)
t_1	10-day moving average of product of t_3 and insolation (ly °C)
t_2	10-day moving average of product of daily minimum air temperature and (1-$e_{w,k}$) (°C)
t_3	Average photoperiod air temperature (°C)
t_4	Range in daily temperature (°C)
t_5	Daily minimum temperature (°C)
t_6	Minimum temperature for net photosynthesis (°C)
t_7	Optimum temperature for net photosynthesis (°C)
t_8	Maximum temperature for net photosynthesis (°C)
t_9	Soil-surface temperature (°C)
t_{10}	Temperature minimum for seed germination (°C)
t_{11}	Soil temperature at 15 cm (°C)
t_{12}	Minimum temperature for crown–shoot flow (°C)
t_{13}	Maximum temperature for frost damage to live shoots (°C)
w_1	Distribution of water-absorbing capacity, normalized (ND)
w_2	Permanent wilting soil-water potential (-bars)
w_3	Temperature and precipitation accumulation (°C cm H$_2$O)
w_4	Daily precipitation (cm H$_2$O)
w_5	Water content of plant canopy
δt	Time step (days)
Ψ	Soil-water potential (-bars)

grasses are assumed to have more roots near the surface than forbs and shrubs. Cactus roots are assumed to be closest to the soil surface. Root distribution is needed to simulate surface-rooted types that are more responsive to small rains than deep-rooted types.

Interspecific differences in the permanent wilting soil-water potential (Slatyer, 1957) are specified via the variable w_2, the permanent wilting soil-water potential for producer group k. The effect of soil-water potential on a producer is calculated in Eq. (3.2):

$$a = \begin{cases} 1 - \dfrac{\psi_j}{w_2}, & \psi_j \geq w_2 \\ 0, & \psi_j < w_2 \end{cases} \tag{3.2}$$

$$e_w = \sum_{j=1}^{n} a w_1(j)$$

where n is the number of simulated soil strata. As all strata approach field capacity, e_w approaches 1.0; if all strata are at or below w_2, e_w is 0.0. This relationship between depth, soil-water content, and root distribution allows differences in wilting and root-depth characteristics of producers to be included. The relationship between soil-water potential and photosynthesis is sigmoid (Brix, 1962; Brouwer, 1963, but a linear relationship seems a better approximation when considering several groups.

Vegetative Phenology. Under lush conditions, vegetative development is slowed to increase carbohydrate assimilation for use as winter reserves for perennials, or seeds in annuals. Sunny, warm weather is assumed to accelerate phenological change, and cool, wet weather slows it. In the desert annual *Plantago insularis,* the earliest flowering and fruiting were at medium water stress, maximum flowering and fruiting were at low water stress, and high water stress accelerated the decline of the population (Klickoff, 1966).

Progression to the next vegetative phenophases is a function of t_1, the current phenophase and e_w [Eq. (3.3)]:

$$f(x,y) = r_1(x)P(x)\alpha(t_1)\beta(e_w)\, \delta t \tag{3.3}$$

where

$$\alpha(t_1) = h[p_x, p_y, 0, 1, t_1] \tag{3.4}$$

and

$$\beta(e_w) = 0.5[\sin(\pi e_w) + 1] \tag{3.5}$$

$f(x,y)$ is the flow from phenophase x to phenophase y, and h is a function defined in Table 3.3. Note that α lies between 0 and 1. If t_1 is less than the threshold value for a particular phenophase or if none of the plant is in phenophase x, flow to the next phenophase is zero. Also, drought and abundant soil water slow vegetative change [Eq. (3.5)], which yields a value of 0.5 near field capacity and permanent wilting and a value of 1 midway between these two limits.

Reproductive Phenology. Flowering begins when the threshold value of t_1 is passed.

$$f(4,5) = r_1(4)P(4)\alpha(t_1)e_w\, \delta t \tag{3.6}$$

In Eq. (3.6) the soil-water potential term, e_w, decreases the rate of change to a reproductive state with decreasing soil-water potential. Thus flowering occurs when heat and carbohydrate reserve requirements are satisfied.

With continued soil drying, plant development proceeds to phenophase 6:

$$f(5,6) = \begin{cases} r_1(5)P(5)\alpha(t_1)\, \delta t,\ e_w \leq 0.8 \\ -\gamma_1(5)P(6)\, \delta t,\ e_w > 0.8 \end{cases} \tag{3.7}$$

However, with an increase in soil-water potential with e_w greater than 0.8, a reverse flow from a predominantly fruiting condition to a flowering condition occurs. The annual forb *Clarkia tembloriensis* resumed flowering, even though near senescence, after an unusually heavy summer rain (Sauer, unpublished data). The perennial grass *Bouteloua gracilis* responded similarly (Sims et al., 1973). The flow from phenophases 5 to 6 is also a function of t_1, indicating dependence on heat and carbohydrate accumulation.

Senescence. Soil water has a major influence on phenological change as the plants progress toward senescence at the end of the growing season. Senescence is an adaptation to unfavorable periods in a plant's life cycle (Leopold, 1961), and thus increasing flow rates to phenophase 7 with increasing drought is reasonable. Flow to phenophase 7 [Eq. (3.8)] is proportional to $1-e_w$ so that with increasing soil drought, senescence occurs sooner.

$$f(6,7) = r_1(6)P(6)\alpha(t_1)(1 - e_w)\, \delta t \tag{3.8}$$

An additional pathway to senescence is simulated by flowing from phenophases 4 to 7:

$$f(4,7) = r_1(4)[P(4) - f(4,5)]\alpha(t_1)(1 - e_w)\, \delta t \tag{3.9}$$

where $\alpha(t_1)$ is defined by Eq. (3.4). This bypass of the reproductive phenophases becomes increasingly significant with increasing drought.

Ending Senescence. The modeled response of both perennial and annual species to the reoccurrence of cool, wet weather is to end senescence. Terminating senescence is simulated by transferring the contents of all phenophases to phenophase 1, winter quiescence. The onset of cool, wet conditions is indicated by a decrease of the moving average of the product of maximum air temperature and soil-water deficit ($1-e_w$) below a species-specific threshold (t_2). The delay of plant response to changing weather is simulated by the 10-day moving average. The delay achieved by the moving average is determined by the rate of change of values averaged.

Because new growth is in phenophase 2, nearly complete (<1 g/m² remaining) shoot removal (grazing) also ends senescence and causes all material to move into phenophase 1.

The flows calculated above are added and subtracted from the appropriate phenophases to update phenophase contents. The mean phenophase for each species, M, is calculated by summing the product of each phenophase index (1,

2, 3, . . . , 7) and its contents (a decimal fraction). The result varies from 1.0 to 7.0 and is the phenological index used elsewhere in the model.

3.3.3 Standing-dead Phenology

Palatability and nutrient values for standing dead are required by consumer sections of ELM. Dead shoots lose forage value through leaching (rainfall) and microbial activity. Because standing-dead forage quality continually decreases through the season, a cumulative sum is used to indicate the extent of change from recently dead material. The cumulative sum is restarted when new standing-dead material is produced as the plants progress through the intermediate vegetative phenophases ($M = 2$–3).

Changes in standing dead from a recently dead condition to leached and bleached straw is represented as follows:

$$M = \begin{cases} p_9 w_3 + 7, \ p_9 w_3 \leqslant 2 \\ 9 \qquad\quad , p_9 w_3 > 2 \end{cases} \tag{3.10}$$

where w_3 is an accumulation of daily values of the product of rainfall (cm) and mean canopy temperatures (°C); p_9 is a constant. The range 7–9 of values of M was chosen to extend the live-plant phenology range and indicates lower nutritive quality of dead forage than live.

3.4 Carbon Submodel

The objective of the carbon submodel is to simulate the carbon dynamics of several producer groups. The response of the submodel is to be biologically reasonable for weather conditions that vary from severe drought to conditions in which water, nitrogen, and phosphorus are nonlimiting. In addition, response to grazing should also show the trend expected from data and experience. Multiyear simulations to evaluate cumulative effects of treatments require that the submodel have an overwintering capacity.

The carbon pathway on the primary producer model is shown in Figure 3.2. The paths that return dead plant material to the atmosphere are simulated in the decomposer section (Chap. 6). Photosynthesis fixes atmospheric carbon in shoots from which it is distributed to shoots, standing dead, crowns, seeds, and roots. Root respiration returns carbon to the atmosphere directly, and root death moves live root carbon to belowground litter. The fall of standing dead moves carbon to aboveground litter. Seed carbon returns to roots and shoots with seed germination. Shoots (leaves and stems), standing dead, crowns, live roots, and seeds are simulated for each producer group. The above- and belowground litter compartments are common to all producers.

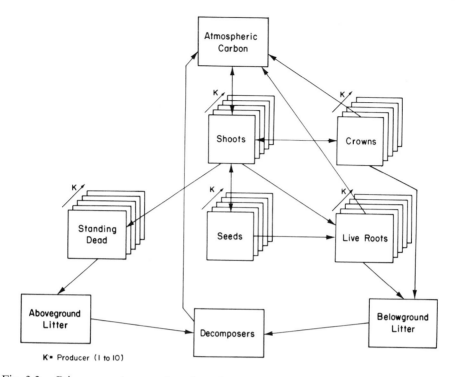

Fig. 3.2.   Primary producer carbon flow diagram (all compartments are species-specific except the dead-root and litter compartments common to all producers)

3.4.1 Gross Photosynthesis

The gross photosynthetic rate g of each producer group is regulated by soil-water potential, photoperiod temperature, nitrogen, phosphorus, intercepted insolation, quantum efficiency, light-saturation level, and phenology:

$$g = 1.032 \min(e_w, e_t, e_N, e_P, e_a)\min(i_1, i)i_2 Q \tag{3.11}$$

The variables e_w, e_t, e_N, e_P, e_a, and i_2 are between 0.0 and 1.0 and dimensionless. The constant 1.032 converts cal/cm² to g C/m² (grams of carbon), based on Loomis and Williams (1964).

The concept of a gross photosynthetic rate limited by available photosynthetically active incident radiation $\min(i_1, i)i_2$ is based largely on Connor (1973) and the calculation of e_w is given in Eq. (3.2).

The factors regulating gross photosynthesis are presented as limiting with the smallest factor having control of the process. This formulation was used because data to support the use of other representations of interactions between all factors were not available.

Temperature.   The temperature used to regulate gross photosynthetic rate is average photoperiod canopy temperature t_3:

$$t_3 = t_4(1 - \cos a)/a + t_5 \qquad (3.12)$$

where a is $l\pi/(l + 4)$, l is photoperiod length in hours, t_5 is daily temperature minimum, and t_4 is daily temperature range. Equation (3.12) was determined by integrating two sine curves joined at extrema to produce the maximum daily air temperature at 1200 and the minimum at 0600.

For each producer a minimum, optimum, and maximum photoperiod temperature for photosynthesis was identified (Mooney et al., 1964; Mooney et al., 1966; Dye et al., 1972; Knievel and Schmer, 1971; Bokhari and Dyer, 1973). The function h, using average photoperiod temperature as the argument, produced an s-shaped curve for e_t that is zero at or below the minimum temperature and 1.0 at or above the optimum.

Nitrogen and Phosphorus.   Young plants are more responsive to mineral nutrition changes than are old plants (Remy, 1938). This is modeled by having phenology and changes in shoot concentrations of nitrogen and phosphorus regulate the effect of N and P on photosynthesis. Both e_N and e_P depend on the phenophase of the producer using a Michaelis–Menton expression and are shown in Figure 3.3 for N and Figure 3.4 for P.

The nutrient submodels determine shoot concentrations of N and P for all producers combined; thus the nitrogen-starvation level (0.007 g N/g dry wt shoot) and the phorphorus-starvation level (0.001 g P/g dry wt shoot) (C. V. Cole, personal communication) do not depend on producer category.

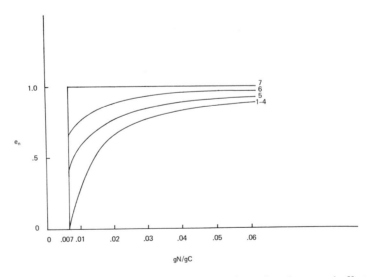

Fig. 3.3.   Relationship between shoot N concentration, phenology, and effect of N

Phenology. The effect of phenology (e_a) varies from 1.0 to 0.0 as the mean phenophase varies from 1.0 to 7.0. For phenologic states up to flowering (phenophase 5), e_a is 1.0; thereafter it decreases linearly to 0.0 at senescence. A value of 0.0 at senescence prevents photosynthesis in dormant plants, and lower values from flowering to senescence reflect the lower photosynthetic efficiency of older shoots.

Insolation. The abiotic section of ELM simulates values for diffuse (cloudy-sky) and direct (clear-sky) photosynthetically active radiation. All diffuse radiation is intercepted. Direct radiation is divided into scattered (0.25) and nonscattered (0.75) fractions (de Wit, 1965). The scattered fraction is divided equally into upward and downward components. Nearly all of the downward component (90%) is absorbed by the soil surface. Of the upward scattered component, 10% escapes from the canopy because of the relatively low leaf area index (<1.0). The insolation potentially available for interception [i of Eq. (3.11)] is thus the sum of the diffuse, nonscattered direct, 10% of the downward scattered, and 10% of the upward scattered radiation.

A distinction between C_3 and C_4 species is the minimum value of insolation required for a maximum rate of photosynthesis [i_1 in Eq. (3.11)]; Ludlow and Wilson, 1971); C_3 species saturate at lower light intensities than do C_4 species. *Bouteloua gracilis,* the C_4 species (Williams and Markley, 1973) representing warm season grasses, has a saturation level of 450 ly. The other noncactus species are C_3 (Williams and Markley, 1973) and were given saturation values of 400 ly. The saturation level for cactus was set high, 500 ly, because succulents are not light-saturated (Mooney, 1972). For comparison, maximum simulated insolation values were approximately 500 ly.

Light is intercepted by both live and dead shoots, but only the fraction of light intercepted by the live portion is avilable for photosynthesis. Insolation interception is calculated from cover and leaf-area index to estimate the fraction (i_2) of

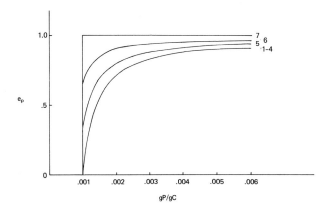

Fig. 3.4.   Relationship between shoot P concentration, phenology, and effect of P

insolation available for photosynthesis and thereby estimate the potential rate of gross photosynthesis.

The basal area b of each species is calculated from shoot weight (live, T, and dead, D)

$$b = \begin{cases} (T + D)/c_2, & (T + D) < c_2 \\ 1 & ,(T + D) \geq c_2 \end{cases} \quad (3.13)$$

as the fraction of 1.0 m^2 covered by a producer. If a producer's basal area exceeds 1.0 m^2, $b = 1$ and further growth is assumed to be upward.

Leaf area (one side, live plus dead) is a linear function of the known weight of one m^2 of leaf area: $(T + D)/c_3$. Typical values of c_3 are 24–48 g C (Knight, 1972; Brown, 1974). The leaf-area index, a, of a producer is the ratio of leaf area to cover. The cover of each species is assumed nonoverlapping in the presence of bare soil. The computation of a under the condition of total cover is described below.

The fraction of incident shortwave radiation intercepted by a producer canopy, i_3, is estimated from Anderson and Denmead (1969):

$$i_3 = 1 - \exp(-Za) \text{ where } \begin{cases} Z = 1.25 \text{ for } a \leq 0.5 \\ Z = 1.1 \text{ for } 0.5 < a < 1 \\ Z = 0.9 \text{ for } a \geq 1 \end{cases} \quad (3.14)$$

The fraction of shortwave radiation incident on the m^2 intercepted by live and dead shoots of a producer is the product of i_3 and b.

Light interception as calculated above applies when shoots do not overlap and compete for light. However, if water and N are added, leaf-area indices can increase 10-fold (Knight, 1973). If the ground surface is covered, incoming insolation is divided among the producers in proportion to their leaf-area index. The fraction of light intercepted by the live shoots, i_2, is calculated from the fraction of live shoots to total shoots.

Quantum Efficiency. The quantum efficiency [Q in Eq. (3.11)] is greater for C$_4$ species than for C$_3$. Maximum values for quantum efficiency range from 0.1 for C$_4$ species to 0.06 for C$_3$ species (Ludlow and Wilson, 1971). The quantum efficiency of cactus species (crassulacean-acid metabolism) is approximately 8% that of C$_4$ species or 0.008 (Sesták et al., 1971).

3.4.2 Shoot Respiration

Shoot respiration rate, a temperature-dependent function of the gross rate of photosynthesis, increases linearly from 0.0 to 0.3 of the gross photosynthetic rate as average photoperiod temperature rises from the minimum to the optimum temperature for net photosynthesis, and increases linearly from 0.3 to 1.0 of the gross rate as the average photoperiod temperature increases to the maximum temperature for net photosynthesis. (Dye et al., 1972). Grazing damage (see below) further increases shoot respiration.

3.4.3 Net Photosynthetic Rate

The net photosynthetic rate of each producer is the difference between the gross photosynthetic rate and the shoot respiration rate. Gross photosynthesis and shoot respiration are modeled separately so that effects of variables such as temperature, soil water, and phenology on net and gross photosynthesis can be differentiated.

3.4.4 Effect of Grazing

The relationship between grazing, growth, and primary production is a difficult problem. Initially, it was assumed that grazing influenced primary production through removal of green tissue. Simulations with grazing showed very little reduction in net production, even at unrealistically high stocking rates. Furthermore, no cumulative effect of grazing was seen after 4 simulated years, even though 7–10% of the daily growth was being removed. Growth and consumption rates were checked and found correct. The impact of grazing is more than just herbage removal.

Shoot respiration was hypothesized to be increased by grazing. Respiration was increased to a maximum of 125% of what it would be without grazing to account for damage from trampling and incomplete removal [Eq. (3.15)].

$$e_g = 1 + 0.25 \ L/(c_4 n) \tag{3.15}$$

where L is live material grazed, c_4 is a constant, and n is the net photosynthesis rate.

In addition to the significant damage caused by trampling (Laycock et al., 1972), herbage removal and trampling increase the fraction of photosynthate translocated from the shoots to the roots [Eq. (3.16)]

$$a_3 = c_5 + (1 - c_5)L/n \tag{3.16}$$

where a_3 regulates carbon flow from shoots and c_5 is a constant. In Eq. (3.16), a_3 varies from a small fraction of photosynthate flowed from the shoots to 1.0 as grazing increases from zero to a level sufficient to cause a "pristine pasture" to become "heavily overgrazed" in less than 3 yr.

Annual producers were treated differently because annuals do not store reserves in roots and crowns for winter. In the case of an annual, only shoot respiration increases with grazing.

3.4.5 Translocation to Roots and Crowns

The fraction of net photosynthate transferred to live roots and crowns determines photosynthetic area and hence, in large part, shoot growth. Transfer from shoots to roots is regulated by seven factors: (a) live-shoot weight, (b) live-root weight, (c) effects of soil-water potential (e_w), (d) nitrogen (e_N), (e) phosphorus (e_P), (f) phenology, and (g) grazing:

$$a_1 = \begin{cases} c_5 e_p \geqslant .1 \\ \\ c_6 e_p < .1 \end{cases}$$

$$a_2 = \begin{cases} c_5 e_n \geqslant .1 \\ \\ c_6 e_n < .1 \end{cases}$$

$$a_4 = \begin{cases} h(4, 7, c_5, c_6, M), \text{ perennial} \\ c_5 \quad \text{ annual} \end{cases}$$

$$a_7 = c_7 T/R$$

$$a_5 = \begin{cases} a_6, a_7 < c_6 \\ \\ a_7, a_7 \geqslant c_6 \end{cases}$$

$$a_6 = \begin{cases} c_5 e_w \geqslant .2 \\ \\ c_6 e_w < .2 \end{cases} \tag{3.17}$$

$$f = \max(a_1 a_2 a_3 a_4 a_5 a_6)$$

where a_3 is defined in Eq. (3.16); h is defined in Table 3.3; f is the fraction of net photosynthesis translocated to the crowns and roots. Variables a_1, a_2, a_4, and a_6 are between c_5 and c_6, a_5 between zero and c_5, and a_3 between c_5 and 1.0. This functional relationship is based on the observations that shoot removal slows root growth until the original shoot:root ratio is regained (Brouwer and de Wit, 1968; Dittmer, 1973) and that decreased water, nitrogen, and phosphorus in the soil result in increased root:shoot ratios in *Lolium perenne* L. and *Trifolium repens* L. (Davidson, 1969). Under ordinary conditions of soil water potential, nitrogen or phosphorus, these variables have little effect. However, as the effect of one or more of these factors decreases below 0.1 (0.2 for e_w), shoot-to-root transfer increases to the maximum value, c_6. Shoot concentrations of N and P are used as indicators of the soil conditions for these computations.

In a simulation of an Arctic tundra (Miller and Tieszen, 1972) roots did not return to pregrowing season weights at the end of the growing season, suggesting the need for some mechanism for increased belowground translocation, such as increased translocation with advancing phenology. In ELM, phenology controls translocation to belowground parts as follows. Vegetative plants transfer the minimum fraction, c_5, to the roots and crowns. As plants mature the fraction of net photosynthate flowing to the roots and crowns increases linearly to the maximum, c_6. Lower translocation rates in spring produce faster shoot growth and a rapid approach to peak leaf area while the increase in transfers with maturity replenish reserves for winter. For annuals phenology does not regulate translocation from shoots to roots and crowns.

The fraction of net photosynthate sent to roots and crowns is determined by the largest of six fractions [Eq. (3.17)]. The largest, rather than the smallest, of the fractions is used because flow to roots is based on conditions that increase, not decrease, root:shoot ratios. The constants c_5, c_6, and c_7 are adjusted so that on the average, 85% of net photosynthesis for a season (Singh and Coleman, 1974) is translocated to the roots and crowns in ungrazed systems. The division of translocated material between the crowns and roots is constant.

All producer groups are assumed to have similar requirements for root surface and storage per unit leaf area. Initial values of roots and crowns in each producer category are important because they partially regulate the dynamics of the producer system. Unfortunately, available initial conditions were for all producers combined. Experimentally, it is difficult to separate roots by producer or determine the use of stored material for early growth. Roots and crowns are assumed to be 60% live and 40% dead (Singh and Coleman, 1974). Total live root and crown carbon is divided among the producer groups according to the fraction of the total live-shoot weight in each producer group.

3.4.6 Seed Production

For the perennials seed production is relatively unimportant. Annuals, however, are dependent on seed production, although seed storage in the soil may reduce the necessity of seed production in a given year (Lewis, 1962; Mott, 1972). Flowering and fruiting shoot weight, and not photosynthetic rate, are used to regulate seed growth. In wheat, for example, the heads are filled with carbohydrate stored in the stems. If soil water is available, the heads remain green and the phytosynthate contributed by the heads is also available for head filling (Asana and Saini, 1958). Carbon flow to seeds is the product of the fraction of shoot weight in phenophases 5 and 6 (flowering and fruiting), the shoot weight, and a rate constant (r_2).

3.4.7 Seed Germination

Seed germination is represented to complete the carbon pathway and simulate annuals. Germination occurs when the soil-water potential of the top two layers (ψ_1 and ψ_2) averages $>$-5 bars. The rate increases with increasing soil surface temperature from t_{10} to an optimum at ($t_{10} + 5°C$) and decreases to zero at ($t_{10} + 10°C$). Seed weight and a rate constant complete the representation [Eq. (3.18)].

$$a_1 = (t_9 - t_{10} - 2.5)/10$$
$$a_2 = (\psi_1 - \psi_2)/2 \qquad\qquad (3.18)$$
$$a_3 = Sr_3(1 + \sin 2\pi a_1)/2$$

The germination rate (g $C/m^2/day$) is zero unless $a_2 \geq -5$ and $t_{10} < t_9 < t_{10} + 10$, in which case it is a_3. Germinating-seed carbon is equally divided between shoots and roots.

3.4.8 Translocation between Crowns and Shoots

Crowns (B) represent carbohydrate storage as in stem bases in grasses, the woody stems and roots in shrubs, and the pads in cacti. The carbon stored in the crowns is used for initial spring growth. Subsequent photosynthesis replenishes crown reserves. Older plants (mean phenophase > 4) are assumed too mature to translocate carbon from crowns to shoots. Early in the growing season, before shoot weight exceeds 2–5% to total live weight, crown carbon flow to shoots (f) is a function of soil-water potential and temperature [Eq. (3.19)]

$$a_1 = [(t_9 + t_{11})/2 - t_{12}]^+$$
$$a_2 = h(0, 10, 0, 1, a_1)$$
$$a_3 = T/(B + T + R) \qquad\qquad (3.19)$$
$$a_4 = \begin{cases} 1, & a_3 < c_8 \\ 0, & a_3 \geq c_8 \end{cases}$$
$$f = Br_4 a_2 a_4$$

where a_2 is an s-shaped curve going from 0 to 1 as a_1 goes from 0 to 10 and $[x]^+$ = $\max(0,x)$.

Increasing soil temperature increases crown metabolic rate and initiates translocation to shoots. The ratio of live shoot to total live weight indicates when photosynthesis alone can supply shoot growth needs. Additionally, this ratio represents the fraction of live carbon available for initial growth while the remaining carbon is nonlabile or reserved for other purposes. The ratio is adjusted for each producer to provide sufficient growth to obtain the observed peak crop. Since peak crop is also regulated by photosynthetic rate, c_8 is only one of several producer characteristics considered.

When the ratio of shoot carbon to total live carbon exceeds c_8, translocation is from the shoots to the crowns. The fraction of net photosynthate translocated to roots and crowns is determined in Eq. (3.17). The fraction c_9 of the carbon translocated from the shoots that goes to the crowns is constant. Carbon flow to the live roots is the remainder of that translocated from shoots. A consequence of this representation during early spring growth is no shoot-to-crown flow which further increases the early growth rate of shoots.

3.4.9 Crown Respiration

Crown respiration rate increases with increasing temperature of the top 15 cm of soil and crown live weight. When soil temperature is below t_{12}, crown respiration is zero. Crown respiration rate is a product of soil temperature [a_1 in Eq. (3.19)], crown weight, and a rate constant r_5. Crown respiration is 3% to 22% of the total soil respiration rate (Clark and Coleman, 1972).

3.4.10 Crown Death

Information on crown death is limited and assumed to follow:

$$f = \begin{cases} r_6B, & \text{perennial} \\ r_6B(1 - e_a), & \text{annual} \end{cases} \tag{3.20}$$

where B is crown weight and f is flow rate from crowns to belowground litter in g C/m²/day. Phenology is a factor in crown death of annuals only and increases crown death with advancing phenologic state.

3.4.11 Shoot Death

Shoot death in young plants is less than in old plants because young plants have fewer dying leaves (Robson, 1973). The dependence of shoot death on phenology is simulated as a linear increase as mean phenophase progresses from 5 (flowering) to 7 (senescence) (see a_3 below). Shoot death due to drought increases linearly as soil-water potential drops below 20% of the permanent wilting soil-water potential (a_2). Frost damage causes 25% of the live-shoot biomass to be transferred to standing dead if the minimum daily temperature falls below t_{13}.

$$a_1 = 1 - e_w$$
$$a_2 = h(0, 0.8, 0.8, 1, a_1)$$
$$a_3 = h(5, 7, 0, 0.2, M)$$
$$f = \begin{cases} Tr(a_2 + a_3) & ,t_5 > t_{13} \\ T/4 & ,t_5 \leqslant t_{13} \end{cases} \tag{3.21}$$

where f is flow from shoots to standing dead (g C/m²/day).

3.4.12 Root Respiration

Root respiration is a product of average soil temperature in the top 60 cm, root weight, and a rate constant r_8. The effect of soil atmosphere is disregarded. Root respiration increases linearly with increased root weight and average soil temperature. The rate constant, r_5, representing the maximum daily respiration rate, adjusts root respiration rate of each producer to achieve reasonable ratios (0.07–0.15) of root respiration to death (Jameson and Dyer, 1973).

3.4.13 Root Death

Roots die at a constant rate per gram root in moist soil, and root death increases linearly as soil-water potential decreases below 10% of field capacity [Eq. (3.22)]. For perennials the root-turnover rate varies from once every 2 yr (Singh and Coleman, unpublished data) to 4 yr (Dahlman and Kucera, 1965).

$$a_1 = 1 - e_w$$
$$a_2 = h(0.75, 1.5, 0.9, 1, a_1)$$
$$a_3 = 1 - e_a$$
$$f = \begin{cases} r_a R a_2, & \text{perennial} \\ r_a R a_2 a_3, & \text{annual} \end{cases} \tag{3.22}$$

In moist soil ($e_w > 0.1$), 75% of r_9 of live roots die each day; at permanent wilting ($e_w = 0.0$), root death rate doubles. For annuals phenology decreases root death in young plants and increases root death in old plants so that roots and shoots die simultaneously. The functions of root growth and death give the decreased root growth at the end of the growing season observed by Taylor et al. (1970) and the loss of roots as soil dries observed by Klepper et al. (1973).

3.4.14 Fall of Standing Dead

Standing dead is hypothesized to be pushed over by precipitation (both rain and snow) and to have its mechanical strength decreased by an increase in water content. The relative contribution of rain c_{10} (0.99) is set greater than that of water content (0.01), and the storm-size distribution (c_{11}, c_{12}) is set at 0.5–0.8 cm (summer storm size) to agree with the available data that show most loss of standing dead is in the spring and summer months. The functional relationship is shown in Eq. (3.23):

$$a_1 = h(c_{11}, c_{12}, o, c_{10}, w_4)$$
$$a_2 = 20,000 \, w_5/(T_t + D_t)$$
$$a_3 = \begin{cases} a_2, & a_2 \leq -c_{10} \\ 1 - c_{10}, & a_2 > 1 - c_{10} \end{cases} \tag{3.23}$$
$$f = r_{10} D(a_1 + a_3)$$

where a_1 is an s-shaped curve increasing from 0 to 0.99 with increasing storm size from 0.5 to 0.8 cm; a_2 is the ratio of water content to dry weight; 20,000 gives a maximum water-holding capacity of the live plus dead of ½ the dry weight; a_3 is the effect of water content on fall of standing dead relative to a_1; and f is the flow (g C/m²/day) of standing-dead carbon to litter.

3.5   Simulation Results

3.5.1 Biomass

Data collected in 1970–1971 at the Pawnee Site from an ungrazed pasture were used to develop and adjust the producer model. In Figure 3.5 simulated dry weight (2.5 × weight of carbon) is compared with 95% confidence intervals of data for the warm-season grass. Modeled live- and dead-shoot values were added for comparability with data.

The simulation fell below the confidence intervals early, and above the confidence intervals late, in the growing season. The simulation of cool-season grasses (not shown) appeared good. Forbs and shrubs (not shown) remained near the upper limit of the confidence intervals, suggesting an excessive production in these categories. The variation in the biomass of cactus (not shown) was much less than those of the other producer categories and appeared as a straight line. Data provided were the sum of live and dead shoots. The separation of these components was required in the model because standing dead does not photosynthesize and, in fact, interferes by shading. Sims et al. (1971) estimated 60% of the total herbage was green. In the model, green shoots were approximately 82% of the peak live crop, not 60%. The higher proportion green was necessary to obtain the photosynthetic capacity required to match the estimated minimum net primary production of 500 g dry weight · m^{-2} · yr $^{-1}$ (Sims and Singh, 1971).

The transfer of standing dead to litter is not well understood. This process is represented as being controlled by large summer rain storms. It is likely that

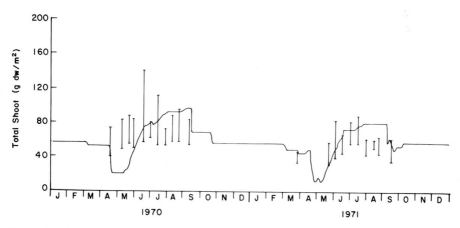

Fig. 3.5.   Comparison of simulation and observed data (live + dead shoots) for warm-season grasses (principally *Bouteloua gracilis*) on the Pawnee Site, 1970–1971 (vertical bars are 95% confidence intervals of observed data)

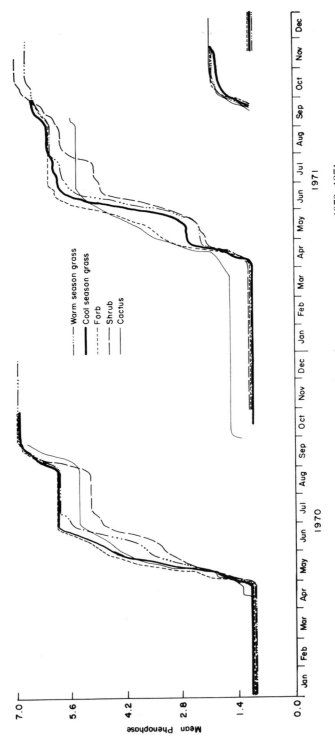

Fig. 3.6. Simulation of mean phenophase for warm-season grasses, 1970–1971

other factors such as wind transferred standing dead to litter. Weakness in the representation of this transfer accounted for simulations of live-plus-dead shoots outside the 95% confidence intervals at the beginning of the growing season.

3.5.2 Phenology

Figure 3.6 shows the mean phenophase of warm-season grasses, and Figure 3.7 shows each phenophase of warm-season grass in 1970. Observed temperature and precipitation were used as driving variables in these simulations. In 1970 simulated development started in late April, slowed in mid-June, continued into August, and at the beginning of September advanced rapidly to phenophase 7 (senescence). The 1971 phenologic progression started in early April, slowed in late April, resumed rapid change in late May, and again slowed in mid-June. In 1971 not all warm-season grass died as occurred in 1970, perhaps because of the wet snowstorm in September 1971. Recall that high soil-water content can reverse the flow to senescence. Warm-season grasses returned to phenophase 1

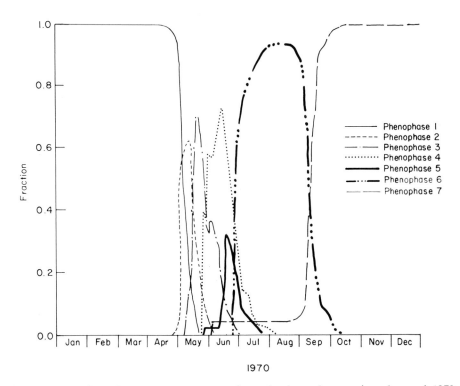

1970

Fig. 3.7.   Fraction of warm-season grasses in each phenophase, using observed 1970 temperature and precipitation of Pawnee National Grassland

later in 1971 than in 1970 because warm 1971 fall temperatures were too high to meet the low-temperature requirements for breaking senescence.

A more detailed look (Fig. 3.7) at the simulation of warm-season grass phenology shows the contents of the 7 phenophases through 1970. Rapid progression through phenophases 2 and 3, the vegetative phenophases, is indicated by the narrow peaks. Phenophase 4 had a longer residence time, whereas phenophase 5 didn't last long, indicating little flowering. Phenophase 6 was quite long and the majority of plants were in this stage from late June to early September, indicating a prolonged seed-ripening period and continued capacity for vegetative response to summer rain. Flow into phenophase 7, senescence, began in early June because soil water was below field capacity, allowing flow from phenophases 4 to 7. Flow into phenophase 7 from phenophase 6 accelerated in late August as temperature and insolation requirements were met. By mid-October all warm-season grass was in phenophase 7. Only before and after the growing season was all warm-season grass in a single phenophase.

3.5.3 Biomass Dynamics

The 1970–1971 dynamics of shoots, crowns, roots, seeds, and standing dead for warm-season grass is shown in Figure 3.8. The observed live-shoot peak at

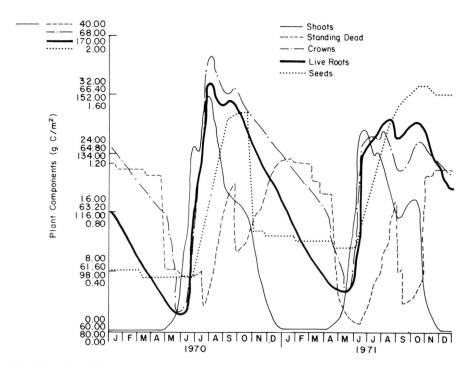

Fig. 3.8.   Simulation of weight change in warm-season grasses (g C/m²) of the five plant components in ELM using 1970–71 observed temperature and precipitation

the Pawnee site occurred in June of both years (Knight, 1972). Simulated warm-season grasses peaked in late July of 1970 and in late June of 1971. The effect of water on production is illustrated by the impact of the late summer 1971 snowfall when live shoots increased shortly before declining to winter values.

The role of crowns was evident early in the growing season when they decreased as carbon flowed to the shoots to provide initial photosynthetic area. Thereafter, the crowns were replenished. More than a single wet growing season might be required to recover from several dry years because crowns need more than one growing season to restore reserves. Similarly, recovery from heavy grazing might require several seasons.

Roots also provide carbon for early growth. However, the production from reserves is not simulated because roots and root reserves are not separated. Roots gain weight rapidly during the growing season, whereafter they decline until the next growing season.

The representation of flow from standing dead to litter facilitates rapid fluctuations in the standing dead. For example, little standing dead was lost during the winter. As growth began, dead shoots decreased rapidly. As green shoots died, standing dead accumulated and remained until the beginning of the next growing season. The transfer of standing dead to litter requires a firmer conceptual base. This seemingly irrelevant aspect of grassland biomass dynamics is important when viewed in an ecosystem context. The quantity of standing dead influences not only forage quality, but rainfall interception and photosynthetic rate.

Seeds increased at the time of flowering and fruiting and decreased during germination. Germination was insignificant during the winter when soil temperatures were low and during the summer when the soil was dry.

The effects of air temperature, soil water, and phenology on warm-season grass photosynthesis are plotted in Figure 3.9, and the effects of nitrogen, phosphorus, and the fraction of incident shortwave radiation intercepted by live warm-season grass shoots are plotted in Figure 3.10. Canopy temperatures before and after the growing season are too low for photosynthesis to occur in warm season grass ($e_t = 0$). During the growing season, however, e_t increased to 1, indicating that the optimum temperature for net photosynthesis was reached. The effect of soil water, e_w, decreased in mid-June to low summer values. Summer rain wetted the soil briefly, as narrow peaks in e_w indicate; hence soil-water stress limited growth for warm-season grass for much of the summer. Phosphorus shoot concentration never limited growth. On the other hand, nitrogen could be limiting, as e_N decreased to less than 0.4. However, because the soil-water effect, e_w, was less than that for nitrogen, it may be assumed that nitrogen limitations did not slow photosynthesis. The fraction of incident solar radiation intercepted by warm-season grasses (i_2) (Fig. 3.10) increased to 33% in late July.

The model provides an opportunity to accumulate flows such as photosynthesis for all producers for a time period to obtain carbon budgets. The annual accumulations of several flows are shown in Table 3.4. Production was less in 1971, probably because 1971 was drier than 1970. The peak live crop was 10.1 g dry weight /m² lower and approximately 1 month later in 1971 than in 1970.

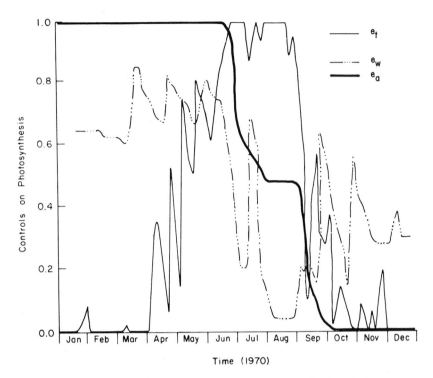

Fig. 3.9.   Seasonal variation in factors that control photosynthetic rate in warm-season grasses

Although respiratory losses were higher in 1971, death losses were lower or the same but did not compensate for the lower photosynthetic rate. Experimental data of this nature would be nearly impossible to obtain. Recognizing the limitations (assumptions) of the model, these data provide valuable information on carbon budgets.

3.5.4 Comparison with 1972 Observed Data

Data from an ungrazed pasture were available for 1972. This 1972 ungrazed pasture is on Manter soil, whereas the 1970–1971 data used to construct and tune the model were taken from a pasture on Ascalon soil approximately 4 km away. These 1972 data were not used in developing or adjusting the model and thus were useful for evaluating the model's performance. Initial values (January 1, 1972) for crowns and roots for each producer were taken as May 15, 1972 values of crowns and May 16, 1972 values of roots since no earlier data were available. Temperature and precipitation data were collected approximately 3.2 km away from the site.

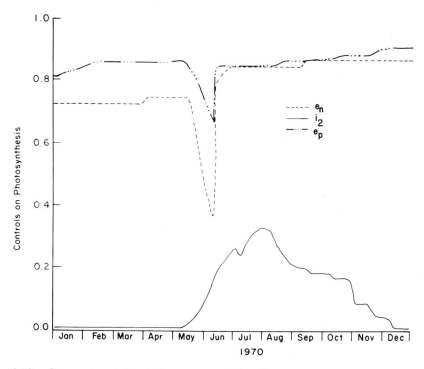

Fig. 3.10.   Seasonal variation in factors controlling photosynthetic rate in warm-season
grasses

Table 3.4   Yearly Sums of Producer Processes for 1970/71 Simulations at the Pawnee
National Grassland

Producer processes	1970	1971
Available total radiation (cal/cm^{-2} yr^{-1})	112,551.0	112,211.0
Intercepted radiation (cal/cm^{-2} yr^{-1})	20,474.0	20,150.0
Net carbon flow to crowns (g dry wt/m^2)	54.5	48.9
Crown death (g dry wt/m^2)	27.3	27.4
Carbon flow to roots (g dry wt/m^2)	364.0	358.8
Crown respiration (g dry wt/m^2)	24.5	28.5
Seed-to-shoot flow	5.3	0.031
Seed-to-root flow	5.3	0.031
Gross photosynthesis	820.5	792.5
Net photosynthesis	557.5	553.0
Root respiration	23.2	25.5
Root death	337.8	327.2
Shoot flow to seeds	5.4	4.6
Shoot death	135.4	140.6
Peak live shoots, date	134.5, (July 25)	124.4, (June 22)
Net production = net photosynthesis − (root + crown respiration)	509.8	499.0

A comparison of 1972 phenology simulation and data is given in Table 3.5. The simulated spring-growth initiation was approximately 1 month earlier, flowering time was approximately 1 wk earlier, and the onset of senescence was slightly earlier than the data indicated. In its present form, the phenology model appears to be adequate, but more data and experimental effort are needed.

Simulated and observed values for standing dead and live shoots of warm season grass are shown in Figure 3.11. Model output did not fall within any of the 95% confidence intervals. More green shoots and less standing dead were simulated than observed. In the remaining four producer categories (not shown) the simulated values of live and dead frequently (74%) passed through the 95% confidence intervals. These discrepancies have several possible explanations. In the 1970–71 ungrazed pastures, the contribution of warm-season grass to the total producer biomass (58% in 1970, 49% in 1971) was greater than in the 1972 pasture (24%). Thus the reserves for warm-season grass growth in the 1972 pasture were large relative to the amount of warm-season grasses, and growth was correspondingly enhanced. Also, species composition of each producer category may be sufficiently different in the 1972 site from the sites used to develop the model that the parameters established were not appropriate. For example, the 1972 site may have a higher proportion of C_4 species in the forb category, in which case a higher quantum efficiency for the forb category would be appropriate for the 1972 pasture.

Another potential source of error was the driving variables. Summer rainfall, mostly convectional with scattered heavy showers, was measured 3.2 km away. The several large rainfall events recorded in 1972 may not have occurred at the sample site.

It would have been desirable to have several years of contrasting weather and grazing-intensity data to use in model development and parameter adjustment. Both 1970 and 1971 were drier than normal. Parameter values may not be appropriate for years of normal or above-normal precipitation.

3.6 Conclusions

Development of this model provided insight into ecosystem functioning that might have been unavailable by any other means. The necessity of representing

Table 3.5. Observed and simulated dates of three phenophases in the ungrazed treatment, 1972

Producer category	Begin growth	Peak flowering	Senescence
Warm-season grasses			
Simulation	mid-April	late June	early October
Data	mid-May	early July	mid-October

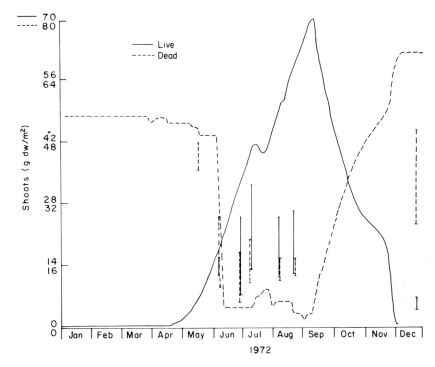

Fig. 3.11. Comparison of simulation and observed data for warm-season grass using
observed 1972 temperature and precipitation at the Pawnee Site

carbon flow and phenologic processes focused attention on some study areas that
would otherwise have been passed over superficially. Of particular importance
were processes critical to determining model and ecosystem response and about
which there was little information.

The transfer of standing dead to litter was important because of the effects of
standing dead on available forage, shade, and precipitation interception. Using
rainfall rate and water content of standing dead to regulate transfer was a partial
answer.

The belowground system functioned as a sink for photosynthate and had a
profound effect on net photosynthetic rate and productivity. The need to estimate the live root and crown weight for each producer category from a total for
all producers compounded the difficulty of estimating the upper and lower limits
for translocation from shoot systems and the rates of translocation from crowns
to shoots for initial growth. The belowground system affected leaf area, which in
turn determined gross photosynthetic rate through light interception. The regulation of flow from the shoots is based on root:shoot ratios, and not translocation of
material belowground. The distinction is important as it relates to net change.
The relationships between shoot/root transfers, the driving variables of soil-
water potential, nitrogen, phosphorus, and the phenology and the relative

weights of shoots and roots must be studied in suitably designed field experiments. The assumption of an exponential distribution of roots with depth may be weak, particularly for very shallow-rooted species. Simulations of some producers (annuals) would have been improved by making root distributions vary through time, thus simulating root growth with depth. More data on root distribution and associated water-absorbing capacity are needed. The belowground system is the least understood element of an ecosystem (Wiegert and Evans, 1964).

The interception of insolation could have been improved by making the constant coefficient of forward scattering and upward and downward reflection a function of canopy characteristics, such as leaf area. This refinement would be particularly significant for those species not light-saturated. The relatively simple representation of seed germination needs improvement, particularly with respect to deserts and annual grasses. The challenge here would be to represent changes in species composition characteristic of annual grasslands (Heady, 1958) and monsoonal climates (Torssel, 1973). The representations of crown and root growth, respiration, and death require elaboration and conceptual and experimental refinements. The functions used here are a first step.

Plant phenology plays a central role in regulating intertrophic relationships and is significant in the control of many intratrophic processes. Phenology is not a primary regulatory variable but a result of interactions between the genetics of a producer and the driving variables. The difficulty of representing these interactions without phenology is that of representing interactions between the genetic makeup and the environment. Such processes as accelerated shoot and root death with plant age, the timing of flowering and fruit set, and the forage value to consumers would be difficult without phenology. Phenology thus merits more study as a separate entity and as part of the ecosystem.

A model is a collection of explicit and implicit hypotheses. Such hypotheses encode our knowledge of the functioning of the grassland ecosystem. The model presented combines hypotheses in a manner that facilitates comparison of results to data and thereby tests the hypotheses. Areas where knowledge about ecosystems is inadequate are thus identified and made the focus of experimental effort.

Acknowledgments. The development of this model would not have been possible without the extensive and sincere assistance of the personnel of the Natural Resource Ecology Laboratory at Colorado State University. A special note of appreciation goes to my colleagues, J. C. Anway, H. W. Hunt, C. F. Rodell, W. J. Parton, R. G. Woodmansee, G. W. Cole, and especially G. S. Innis. The manuscript was improved by the suggestions of D. Botkin and B. Torssell.

This chapter reports on work supported in part by National Science Foundation Grants GB-31862X, GB-31862X2, GB-41233X, and BMS73-02027 AO2 to the Grassland Biome, U.S. International Biological Program, for "Analysis of Structure, Function, and Utilization of Grassland Ecosystems."

References

Anderson, M. C., Denmead, O. T.: Shortwave radiation on inclined surfaces in model plant communities. Agron. J. **61**, 867–872 (1969)

Asana, R. D., Saini, A. D.: Studies in physiological analysis of yield. IV. The influence of soil drought on grain development, photosynthetic surface, and water content of wheat. Physiol. Plant. **11**, 666–674 (1958)

Blaisdell, J. P.: Seasonal development and yield of native plants on the Upper Snake River plains and their relation to certain climatic factors. USDA Tech. Bull. 1190, 1958, 68 pp.

Bokhari, U. G., Dyer, M. I.: The effects of environmental stresses on growth of blue grama *(Bouteloua gracilis)* in environmental control growth chambers. US/IBP Grassland Biome Tech. Rep. No. 227. Fort Collins: Colorado State Univ., 1973, 59 pp.

Botkin, D. B.: Prediction of net photosynthesis of trees from light intensity and temperature. Ecology **50**, 854–858 (1969)

Botkin, D. B., Janak, J. F., Wallis, J. R.: Some ecological consequences of a computer model of forest growth. J. Ecol. **60**, 849–872 (1972)

Bridge, D. W.: A simulation model approach for relating effective climate to winter wheat yields on the great plains. Agr. Met. **17**, 185–194 (1976)

Bridges, K. W., Wilcott, C., Westoby, M., Kirkert, R., Wilkin, D.: Nature: A guide to ecosystem modeling. Presented at Ecosystem Modeling Symposium, 1972 AIBS Meetings, Aug. 26–Sept. 1, 1972. Univ. Minnesota, Minneapolis, 1972 (mimeo.)

Brix, H.: The effect of water stress on the rates of photosynthesis and respiration in tomato plant and loglolly pine seedlings. Physiol. Plant. **15**, 10–20 (1962)

Brouwer, R.: The influence of the suction tension of the nutrient solutions on growth, transpiration, and diffusion pressure deficit of bean leaves *(Phaseolus vulgaris)*. Acta Bot. Neer. **12**, 248–261 (1963); in: Plant Soil Water Relationships: A Modern Synthesis. Kramer, P. J. (ed.). New York: McGraw-Hill, 1969, 482 pp.

Brouwer, R., de Wit, C. T.: A simulation model of plant growth with special attention to root growth and its consequences. *In* Proc. 15th Eastern School of Agricultural Science, Univ. Nottingham. Whittington, W. J. (ed.). London: Butterworths, 1968, pp. 224–242

Brown, L.: Photosynthesis of two important grasses of the shortgrass prairie as affected by several ecological variables. Ph.D. thesis, Colorado State Univ., Fort Collins, 1974, 191 pp.

Caprio, J.: The solar-thermal unit theory in relation to plant development and potential evapotranspiration. Montana Agr. Exp. Sta. Circ. 251. Bozeman: Montana State Univ., 1971, 12 pp.

Clark, F. E., Coleman, D. C.: Secondary productivity below ground in Pawnee Grassland, 1971. US/IBP Grassland Biome Tech. Rep. No. 169. Fort Collins: Colorado State Univ., 1972, 23 pp.

Cline, J. F., Uresk, D. W., Rickard, W. H.: Comparison of soil water used by a sagebrush-bunchgrass and cheatgrass community. J. Range Mgmt., **30**, 199–202 (1977)

Cline, M. G.: Effect of temperature and light intensity on growth of a shrub. Ecology **47**, 782–795 (1966)

Connor, D.: GROMAX: A potential productivity routine for a total grassland ecosystem model. US/IBP Grassland Biome Tech. Rep. No. 208. Fort Collins: Colorado State Univ., 1973, 27 pp.

Dahlman, R. C., Kucera, C. L.: Root productivity and turnover in native prairie. Ecology **46**, 84–89. (1965)

Davidson, R. L.: Effects of soil nutrients and moisture on root/shoot ratios in *Lolium perenne* L. and *Trifolium repens* L. Ann. Bot. **33**, 571–577 (1969)

de Wit, C. T.: Photosynthesis of leaf canopies. Agr. Res. Rep. No. 663. Wageningen, the Netherlands: Centre for Agr. Publ. Docu., 1965, 57 pp.

de Wit, C. T., Brouwer, R., Penning de Vries, F. T. W.: The simulation of photosynthetic systems. *In* Prediction and Measurement of Photosynthetc Productivity: Proc. 1969 IBP/PP Technical Meeting. Setlik, I. (ed.). Wageningen, the Netherlands: Centre for Agr. Publ. Docu., 1970, pp. 47–70

Dickinson, C. E., and Dodd, J. L.: Phenological pattern in the shortgrass prairie. Am. Midland Naturalist **96**, 367–378 (1976)

Dickinson, C. T., Baker, C. V.: Pawnee Site field plant list. US/IBP Grassland Biome Tech. Rep. No. 139. Fort Collins: Colorado State Univ., 1972, 44 pp.

Dittmer, H. J.: Clipping effects on Bermuda grass biomass. Ecology **54**, 217–219 (1973)

Duncan, W. G., Loomis, R. S., Williams, W. A., Hanau, R.: A model for simulating photosynthesis in plant communities. Hilgardia **38**, 181–205 (1967)

Dye, A. J., Brown, L. F., Trlica, M. J.: Carbon dioxide exchange of blue grama as influenced by several ecological parameters, 1971. US/IBP Grassland Biome Tech. Rep. No. 181. Fort Collins: Colorado State Univ., 1972, 44 pp.

Ellis, J. E.: A computer analysis of fawn survival in the prong-horn antelope. Ph.D. thesis, Univ. California, Davis, 1970, 70 pp.

Heady, H. F.: Vegetational changes in the California annual type. Ecology **39**, 402–416 (1958)

Jackson, M. T.: Effects of microclimate on spring flowering phenology. Ecology **47**, 407–415 (1966)

Jameson, D. A., Dyer, M. I.: Process studies workshop report. US/IBP Grassland Biome Tech. Rep. No. 220. Fort Collins: Colorado State Univ., 1973, 444 pp.

Klepper, B., Taylor, H. M., Huck, M. G., Fiscus, E. L.: 1973. Water relations and growth of cotton in drying soil. Agron. J. **65**, 307–310 (1973)

Klikoff, L. G.: Competitive response to moisture stress of a winter annual of the Sonoran Desert. Am. Midland Naturalist **75**, 383–391 (1966)

Knievel, D. P., Schmer, D. A.: Preliminary results of growth characteristics of buffalo grass, blue grama, and western wheatgrass and methodology for translocation studies using ^{14}C as a tracer. US/IBP Grassland Biome Tech. Rep. No. 86. Fort Collins: Colorado State Univ., 1971, 28 pp.

Knight, D. H.: Leaf area dynamics on the Pawnee Grassland, 1970–1971. US/IBP Grassland Biome Tech. Rep. No. 164. Fort Collins: Colorado State Univ., 1972, 22 pp.

Knight, D. H.: Leaf area dynamics of a shortgrass prairie in Colorado. Ecology **54**, 891–896 (1973)

Laycock, W. A., Buchanan, H., Krueger, W. C.: Three methods of determining diet, utilization, and trampling damage on sheep ranges. J. Range Mgmt. **25**, 352–356 (1972)

Lemon, E., Stewart, D. W., Shawcroft, R. W.: The sun's work in a cornfield. Science **174**, 371–378 (1971)

Leopold, A. C.: Senescence in plant development. Science **134**, 1727–1732 (1961)

Lewis, H.: Catastrophic selection as a factor in speciation. Ecology **16**, 257–271 (1962)

Lindsey, A. A., Newman, J. D.: The use of official weather data in spring time temprature analysis of an Indiana phenology record. Ecology **37**, 812–823 (1956)

Loomis, R. S., Williams, W. A.: Maximum crop productivity: An estimate. Crop Sci. **3**, 67–72 (1964)

Ludlow, M., Wilson, G.: Photosynthesis of tropical pasture plant. I. Illuminance, carbon dioxide concentration, leaf temperature, and leaf-air vapour pressure difference. Aust. J. Biol. Sci. **24**, 449–470 (1971)

Miller, P. C., Tieszen, L.: A preliminary model of processes affecting primary production in the Arctic tundra. Arctic Alpine Res. **4**, 1–18 (1972)

Mooney, H. A.: The carbon balance of plants. Annu. Rev. Ecol. Syst. **3**, 315–346 (1972)

Mooney, H. A., West, M., Brayton, R.: Field measurements of the metabolic responses of bristlecone pine and big sagebrush in the White Mountains of California. Bot. Gaz. **127**, 105–113 (1966)

Mooney, H. A., Wright, R. D., Strain, B. P.: The gas exchange capacity of plants in relation to vegetation zonation in the White Mountains of California. Am. Midland Naturalist **72**, 281–297 (1964)

Mott, J. J.: Germination studies on some annual species from an arid region of western Australia. J. Ecol. **60**, 293–304 (1972)

Mueggler, W. F.: Plant development and yield on mountain grasslands in southwestern Montana. U.S. For. Serv. Res. Paper No. INT-124, 1972, 20 pp.

Murphy, C. E., Jr., Knoerr, K. R.: Modeling the energy balance processes of natural ecosystems. Oak Ridge, Tennessee: Oak Ridge National Laboratory Paper No. EDFB-IBP 72-10, 1972.

Remy, T.: Fertilization and its relationship to the course of nutrient absorption by plants. Soil Sci. **48**, 187–209 (1938)

Robson, M. J.: The growth and development of simulated swards of perennial ryegrass. I. Leaf growth and dry weight change as related to the ceiling yield of a seeding sward. Ann. Bot. **37**, 487–500 (1973)

Rose, C. W., Begg, J. R., Byrne, G. F., Torssell, B. W. R., Goncz, J. H.: A simulation model of growth—field development relationships for Townsville stylo (*Stylosanthes humilis* H.B.K.) pasture. Agr. Meteorol. **10**, 161–183 (1972)

Rosenzweig, M. L.: Net primary productivity of terrestrial communities: Prediction from climatological data. Am. Naturalist **102**, 67–74 (1968)

Sauer, R. H.: The relationship of cumulative sums and moving averages of temperature to reproductive phenology in *Clarkia*. Am. Midland Naturalist **95**, 144–158 (1976)

Sesták, Z., Catský, J., Jarvis, P.: Plant Photosynthesis Production: A Manual of Methods. The Hague, The Netherlands, 818 pp. As cited in: W. Junk, 1971, Mooney, H. A.: The carbon balance of plants. Annu. Rev. Ecol. Syst. **3**, 315–346 (1972)

Sims, P. L., Lang'at, R. K., Hyder, D. N.: Developmental morphology of blue grama and sand blue stem. J. Range Mgmt. **26**, 340–344 (1973)

Sims, P. L., Singh, J. S.: Herbage dynamics and net primary production in certain ungrazed and grazed grasslands in North America. In: Preliminary Analysis of Structure and Function in Grasslands. French, N. R. (ed.). Range Sci. Depart. Sci. Ser. No. 10, 1971, pp. 59–124.

Sims, P. L., Uresk, D. W., Bartos, D. L., Lauenroth, W. K.: Herbage dynamics on the Pawnee Site: Aboveground and belowground herbage dynamics on the four grazing intensity treatments; and preliminary sampling on the ecosystem stress site. US/IBP Grassland Biome Tech. Rep. No. 99. Fort Collins: Colorado State Univ., 1971, 95 pp.

Singh, J. S., Coleman, D. C.: Distribution of photoassimilated ^{14}C in the root system of a shortgrass prairie. J. Ecol. **62**, 359–365 (1974)

Slatyer, R. O.: The significance of the permanent wilting percentage in studies of plant and soil water relationships. Bot. Rev. **23**, 586–636 (1957)

Snow, D. W.: Fruiting seasons and bird breeding seasons in the New World tropics. J. Ecol. **56**, 5–6 (1968)

Taylor, H. M., Huck, M. G., Klepper, B., Lind, Z. F.: Measurement of soil-grown roots in a rhizotron. Agron. J. **2**, 807–809 (1970)

Torssell, B. W. R.: Patterns and processes in the Townsville stylo-annual grass pasture ecosystem. J. Appl. Ecol. **10**, 463–478 (1973)

Weaver, J. E.: Classification of root systems of forbs of grassland and a consideration of their significance. Ecology **39**, 393–401 (1958)

Wiegert, R. G., Evans, F. C.: Primary production and the disappearance of dead vegetation on an old field in southeastern Michigan. Ecology **45**, 49–63 (1964)

Williams, G. J., III, Markley, J. L.: The photosynthetic pathway type of North American shortgrass prairie species and some ecological implications. US/IBP Grassland Biome Tech. Rep. No. 229. Fort Collins: Colorado State Univ., 1973, 20 pp.

Zahner, R., Stage, A. R.: A procedure for calculating daily moisture stress and its utility in regression of tree growth on weather. Ecology **47**, 64–73 (1966)

4. A Mammalian Consumer Model for Grasslands

JERRY C. ANWAY

Abstract

This chapter describes a computer-simulation model designed to represent flows of carbon and energy and resultant weight changes in mammalian consumers in a shortgrass prairie ecosystem. One model is used to treat cohorts of mammalian species ranging from mice to cattle, whether carnivorous, omnivorous, or herbivorous. Twenty input parameters describe individual and population characteristics of each mammal simulated. Output variables include numbers and individual weight, food intake, and animal waste products.

4.1 Introduction

The model presented here is the mammalian consumer component of ELM. It was designed to mechanistically represent the dynamics of any given grassland mammal as well as the interconnections of that consumer with the other components. The model is constructed around the basic concept that the state or condition of any mammal is a consequence of the history of the balance between food intake, defecation, urination, and respiration.

The model is a hypothesis about the relationships and interactions of such biological mechanisms as diet selection, metabolism and energy balance, and natality and mortality in their effects on the condition of an average animal within each cohort being simulated. It is a canonical model in the mathematical sense (Courant and Hilbert, 1962) that for each time step it considers all mammalian consumers through one code formulation via different initial conditions, parameters, and environmental factors.

The canonical concept in biology is not new, as most of the basic mechanisms of biology have been developed considering one or more species and refined by application to several species. An example of this is the formulation for basal metabolism ($70w^{0.75}$) discussed by Kleiber (1961). There are many formulations for resting metabolism, utilizing a variety of parameters (Zar, 1973), but the basic formulation remains an equation of the form aw^b where w is weight (Kleiber, 1961; Brody, 1945; Maynard and Loosli, 1962; Gessaman, 1973; Grodzinski and Gorecki, 1967; Lamprey, 1964, Crampton and Lloyd, 1959; Lasiewski and Dawson, 1967). Another example of canonical concept in biology is the general growth curve of Brody (1945).

A number of models have been developed which consider one or more of the mechanisms of energy balance and population dynamics and a few (Smith and Williams, 1973; Timin, 1973; Peden, 1972; Peden et al., 1974; Davis, 1967; Amidon and Akin, 1968; Rice et al., 1974) consider many of them for one or more species. The present model combines those mechanisms appropriate to inter- and intraseasonal biomass dynamics for any mammal or combination of mammals in a grassland ecosystem under the varied environmental conditions encountered for a period of up to 10 years. Biomass dynamics is represented via carbon flow.

Many of the mechanisms and relationships among mechanisms incorporated were those discussed during the US/IBP Grassland Biome Process Studies Workshops held in early 1972 (Jameson and Dyer, 1973) and the Grassland–Tundra International Modeling Workshop held August 14–26, 1972.

4.2 Model Development and Overview

In any biological system two factors change through time: (a) the state variables or system components and (b) flows of information and material (Forrester, 1961). While these two interact to give the dynamics of the state of the system, the agents of action are the flows.

In developing a word model and a box-and-arrow diagram the assumption was made that the two principal processes by which mammalian consumers affect grasslands are intake (via diet selection) and elimination. The principal control on these processes are animal numbers and types and metabolism or energy balance that is influenced by air temperature, animal weight, wastes (gas, urine, and feces), activity, reproductive state, population density, animal phenology, hunger, potential intake amount, food availability or accessibility, preference, and digestibility of foods (see Table 4.1).

The impact of any population on an ecosystem occurs at three levels: (a) organisms, (b) populations, and (c) communities. Whereas all three interact, biologic mechanisms act on the population through the individuals making up that population. Thus the biologic mechanisms simulated are calculated in terms of an individual considered to be the mean for a cohort. Each character is assumed to be normally distributed within a generation. Where appropriate, the influence of the population on other populations is included.

The calculations of metabolic energy requirement, diet selection, food utilization, and biologic index of Table 4.1 are for the average individual, whereas mortality and natality are distributed over the population. Each of these calculations is discussed in Section 4.3.

Figure 4.1 illustrates the principal components of the hypothesized system relationships. Food sources include up to 15 food categories (e.g., live warm-season grass, grasshoppers, mice), which are identified by way of the input parameters for each consumer type considered. The animal waste products are feces, urine, methane, CO_2, and dead animal material. Flows to secondary consumers are considered where appropriate. Flows to and from the source and

Table 4.1. Outline of major calculations (each computation is specific to the consumer category)

Food required to meet metabolic energy requirement (grams food carbon per day); F

$$F = 70 w^{0.75} h_i a c_1 / E_f$$

Diet selection
   Preference per food category; p_i

$$a_i = X_i (1.01 - i_b) i_p$$
$$p_i = a_i / \Sigma a_i$$

Intake per food category; F_i

$$F_i = (F d^{-1} + H + F_0) e_w e_c e_a e_d p_i$$

Secondary production increment; s

$$s = t - f_c - g - u - F - F_0$$

Biological index; b_i

$$b_i = \begin{cases} \min (i_a, i_w) & \text{if } i_a \leqslant 0.7 \text{ or } i_w \leqslant 0.7 \\ (i_a + i_w)/2 & \text{otherwise} \end{cases}$$

Mortality; M

$$M = (M_A + M_p + M_s + M_0) \, dt$$

Average adult weight; w

$$w_{t+dt} = (w_t N_t + s - M w_t)/N_{t+dt}$$

Natality; N_0

$$N_0 = N_m f_f N_e b_i$$

Birth weight; w_b

$$w_b = 0.053 \, w b_i$$

---

sink include human manipulation, birth, immigration, and emigration. The individual and population characteristics of each mammal considered are described by the input parameters with their values and sources shown in Table 4.2. The model is structured to consider up to 10 consumer types. Values in Table 4.2 are for the seven consumers (cow, jackrabbit, coyote, grasshopper mouse, deer mouse, ground squirrel, and kangaroo rat) for which simulation results are discussed.

In Table 4.3 the symbols used in this description are defined and their units given. Many of these factors vary with consumer category and cohort and the

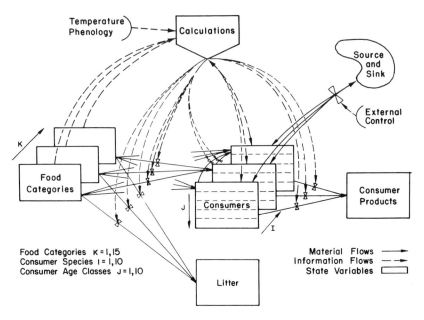

Fig. 4.1.   Information flows, material flows, and state variables considered in the model
(calculations outlined in greater detail in Table 4.1)

computations below are carried out for each separately. To facilitate presentation these extra subscripts (for consumer category and cohort) are omitted.

4.3 Biological Components

The components outlined in Table 4.1 were selected after discussions with
range and animal scientists and numerous simulation exercises. The calculations
are executed each time step for each cohort within each consumer type. In
addition, producer–consumer and consumer–decomposer interpopulation
dynamics are determined by flows between submodels. SIMCOMP (Gustafson
and Innis, 1973) accounts for the material flowing per time step and updates the
contents of each compartment for each time step. A variety of time steps from
0.5 to 3 days and 1-, 2-, and 3-year simulations have been carried out.

4.3.1 Energy

The daily food requirement of an animal is a basic factor controlling behavior.
It was considered to be a function of basal metabolism, heat increment, activity,
and energy content of food. This requirement is converted to carbon amounts as
carbon is the material represented in the model. Basal metabolism under thermo-
neutral conditions is computed $70w^{0.75}$ (Crampton and Lloyd, 1959; Maynard and

Loosli, 1962). Others (e.g., Brody, 1945) prefer $70w^{0.73}$, but considering the level of resolution of this model, such differences are insignificant.

An amount of heat above basal metabolism is required for the calorigenic effect of food (heat loss of digestion) and the effect of external air temperature (cost of thermoregulation) on heat production. The curve used (Fig. 4.2) is based on one from Kleiber (1961), where basal metabolism was multiplied by 1.42 to account for heat production by fed animals and increased from this level as air temperature decreased below 13°C. If this model were used for animals outside the temperate zone, the simulated response to cold might be inappropriate (Scholander et al., 1950). Mean daily air temperatures are supplied by the abiotic submodel. The heat-increment fraction h_i multiplied by basal metabolism represents the daily maintenance energy requirements per animal.

An activity factor, a, further adjusts maintenance energy to obtain total daily energy requirements. This factor is 2 for herbivores and 1.4 for carnivores and omnivores (Crampton and Lloyd, 1959) in the absence of data to the contrary. Literature values for this activity factor range from 1.3 to 4.8 (Crampton and Harris, 1969; Moen, 1973; Graham, 1964, Clapperton, 1961; Van Dyne and Van Horn, 1965; Johnson and Groepper, 1970).

To determine the amount of food carbon in kg/day needed to meet the energy requirement, the energy requirement is divided by the energy content of food (E_f, an input parameter) and multiplied by the carbon fraction in food (c_1). The results of this calculation

$$F_i = 70 \ w^{0.75} \ h_i a c_1 / E_f \tag{4.1}$$

are multiplied by the time step to determine the amount of food carbon needed to a given time step.

4.3.2 Diet

One of the more important and variable points of impact of consumers on a grassland system is forage removal (Van Dyne, 1969). As the seasons progress, the quantity of any food category selected may affect subsequent primary production. Diet analyses show that consumers prefer certain species (food categories) and prefer early phenologic stages within these categories (Wallace, et al., 1972; Flake, 1973; Vavra, 1972, Vavra et. al., 1973; A. Johanningsmeier, personal communication). This preference for food type is calculated as a function of food quantity, bilogical index, and palatability index.

The quantity of each food category present (X_i) is the amount of live and standing dead shoots, and seeds from the producer submodel and the amount of grasshoppers and small mammals from the consumer models.

Food quality has a strong influence on diet selection. To incorporate this effect, a food biological index, i_b, is determined for each food category (from 0.0 to 1.0). The food biological index for animal types consumed is an average of the biological index (discussed later), weighted by the number of individuals in each

Table 4.2. Input parameters

Organism	Parameter	Value	Source
Cow (*Bos taurus*)	Activity factor (a)	2.0	Crampton and Lloyd (1959)
	Number of reproductions per year	1.02	Altman and Dittmer (1964)
	Number of offspring expected at birth (N_e)	1.0	Personal observation
	Life span in days (l_m)	5,450.0	Altman and Dittmer (1964)
	Environmental threshold for reproduction	5,000.0	Experimental simulations
	Adult body weight (w_a)	58.9	J. D. Wallace, personal communication
	Model body weight (initial g C)	57.0	Dyck and Bement (1972)
	Minimal feeding range (A)	8,000.0	Dyck and Bement (1972)
	Fraction or number of animals/m^2	0.0	Not applicable
	Fraction of food lost as urine	0.06	Cook (1970)
	Fraction of food lost as methane	0.06	Cook (1970)
	Caloric content of food (E_f)	4.5	Beck (1969)
	Minimal food amount (F_m)	10.0	Dyck and Bement (1972)
	Digestibility (d)	See discussion	Personal estimate
	Fraction of food wasted	0.75	Personal estimate and experimental simulations
	Innate preference for food sources	0.0–1.0	Personal estimate and experimental simulations
	Weight-deviation margin (t_w)	0.2	Personal estimate and experimental simulations
	Nominal mortality rate (M_0)	0.0	Not applicable
	Migration time		
	into system	0.0	Not applicable
	out of system	0.0	Not applicable
	Other[a]		
Coyote (*Canis latrans*)	Activity factor	1.4	Crampton and Lloyd (1959)
	Number of reproductions per year	1.0	Lechleitner (1969)
	Number of offspring expected at birth	4.0	Haskell and Reynolds (1947)
	Life span in days	1,800.0	Rogers (1965)
	Environmental threshold for reproduction	5,000.0	Experimental simulations
	Adult body weight (initial g C)	1.61	Lechleitner (1969)
	Model body weight (initial g C)	1.4	Personal estimate

Parameter	Value	Source
Minimal feed range (m²)	45,455,000.0	L. F. Paur and N. R. French, personal communication
Fraction or number of animals/m²	0.000000022	L. F. Paur and N. R. French, personal communication
Fraction of food lost as urine	0.03	S. Smith, personal communication
Fraction of food lost as methane	0.01	S. Smith, personal communication
Caloric content of food (kcal/g)	5.1	Personal estimate
Minimum food amount (g/m²)	0.013	Personal estimate
Digestibility	0.8	Personal estimate
Fraction of food wasted	0.01	Personal estimate
Innate preference for food sources	0.0–1.0	Personal estimate and experimental simulations
Weight-deviation margin	0.11	Personal estimate
Nominal mortality rate	0.0015	
Migration time		
into system	0.0	Not applicable
out of system	0.0	Not applicable
Other[a]		
Activity factor	2.0	Crampton and Lloyd (1959)
Jackrabbit (*Lepus californicus* and *Lepus Townsendii*)		
Number of reproductions per year	1.5	French, et al. (1965)
Number of offspring expected at birth	5.0	Personal estimate
Life span in days	1,440.0	Altman and Dittmer (1964)
Environmental threshold for reproduction	6,000.0	Experimental simulations
Adult body weight (initial g C)	0.42	Hansen et al. (1969)
Model body weight (initial g C)	0.36	Hansen et al. (1969)
Minimal feeding range (m²)	80,000.0	Donoho (1971)
Fraction or number of animals/m²	0.0000155	Donoho (1971)
Fraction of food lost as urine	0.02	S. Smith, personal communication
Fraction of food lost as methane	0.02	S. Smith, personal communication
Caloric content of food (kcal/g)	4.5	Personal estimate
Minimum food amount (g/m²)	3.5	Personal estimate
Digestibility	0.6	Personal estimate
Fraction of food wasted	0.01	Personal estimate
Innate preference for food sources	0.0–1.0	Personal estimate and experimental simulations

Table 4.2. Input parameters (continued)

Organism	Parameter	Value	Source
	Weight-deviation margin	0.11	Personal estimate
	Nominal mortality rate	0.0029	Experimental simulations
	Migration time		
	into system	0.0	Not applicable
	out of system	0.0	Not applicable
	Other[a]		
Grasshopper mouse (*Onychomys leucogaster*)	Activity factor	1.4	Crampton and Lloyd (1959)
	Number of reproductions per year	2.5	Lechleitner (1969)
	Number of offspring expected at birth	4.0	Lechleitner (1969)
	Life span in days	1,200.0	Altman and Dittmer (1964)
	Environmental threshold for reproduction	5,000.0	Experimental simulations
	Adult body weight (initial g C)	0.0036	Grant (1972)
	Model body weight (initial g C)	0.003	Grant (1972)
	Minimal feeding range (m²)	2,400.0	Grant (1972)
	Fraction or number of animals/m²	0.00041	Grant (1972)
	Fraction of food lost as urine	0.05	S. Smith, personal communication
	Fraction of food lost as methane	0.02	S. Smith, personal communication
	Caloric content of food (kcal/g)	5.7	Johnson and Groepper (1970)
	Minimum food amount (g/m²)	3.5	Flake (1973)
	Digestibility	0.87	Johnson and Groepper (1970)
	Fraction of food wasted	0.07	Personal estimate
	Innate preference for food sources	0.0–1.0	Personal estimate and experimental simulations
	Weight-deviation margin	0.03	Personal estimate
	Nominal mortality rate	0.0035	Experimental simulations
	Migration time		
	into system	0.0	Not applicable
	out of system	0.0	Not applicable
	Other[a]		
Deer mouse (*Peromyscus maniculatus*)	Activity factor	1.4	Crampton and Lloyd (1959)
	Number of reproductions per year	4.0	Grant (1972)
	Number of offspring expected at birth	4.0	Lechleitner (1969)

Life span in days	600.0	Personal estimate
Environmental threshold for reproduction	4,800.0	Experimental simulations
Adult body weight (initial g C)	0.0025	Grant (1972)
Model body weight (initial g C)	0.0021	Grant (1972)
Minimal feeding range (m²)	4,000.0	Grant (1972)
Fraction or number of animals/m²	0.00024	Grant (1972)
Fraction of food lost as urine	0.05	Personal estimate
Fraction of food lost as methane	0.02	Personal estimate
Caloric content of food (kcal/g)	5.1	Golley (1961)
Minimum food amount (g/m²)	3.5	US/IBP data
Digestibility	0.8	Johnson and Groepper (1970)
Fraction of food wasted	0.07	Personal estimate
Innate preference for food sources	0.0–1.0	Personal estimate and experimental simulations
Weight deviation margin	0.027	Experimental simulations
Nominal mortality rate	0.007	Experimental simulations
Migration time		
into system	0.0	Not applicable
out of system	0.0	Not applicable
Other[a]		
Activity factor	2.0	Crampton and Lloyd (1959)
Thirteen-lined ground squirrel (*Spermophilus tridecemlineatus*)		
Number of reproductions per year	1.0	Lechleitner (1969)
Number of offspring expected at birth	8.0	Lechleitner (1969)
Life span in days	1,080.0	Personal estimate
Environmental threshold for reproduction	4,800.0	Experimental simulations
Adult body weight (initial g C)	0.018	Grant (1972)
Model body weight (initial g C)	0.015	Grant (1972)
Minimal feeding range (m²)	7,500.0	Grant (1972)
Fraction or number of animals/m²	0.00015	Grant (1972)
Fraction of food lost as urine	0.05	S. Smith, personal communication
Fraction of food lost as methane	0.02	S. Smith, personal communication
Caloric content of food (kcal/g)	5.5	Golley (1961)
Minimum food amount (g/m²)	3.5	US/IBP data
Digestibility	0.8	Johnson and Groepper (1970)
Fraction of food wasted	0.07	Personal estimate
Innate preference for food sources	0.0–1.0	Experimental simulations
Weight-deviation margin	0.035	Experimental simulations

Table 4.2. Input parameters (continued)

Organism	Parameter	Value	Source
	Nominal mortality rate	0.0038	Experimental simulations
	Migration time		
	into system	Day 100	Lechleitner (1969)
	out of system	Day 300	Lechleitner (1969)
	Other[a]		
Ord's kangaroo rat (Dipodomys ordii)	Activity factor	2.0	Crampton and Lloyd (1959)
	Number of reproductions per year	1.0	Lechleitner (1969)
	Number of offspring expected at birth	3.0	Lechleitner (1969)
	Life span in days	800.0	Personal estimate
	Environmental threshold for reproduction	4,000.0	Experimental simulations
	Adult body weight (initial g C)	0.0072	Grant (1972)
	Model body weight (initial g C)	0.0058	Grant (1972)
	Minimal feeding range (m^2)	8,000.0	Grant (1972)
	Fraction or number of animals/m^2	0.0001	Grant (1972)
	Fraction of food lost as urine	0.01	S. Smith, personal communication
	Fraction of food lost as methane	0.0	S. Smith, personal communication
	Caloric content of food (kcal/g)	5.5	Golley (1961)
	Minimum food amount (g/m^2)	3.5	US/IBP data
	Digestibility	0.8	Johnson and Groepper (1970)
	Fraction of food wasted	0.20	Personal estimate
	Innate preference for food sources	0.0–1.0	Experimental simulations
	Weight-deviation margin	0.03	Experimental simulations
	Nominal mortality rate	0.0038	Experimental simulations
	Migration time		
	into system	0.0	Not applicable
	out of system	0.0	Not applicale
	Other[a]		

[a]These are numbers assigned by the user for bookkeeping purposes in the model. They are the state-variable category number, the number of consumers considered, an array of food sources by state variable number, number of food categories, and animal type.

Table 4.3. Definitions of terms used in this chapter

Symbol	Definition
A	Feeding area for an individual (m²/individual)
a	Activity factor (ND)
a_s	Age, simulated (days)
c_1	Carbon content of food = 0.4 (gC/g dry wt)
d	Digestibility (ND)
dt	Time step (days)
e_a	Effect of age of consumption (ND)
e_c	Effect of density on consumption (ND)
e_d	Effect of digestibility on consumption (ND)
E_f	Energy content of food (kcal/g dry wt)
e_w	Effect of weight deviation on consumption (ND)
f_c	Feces (g/day)
f_f	Fraction female (ND)
F_i	Food required to meet metabolic energy demand (g/individual day^{-1})
F_m	Food density below which it is difficult for a consumer to find food (g/m²)
F_o	Food needs of offspring (kg/individual day^{-1})
F_t	Total food available (g/m²)
g	Gas (g/day)
G_i	Individual intake of food category i (kg/individual day^{-1})
H	Food needs carried forward from previous time step (kg/individual day^{-1})
h_i	Heat increment (ND)
i_a	Index of relative age (ND)
i_b	Food biological index (ND)
i_p	Food palatability index (ND)
i_w	Index of relative weight (ND)
M_a	Mortality from old age (individuals/day)
M_o	Mortality from other causes (individuals/day)
M_p	Mortality from predation (individuals/day)
M_s	Mortality from starvation (individuals/day)
N	Density of animals (individuals/m²)
N_e	Number of offspring expected per mature female (ND)
N_m	Density of mature individuals (individuals/m²)
N_o	Density of offspring (individuals/m²)
p_i	Preference for food of category i (ND)
p_s	Proportion starving (ND)
s	Secondary production increment (g/day)
t	Total intake (g/day)
u	Urine (g/day)
w	Weight (kg/individual)
w_a	Adult weight (kg/individual)
w_b	Birth weight (kg/individual)
w_d	Relative weight deviation = $(w - w_e) / w_e$ (ND)
w_e	Expected weight (kg/individual)
w_s	Weight of dying individuals (kg/individual)

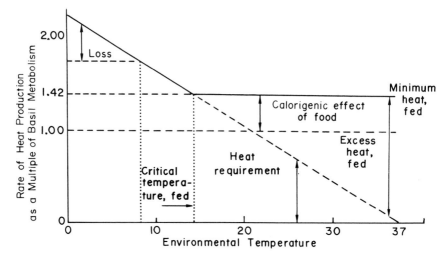

Fig. 4.2.   Heat increment of animal vs. environmental temperature [from Kleiber (1961), p. 274]

cohort. The food biological index for plants is the phenological index from the producer submodel converted to a scale of 0.0 to 1.0. Plants in late phenological stages and adult animals near their expected weight have biological indices near 1. Preference involves $1.01-i_b$, so that plants and animals of low biological index are preferred [Eq. (4.2)]. The 0.01 accommodates a positive preference even for large animals and plants in late phenological stages.

The acceptability by the consumer of the foods available also influences preference. For example, some foods will be eaten even if quantity or quality are relatively low, while others are acceptable only if quantity or quality are very high. Thus a palatability index, i_p, for each food category as a relative index of food acceptability is included (see Table 4.2) for each consumer. The palatability index is scaled from 0.0 to 1.0, with foods that are nearly always acceptable assigned a value of 1.0 and those seldom acceptable, a palatability index near zero.

$$a_i = X_i(1.01 - i_b)ip$$
$$p_i = a_i/\Sigma a_i \qquad\qquad (4.2)$$

4.3.3 Daily Intake

Daily food intake per animal per food category (G_i) is

$$G_i = (Fd^{-1} + H + F_0)e_w e_c e_a e_d p_i \qquad\qquad (4.3)$$

where F is the food required [Eq. (4.1)]; d is the digestibility of food; H is the hunger; F_0 is the food needs of offspring; e_w is the influence of weight deviation; e_c is the density-consumption factor; e_a is the age effect; e_d is the digestibility effect for large herbivores; and p_i is the preference as calculated above. If G_i exceeds the amount present in a category, the excess is taken from the other food categories.

Digestibility is used to account for the indegestible fraction of intake lost as feces. Digestibility of intake for large herbivores is an important control on energy balance and changes relative to maturity (phenology) of the herbage consumed. Nylon-bag trials by Wallace et al. (1972) and Rauzi et al. (1969) showed that digestibility for cattle and sheep varied with the phenological stage of the herbage (Fig. 4.3). The curve used in the model to determine digestibility of any food category as a function of its phenological index is also shown in Figure 4.3. The value of digestibility is an average of the digestibility of each food category weighted by the amount consumed per category. Digestibility of intake for carnivores and omnivores is less variable and is supplied as an input parameter (see Table 4.2).

The value for H represents the difference between desired intake and available food if available food was less than desired on the previous day; otherwise, this value is zero. The value for H in any given time step is limited to 10% of the current food requirements.

Food needs by offspring are determined from similar calculations. These needs are added to parental needs in proportion to the offspring's dependence. The offspring's period of dependence or nursing period is 0.04 of its life span. Four percent of the life span may not be exactly correct for any given mammal, but it approximates the values given in several life tables (Altman and Dittmer, 1964; French and Grant, personal communication; Lechleitner, 1969). Carnivores, omnivores, and small herbivores are completely dependent for 95% of this period with increasing independence for the last 5%. The degree of dependence

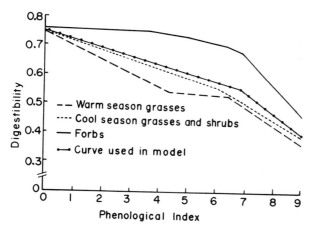

Fig. 4.3. Digestibility for large herbivores as a function of phenological index

for large herbivores follows the curve in Figure 4.4. During the period of partial independence, the fraction of dependence is the proportion of offspring's total intake obtained from the parent. The amount added to parental intake is the sum of food needed by all dependent offspring of that parent. Under normal conditions this allows offspring to achieve approximately 60% of adult weight by weaning time. This weaning weight compares favorably with observations (Moen, 1973; Altman and Dittmer, 1964).

The expected weight of an animal is computed as

$$w_e = w_a(1 - \exp(-9a_s/1_m)) \tag{4.4}$$

where w_a, a_s, and 1_m are adult weight, age and life span, respectively. It was found that by assigning a birth age of $0.006 \ 1_m$, the above formula gave an expected weight within 10% of observed birth weights (Altman and Dittmer, 1964; Lechleitner, 1969).

The effect of weight deviation, on daily intake, e_w, is computed as

$$e_w = 1 - r_i f(w_d) \tag{4.5}$$

where

$$f(w_d) = \begin{cases} w_d \text{ if } |w_d| \leq 0.2 \\ \pm 0.2 \text{ if } w_d | > 0.2 \end{cases}$$

w_d is relative weight deviation $[(w - w_e)/w_e]$, and r is a parameter (set to 2.0)

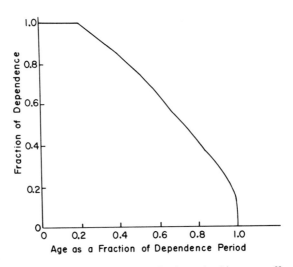

Fig. 4.4.   Degree of dependence for large herbivorous offspring

determining the effect of the weight deviation on intake. Relative weight deviation and expected weight are discussed later.

Various factors have been proposed and discussed as limiting factors for species (Andrewartha, 1957; Main et al., 1959; Wagner and Stoddart, 1972; Hairston et al., 1960; Watt, 1968, MacArthur, 1958). Most of these relate to one or more of food, density, and weather. One climatologic aspect, namely, temperature has been discussed earlier as it influences metabolism. The influence of the remaining aspects of weather may be related to density, since the availability of shelter from rain or cold temperatures may decrease as density increases (Emlen, 1973).

Food as a limiting factor is periodic in its influence. Generally, during the growing season and early fall, food is abundant while during late winter and early spring prey and producer species are at their lowest. On an annual basis and under average conditions predators harvest less than 20% of the prey populations (Kolenosky, 1972; French et al., 1967; Southern and Lowe, 1968; Kemp and Keith, 1970). Even large herbivores (cattle) grazing an area heavily remove less than 20% of total herbage production (US/IBP Grassland Biome data). But scarcity of food may influence intake, especially if animal density is high. The impact of density and relative food availability F_r on intake was implemented as

$$F_r = F_t/(NF_m A) \tag{4.6}$$
$$e_c = 1 - \exp(-10F_r) \tag{4.7}$$

where F_t is food; N is animal density; and F_m is minimum food density per animal; A is minimum feeding area per animal; and $F_m A$ is the minimum food per animal below which hunger occurs. Thus as consumer density increases and/or food decreases, relative food available F_r decreases. This decrease is gradual until F_r drops below 0.2 below which intake is sharply decreased (Fig. 4.5).

A young animal's food intake is affected by its growth. To simulate this, the following function represents the age effect:

$$e_a = 1 + \exp(-18a_s/1_m) \tag{4.8}$$

The effect of digestibility, e_d, on large herbivore intake varies linearly from 0.85 to 1.25 as digestibility increases from 0.45 to 0.70. According to Blaxter et al. (1966) and Hodgson and Wilkinson (1968), this relationship is implemented only for large herbivores. For carnivores and omnivores it is assumed that digestibility is constant.

When desired intake per food category has been determined, it is compared with the amount available. If the amount available is less than the amount desired for any category, the lack is supplemented from other categories in preference order, if possible. Any remaining unfulfilled desire is carried to the next time step as hunger, H.

Not all food that is damaged or destroyed is ingested. With some mammals a proportion is dropped and lost, while some practice hoarding (Emlen, 1973) and

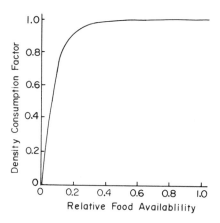

Fig. 4.5.   Density-consumption factor vs. relative food available

some cannot consume all of a large prey item before decomposition occurs. Additionally, large herbivores damage aboveground plant parts by trampling (Laycock et al., 1972; Laycock and Harniss, 1974). Flows representing these effects are considered to be proportional to feeding activity and are implemented as a fraction (input parameter) of the food ingested from each category.

4.3.4 Secondary Production

Gain or loss of weight is a function of total intake and material lost via gas, urine, and feces (Table 4.1). The amount of fecal material is the amount of nondigestible intake, whereas gas and urine losses are a fraction (input parameters) of the digestible intake (Cook, 1970).

To determine food utilization in respiration, materials lost as feces, urine, gas, activity, and maintenance are subtracted from total intake. If the remaining value is positive, it is multiplied by 0.4 to obtain weight gain (anabolism is approximately 40% efficient). The other 60% is added to F_j to give total respiration. Should intake be less than F_j (indicating a loss in weight), F_j minus intake is divided by 0.8 to obtain total weight loss, since catabolism is approximately 80% efficient (Cook 1970, K. Hutchinson, personal communication). Should the animal have dependent offspring, the food needs of the offspring are supplied before gain calculations are made.

4.3.5 Biological Index

Among the factors controlling the response of an animal to its environment are its general health and phenology. To include this as a determinant of bioenergetics and demography, a biological index is determined for each consumer category and cohort, considering age relative to life span and weight relative to that expected for its age (Table 4.1). This biological index is used to influence reproduction, energy balance, diet selection, susceptibility to predation, and other functions.

An age- or maturity-index value is determined as a function of relative age (simulated age divided by average life span) (Fig. 4.6). Thus, the very young and the very old have a low age index while mature individuals have an age index near 1. This curve is based on the premise that most mammals are sexually mature at 25% of their live span.

A weight index value (Fig. 4.7) is determined as a function of the relative weight deviation of the animal via w_d. The expected weight function [Eq. (4.4)] incorporates the premise that mammals (generally) reach 93% of their adult weight by the time they reach 25% of their life span. The biological index is defined as the smaller of age or weight index if either of these values is less than 0.7. Should both be above 0.7, then the arithmetic mean of the two is used, since at high levels the age and weight of an animal are assumed to interact in determining condition. At low levels, Liebig's (1849) law of the minimum is thought more appropriate, and so the minimum (limiting?) value is used.

4.3.6 Mortality

With each time step the effect of mortality on population weight and numbers is calculated (Table 4.1). Sources of mortality are classified in four categories: old age, starvation, predation, and other causes. Death from old age occurs when the

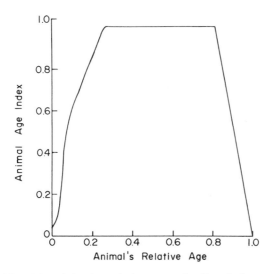

Fig. 4.6.   Animal age index per animal's relative age

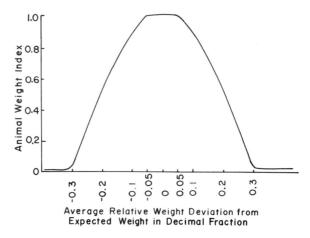

Fig. 4.7.   Animal weight index per its deviations from expected weight

relative age reaches 1.0. At this time the remaining members in that cohort are transformed to the dead category.

Predation loss is determined by summing all flows from each food category. If the food category is animal, the total flow is treated in the next time step as mortality. Predation mortality is distributed among cohorts in proportion to the reciprocal of cohort's biological index.

Starvation rate is a function of two factors: (a) the deviation of modeled weight from expected weight and (b) the starvation level for the population. This series of calculations considers the weight distribution in a natural population to be Gaussian. When the weight deviation w_d falls below a threshold (t_w), a starving proportion of the population (Fig. 4.8) is calculated. The weight deviation below which starvation mortality occurs is $3t_w$. For example, if a small mammal would starve when its weight was 70% of the expected weight, then t_w was 0.1. The proportion starving, s, is calculated as

$$ s = \frac{w_e - w}{\sigma} \, n(w_s ; w, \sigma) $$

where $\sigma = w_e t_w$ and $w_s = (w_e + w)/2 - 3\sigma$. The independent variable w_s in the normal distribution function is chosen as the average of $w_e - 3\sigma$ and $w - 3\sigma$, an approximation for the average weight of animals below the starvation threshold. The product sN_j approximates the number of individuals falling below the threshold of expected weight variation, and that number of animals is removed as starved.

Mortality from causes other than old age, starvation, and predation is supplied as a relative rate (input parameter) for each consumer and includes death from disease, competition, or predation not otherwise included in simulation. If the mean weight is less than the expected weight, the weight of the individuals removed by predation, starvation, and other causes is w_s. After numbers and weight are reduced by mortality, the carbon remaining in the cohort is then divided by the population to determine average carbon weight in this cohort.

4.3.7 Natality

Natality (Table 4.1) is contingent on a number of factors, such as environment, biological index, and frequency of reproduction.

The environmental influence is simulated using a 20-day running average of the product of sunlight, average air temperature, and precipitation. This procedure acknowledges the effect of photoperiod, temperature, and moisture on the reproductive success of many mammals. As the year progresses, this running average first increases and then decreases. A threshold for each consumer determines whether environmental conditions are suitable for reproduction. For aseasonal reproducers the threshold value is set so low that it has no impact.

A parental biological index greater than 0.83 is necessary for reproduction to occur. This simulates the influence of parental health and maturity on reproduc-

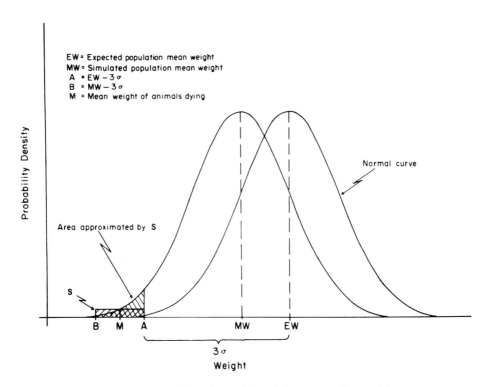

Fig. 4.8. Proportion of population below starvation weight.

tion. The number of reproductions per year is divided into half a year to determine the interval between each, assuming that reproduction occurs during only half of the year. If environment, biological index, and reproductive interval permit, reproduction can occur. For those animals reproducing in the spring the first reproduction occurs as soon as the thresholds for environment and biological index are met.

At each reproduction the number of offspring produced (N_0) is calculated as

$$N_0 = N_m f_f N_e b_i \qquad (4.10)$$

where N_m is the number of sexually mature animals; f_f is the fraction of females in population; N_e is the number of offspring expected; and b_i is the parental biological index. The parental biological index is included here since it reflects parental condition as influenced by age and weight.

The birth weight of the offspring is computed as the expected weight times the parental biological index. The expected weight is 5.3% of adult weight. The reported values (Altman and Dittmer, 1964) are 3.0% to 8.0% for mammals and birds, but for the mammals considered here 5.3% gives birth weights within 20% of observed values (Altman and Dittmer, 1964; Lechleitner, 1969).

4.4 Results and Discussion

For each consumer 20 initial values and parameters must be supplied (see Table 4.2). The goal was to describe a consumer with a minimal number of parameters to meet the objectives of the model. While the inclusion of more parameters and mechanisms might have increased the resolution of the model, there were two reasons for not doing so: (a) in modeling large systems, time and computer space soon limit the level of resolution and (b) the increased resolution would be spurious if those parameters could not be measured experimentally (Von Bertalanffy, 1968).

The model flows and state variables comparable with data are discussed in this section. Where data was lacking, the biological reasonableness of the representation is discussed.

4.4.1 Energy

The metabolic energy requirement as simulated under the variety of conditions in the model has not been determined experimentally. However, each of the components of this calculation has been well documented in the literature. Comparison of food-intake data and simulated values can indicate the accuracy of the metabolic energy-requirement calculations. Data on mammalian intake under field conditions are only available for cattle, and this comparison is made and discussed later.

4.4.2 Diet

Data on diet composition were available for some mammals under conditions similar to those simulated in the model. Model output and data values for these are given in Tables 4.4–4.6. In Tables 4.4 and 4.5, the producer and abiotic submodels for the Pawnee Site were used to simulate food available. The results in Table 4.6 use field data for amounts of food available. The model results and field data in the small mammal diets of Table 4.4 were for 1971. The simulated cattle diets shown in Table 4.5 were for 1970–1971. Although the model was developed and tuned on the 1970–1971 data for population dynamics, the 1971 diet-composition data were not used. Thus the 1971 data shown in Tables 4.4 and 4.5 were not used in constructing and tuning the model. The bimonthly average percentage intake for *Onychomys leucogaster* (grasshopper mouse) on the Pawnee Site from May 1969 to April 1970 (Flake, 1973) and model output from a simulation using field-determined food availability are shown in Table 4.6.

The two sets of small mammal results are similar except for intake values of Coleoptera, grasshoppers, and fall seeds. An explanation for this discrepancy is that the model used total live carbon for these food categories and that the amount of live Coleoptera and grasshoppers in the fall is relatively low (Van Horn, 1972). However, their availability as frost-killed corpses and a large weight per individual insect would account for the intake apparent in the data. As corpse grazing was not considered in formulation of the model, the model results indicate low amounts in these categories corresponding to modeled live insects and larger amounts of seed that was the next most available and desired food.

Diet selection data for small mammals for 1971 was not available in the same format as 1970; thus Table 4.4 presents only the three food categories of arthropods, seeds, and other plant material for the times and animals available.

Table 4.4. Average percentage intake by food category for three small mammals (D column = US/IBP data, 1971; M column = model results)

Food category	March		May		June		August	
	D	M	D	M	D	M	D	M
Onychomys leucogaster								
Arthropods	100	97	98	92	95	88	98	92
Seeds	0	2	0	1	0	1	0	2
Plant materials	0	1	2	7	5	11	2	6
Peromyscus maniculatus								
Arthropods	70	69	59	26	79	21	57	37
Seeds	20	30	0	3	1	2	0	5
Plant materials	10	1	41	71	20	77	43	58
Spermophilus tridecemlineatus								
Arthropods	—	—	51	34	37	41	—	—
Seeds	—	—	28	48	38	38	—	—
Plant materials	—	—	21	18	25	21	—	—

Table 4.5. Average percentage intake by food category for cattle under two grazing conditions for 2 yrs (D column = 1969–1971 data; M column = model results)

Food category	December 1969		June 1970		July 1970		August 1970		June 1971		August 1971		October 1971	
	D	M	D	M	D	M	D	M	D	M	D	M	D	M
Light grazing														
Cool-season grasses	10	20 −	26	9 +	20	7 +	17	8 +	25	7 +	13	5 +	26	9 +
Warm-season grasses	62	58 +	56	68 −	56	72 −	56	66 −	61	74 −	76	79 −	72	80 −
Forbs	23	12 +	16	21 −	22	19 +	25	24 0	13	17 −	10	14 −	1	10 −
Shrubs	15	10 +	2	2 0	2	2 0	2	2 0	1	2 0	1	2 0	1	1 0
Heavy grazing														
Cool-season grasses	7	4 +	24	9 +	18	12 +	12	5 +	17	6 +	11	5 +	10	20 −
Warm-season	43	60 −	54	62 −	56	72 −	61	74 −	66	72 −	81	80 0	61	40 +
Forbs	40	11 +	19	24 −	24	13 +	25	9 +	15	18 −	2	10 −	2	16 +
Shrubs	10	25 −	3	5 0	2	3 0	2	12 −	2	4 0	6	5 0	27	24 0

Table 4.6. Bimonthly averages of percentage intake by food category for *Onychomys leucogaster* [D column = May 1969–April 1970 data (Flake, 1973); M column = model results]

Food categories	Jan–Feb		Mar–Apr		May–Jun		Jul–Aug		Sep–Oct		Nov–Dec	
	D	M	D	M	D	M	D	M	D	M	D	M
Warm-season grasses	4	3	2	1	3	6	7	12	7	4	4	13
Cool-season grasses	6	2	7	2	1	6	1	1	1	2	4	6
Forbs	9	3	3	1	6	6	7	8	7	6	8	13
Seeds	25	24	15	15	3	0	4	7	12	12	16	45
Spiders	3	3	3	2	3	1	3	1	3	1	4	1
Leafhoppers	6	3	6	4	0	2	0	2	1	1	2	1
Lepidoptera larvae	4	5	14	16	18	15	1	9	2	6	1	1
Coleoptera	30	28	27	19	54	44	56	40	40	39	38	5
Grasshoppers	12	29	23	40	12	20	21	20	27	29	23	15

The model results (M column) are comparable with data (D column) for *Onychomys leucogaster* (grasshopper mouse) and *Spermophilus tridecemlineatus* (13-lined ground squirrel) but are disproportionately high in the plant-material category for *Peromyscus maniculatus* (deer mouse) in the summer. The model only allowed consumers to eat live plant parts that were nearly nonexistent in March and abundant in the summer. The inclusion of standing dead as a food category and a reduced palatability index for live plant parts would more appropriately simulate the feeding impact of *Peromyscus* on the system.

The cattle-diet data shown in Table 4.5 incorporate results for two grazing regimes over a 2-yr period. The light grazing involved 12 animals per half section (average initial weight of 218 kg); heavy grazing was 30 animals per half section (average intitial weight of 213 kg). More data were available on cattle than any other mammal; thus more comparisons between data and simulation results were made for this group.

Warm-season grasses dominate the cattle diet in both model and data for all dates and treatments. Except for December 1969 and August and October 1971, the predicted value exceeds the measured value. Since the values presented are percentages, this common overestimate of the most important item should be compensated for by a frequent underestimate of the less important items. This is seen in that cool-season grasses are commonly underestimated in the diet. This is partly a result of a simulated cover response of forbs and cool-season grass to grazing. Details of the physiological effect of grazing on native plants were not known and could only be approximated in the producer simulation. It resulted that the availability of cool-season grasses and forbs was commonly underestimated and this is reflected in the simulated diet selection.

Greater accuracy and flexibility of the diet-selection calculations could be achieved by incorporating a selectivity factor (Ellis et al., 1976). This factor would account for the observation that consumers may ingest foods on the basis of density (nonselective feeding) rather than on a preference-density basis (selective feeding utilized here). Ivlev (1961) and Holling (1965) observed that diet discrimination increased as the animal became satiated. Rice et al. (1971) noted that antelope, being smaller with lower population density than cattle and having a relatively greater abundance of available food, selected diets that departed radically from the proportion in which these foods occurred in the environment, while cattle diets followed availability more closely.

4.4.3 Secondary Production

Comparison of intake, gain, and feed conversion rates shown in Table 4.7 provides circumstantial evidence that the metabolic energy requirement and food-utilization functions applied to cattle are reasonable. Both the observed and modeled values involved approximations for which minor changes could readily explain the differences obtained. For example, the moisture content used by Dyck and Bement (1972) to estimate intake was based on hand-picked samples collected by observers moving with the grazing herd. The forage moisture

Table 4.7. Cattle herbage intake, gain, and feed-conversion rate for three grazing periods in 1970–1972 (E = estimated from data and tables; M = model results; intake and gain are mean values in kg/head, conversion rate = kg intake/kg gain)

	June 15–September 3, 1970		June 2–September 21, 1971		May 23–October 10, 1972	
	E	M	E	M	E	M
Heavy grazing						
Intake	510.0	504.0	704.1	780.3	986.2	955.3
Gain	40.6	38.5	54.4	58.3	78.8	70.8
Conversion rate	12.6	13.1	12.9	13.4	12.5	13.5
Light grazing						
Intake	585.8	568.2	780.8	776.5	1089.5	1114.7
Gain	52.7	43.1	89.8	57.5	115.3	81.7
Conversion rate	11.1	13.2	8.7	13.5	9.5	13.6

content, mean daily temperature, and water intake were used with the table entitled "Feed-intake Rates of European Cattle in Pounds of Herbage Dry Matter Eaten per Gallon of Water Drunk" in Hyder et al. (1966) to estimate intake. Within the model digestibility is an important control on gain. The digestibility curve used is an average of curves from Wallace et al. (1972) keyed to the producer section by phenological stage. The 1970–1971 results in Table 4.6 should have been good since data for these years were used in model development. On the other hand, the intake data for 1970–1972, the gain for 1972, and the resultant conversion rates were not used in model development.

The conversion rate for heavy grazing (Table 4.7) indicates a slightly higher intake per unit gain than was estimated from data. The most likely source of this discrepancy is the digestibility curve employed. This is especially true for standing dead as the 1972 data-collection period included late September and early October. These times were not included in previous sampling; hence estimates of digestibility were probably low.

The comparison of conversion rates under light grazing is even more striking. The total intake values were similar, but data indicated a greater gain and lower conversion rate relative to the model. This probably results from the ability of cattle in lower densities to select a more digestible diet. The amount of food available, relative to the food requirement, is higher under light grazing than under heavy, thus allowing greater opportunity for selectivity. This further corroborates the need for a selectivity factor in diet selection.

4.4.4 Biological Index

Figure 4.9 illustrates the fluctuations in biological index for populations of adults and growing offspring of a long-lived mammal (coyote) during 1972, and

Figure 4.10 illustrates the index during 1972 for a short-lived mammal. For the coyote the index remained relatively constant or decreased gradually during the winter as a result of limited food and adverse environmental conditions. When the young were born, the index for adults dropped and then gradually returned to normal as the litter matured. The index for the young started near 0.2 at birth, dropped to nearly 0 in the first week of life, and then climbed to adult levels as the young matured and the weak were removed by mortality. The index is scaled such that under normal conditions, when an animal reaches 25% of its life span or 93% of its potential adult weight, the index is 0.83. This level (0.83) was chosen as the minimum for reproduction to occur. The coyote young reach this point in the fall, but because of environmental conditions, do not reproduce until the next spring. This agrees with the observation of Knowlton (1972) and Dunbar (1973).

Peromyscus (Fig. 4.10) adult populations reproduced three times with the first litter reaching maturity early enough in the year to reproduce also. Table 4.8 gives literature values and simulated values of age in days at which the mammals considered here reach sexual maturity.

4.4.5 Natality

Modeled litter sizes for 1972 are shown in Table 4.9. These values are in general lower than the literature values but represent the population mean rather than the mean of litter-bearing females. Frequency of reproduction is related to

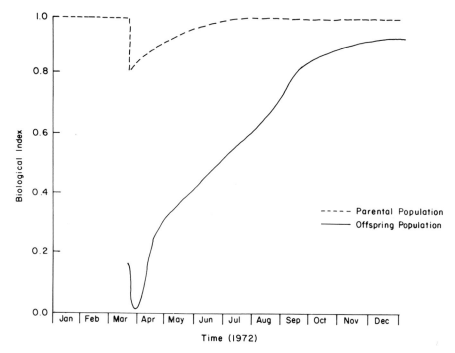

Fig. 4.9.   Biological index for *Canis latrans* populations in 1972 simulation

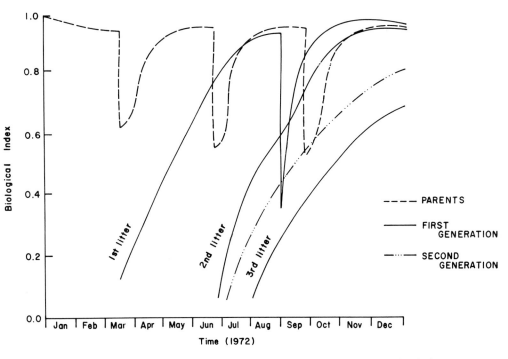

Fig. 4.10.   Biological index for *Peromyscus* populations in 1972 simulation

maturity and health (biological index) and environmental conditions. Environmentally, 1972 was an approximately average year.

4.4.6 Mortality

Although mortality affects populations throughout the year, there are periods of higher mortality during reproduction and in winter. Figure 4.11 shows the number of animals per hectare for each cohort simulated for *Peromyscus mani-*

Table 4.8.   Age at sexual maturity in days [D = data from Lechleitner (1969) and French et al. (1965); M = model biological index = 0.83]

Species	D	M
Canis latrans	210	185
Dipodomys ordii	42	50
Lepus californicus	160	144
Onychomys leucogaster	90	95
Peromyscus maniculatus	50	48
Spermophilus tridecemlineatus	72	90

Table 4.9.   Reproductive frequency and litter size per female [D = data from Lechleitner
(1969) and Haskell and Reynolds (1947); M = model simulation results, 1972]

Species	Reproductive frequency		Litter size	
	D	M	D	M
Canis latrans	1	1	4– 6 $\bar{X} = 5$	4.0
Dipodomys ordii	1	1	2– 4 $\bar{X} = 3$	2.97
Lepus californicus	1–2	2	1– 5 $\bar{X} = 2.3$	2.1
Onychomys leucogaster	1–3	2	2– 6 $\bar{X} = 4$	3.9
Peromyscus maniculatus	2–4	3	1– 9 $\bar{X} = 5$	3.9
Spermophilus tridecemlineatus	1	1	5–12 $\bar{X} = 7$	7.8

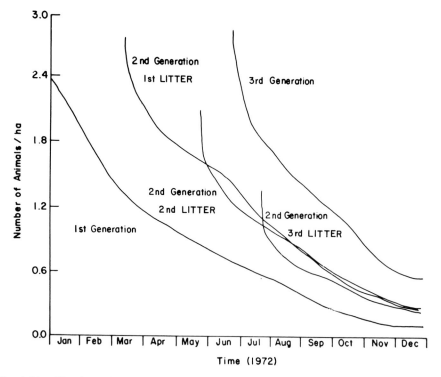

Fig. 4.11.   Number of animals surviving for five cohorts of *Peromyscus* simulation in
1972

culatus in 1972. The greatest decrease (mortality) occurred during the first few weeks of life, with the next greatest decrease occurring from January through mid-April.

4.4.7 Population Dynamics

The separate influences of mortality, natality, and growth on population are important, but it is the combined effect that determines population dynamics and can best be compared with observed values. The population dynamics for 1972 for the mammals simulated are shown in Figures 4.12–4.17. The distinction of grazing treatment is made for the four small mammals, as the data and model indicate differences of dynamics under these two treatments.

The modeled response of *Onychomys* (Fig. 4.12) for both light and heavy grazing was generally higher than data means but fell within the 95% confidence intervals for all but two points. *Peromyscus* simulated dynamics (Fig. 4.13) fit the confidence intervals poorly, while *Spermophilus* (Fig. 4.14) and *Dipodomys* (Fig. 4.15) fell within the confidence intervals at all points. The wide confidence bands indicate the difficulty of estimating these populations. Data for jackrabbits were not available for 1972. Comparison with 1970 and 1971 census data for jackrabbits on the Pawnee Site indicated the low point of the curve to be

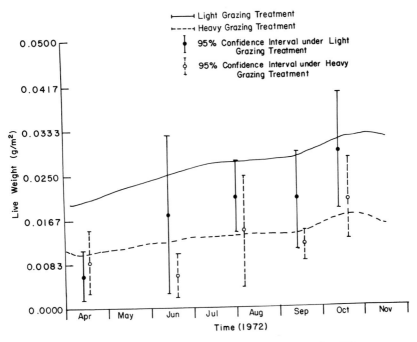

Fig. 4.12.   Biomass of *Onychomys leucogaster* in 1972

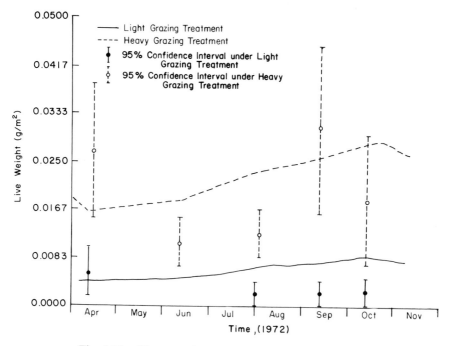

Fig. 4.13. Biomass of *Peromyscus maniculatus* in 1972

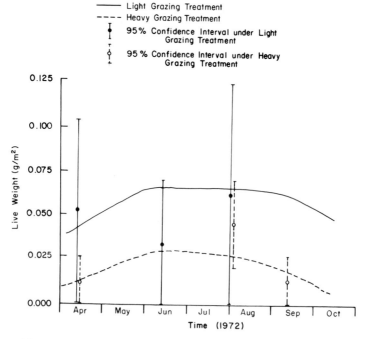

Fig. 4.14. Biomass of *Spermophilus tridecemlineatus* in 1972

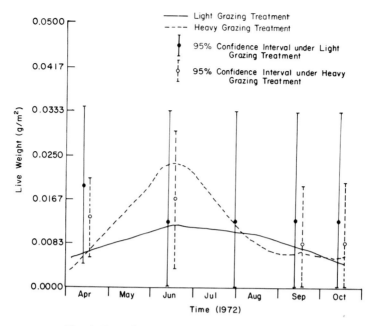

Fig. 4.15.   Biomass of *Dipodomys ordii* in 1972

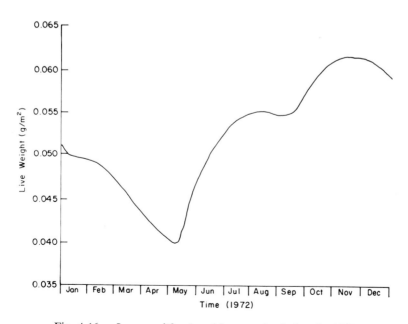

Fig. 4.16.   *Lepus californicus* biomass simulation for 1972

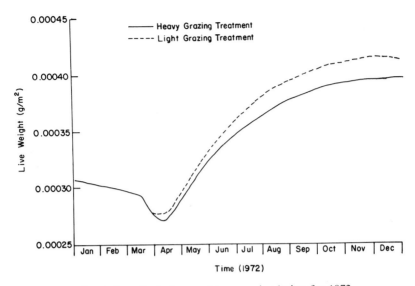

Fig. 4.17. *Canis latrans* biomass simulation for 1972

appropriate, but the simulated value (Fig. 4.16) should have reached 0.11 g/ha instead of 0.062.

The coyote biomass for 1972 (Fig. 4.17) showed a net increase, more so in the light grazing treatment than the heavy. This was a response to the increase in small mammal and rabbit populations during the year. The distinctions between light and heavy grazing are not realistic because coyotes are too far-ranging compared to the size of the grazing-treatment pastures. The results do, however, illustrate the response of the modeled animals to increased food supply. Data for coyote biomass at the Pawnee Site were not available. Initial values were based on data from an aerial survey by the Department of Fishery and Wildlife Biology of CSU covering an area that included the Pawnee Site. The shape of the curve (Fig. 4.17) is similar to that from other areas. Knowlton (1972) reports that as a result of "dispersal and mortality, populations normally attain their lowest levels just prior to the whelping season." A plot of animal numbers would show that the weight density increase (Fig. 4.17) during May to October was due to growth and fattening because mortality steadily reduced population size during this period.

4.5 General Discussion

System response of the mammalian consumers on the shortgrass prairie during 1970–1972 is summarized in Tables 4.10–4.13. Table 4.10 shows the amount of primary producer biomass consumed and wasted per year by mammals during the 1970, 1971, and 1972 simulations. The values in Table 4.10 may seem high, but they include the effects of trampling and wasting of food. For

Table 4.10. Total consumption and wastage of primary producer biomass by mammals (g dry wt · m^{-2} · yr^{-1})

Grazing	1970	1971	1972
None	1.6	1.7	1.9
Light	20.0	20.2	21.0
Heavy	47.5	46.0	46.0

cattle this was simulated by having 0.75 g of food trampled or wasted for each gram ingested. For small mammals the amount wasted ranged within 10–20% of intake, depending on food habits.

Table 4.11 summarizes the simulated secondary production for 1970, 1971, and 1972 under the three grazing treatments. This table illustrates the response of the model to the improved environment of 1972 (1970 and 1971 were dry years) and the effect of the reduced number of animals in the heavy-grazed treatment.

The conversion rates in Table 4.12 were determined by dividing the total food harvested or wasted (Table 4.10) in g/m^2 dry wt by the total secondary production in g/m^2. These values seem high when compared with the conversion rates of cattle in Table 4.7, but they do illustrate the impact of mammals on the system. If one considered only the food consumed and disregarded wastage, the conversion rates shown in Table 4.13 would obtain (cf. Table 4.7).

The dynamics of the mammals simulated were generally comparable with field data. While some detail was sacrificed in the (canonical) modeling approach, the accuracy of this submodel was commensurate with the other model components. Thus for large-scale models the canonical approach seems to be a realistic way to simulate a variety of animals with a minimum of coding.

Throughout Section 4.4 model results were compared with data; this is a form of validation. It is the opinion of the present author, however, that at least five years of reasonably representative data are needed to build and adjust such a model. Preliminary adaptations indicate that those systems of a more variable nature such as a desert grassland may require ≥ 7 yr of data. The present model used only 2 yr of data (1970–1971) collected specifically for its development.

Table 4.11. Simulated secondary production (g wet wt · m^{-2} yr^{-1})

Grazing	1970	1971	1972
None	0.12	0.13	0.19
Light 12, 12, 12[a]	0.71	0.86	0.95
Heavy 35, 30, 27[a]	1.92	2.08	1.89

[a] Animals/half section during 1970, 1971, and 1972, respectively. Average initial weight of 200 kg.

Table 4.12. System conversion rate[a] for simulated secondary production

Grazing	1970	1971	1972
None	13.3	13.1	10.0
Light	28.1	23.5	22.1
Heavy	24.5	22.1	24.3

[a] $\dfrac{\text{Dry wt food harvested or wasted } (g/m^2)(\text{Table 4.10})}{\text{Mammal production } (g/m^2)(\text{Table 4.11})}$

Table 4.13. Metabolic conversion rate[a] for simulated
secondary production

Grazing	1970	1971	1972
None	12.5	12.3	9.5
Light	15.1	13.4	12.6
Heavy	14.1	12.6	13.9

[a] $\dfrac{\text{Dry wt food intake } (g/m^2)}{\text{Secondary production } (g/m^2)}$

These 2 yr were similar in having slightly subnormal precipitation. Validation
was accomplished using data from 1972, which was a wetter year with a different
rainfall pattern. Thus while the results of the simulation were encouraging, their
accuracy could have been much improved if data spanning a greater range of
conditions had been available during implementation.

Acknowledgments.   The author wishes to acknowledge contributions of concepts, data, and relationships by participants in the mammalian sections of the
Process Studies Workshops Numbers 1, 2, and 3 held January, February, and
March, respectively, of 1972 and the Grassland–Tundra International Modeling
Workshop, August 14–26, 1972, both held at Colorado State University. In
addition, may special thanks to all those cited for personal communication within
the chapter.
   This chapter reports on work supported in part by National Science Foundation Grants GB-31862X, GB-31862X2, GB-41233X, and BMS73-02027 A02 to the
Grassland Biome, U.S. International Biological Program, for "Analysis of Structure, Function, and Utilization of Grassland Ecosystems."

References

Altman, P. L., Dittmer, D. S.: Biology Data Book. Washington, D.C.: Federation of American Society of Experimental Biology, 1964, 633 pp.

Amidon, E. L., Akin, G. S.: Dynamic programming to determine optimum levels of growing stock. Forest Sci. **14**, 287–291 (1968)

Andrewartha, H. G.: The use of conceptual models in population ecology. Cold Spring Harbor Symp. Quant. Biol. **22**, 219–236 (1957)

Beck, R.: Diets of steers in southeastern Colorado. Ph.D. thesis, Colorado State Univ., Fort Collins, 1969, 53 pp.

Blaxter, K. L., Wainman, F. W., Davidson, J. L.: The voluntary intake of food by sheep and cattle in relation to their energy requirements for maintenance. Anim. Prod. **8**, 75–83 (1966)

Brody, S: Bioenergetics and Growth. New York: Reinhold, 1945, 1023 pp.

Clapperton, J. L.: The energy expenditure of sheep walking on the level and on gradients. Proc. Nutr. Soc. **20**, 31–32 (1961)

Cook, C. W.: Energy budget of the range and range livestock. Fort Collins: Colorado State Univ. Exp. Sta. Bull. TB109, 1970, 28 pp.

Courant, R., Hilbert, D.: Methods of Mathematical Physics, Vol. II. New York: Wiley-Interscience, 1962, pp. 106–107.

Crampton, E. W., Harris, L. E.: Applied Animal Nutrition. San Francisco: Freeman, 1969, 753 pp.

Crampton, E. W., Lloyd, L. E.: Fundamentals of Nutrition. San Francisco: Freeman, 1959, 494 pp.

Davis, L. S.: Dynamic programming for deer management planning. J. Wildl. Mgmt. **31**, 667–679 (1967)

Donoho, H. S.: Dispersion and dispersal of white-tailed and black-tailed jackrabbits, Pawnee National Grassland. US/IBP Grassland Biome Tech. Rep. No. 96. Fort Collins: Colorado State Univ., 1971, 52 pp.

Dunbar, M. R.: Seasonal changes in testis morphology and spermatogenesis in adult and young-of-the-year coyotes (*Canis latrans*). M.S. thesis, Oklahoma State Univ., Stillwater, 1973, 24 pp. (mimeo.)

Dyck, G. W., Bement, R. E.: Herbage growth rate, forage intake, and forage quality in 1971 on heavily- and lightly-grazed blue grama pastures. US/IBP Grassland Biome Tech. Rep. No. 182. Fort Collins: Colorado State Univ., 1972, 17 pp.

Ellis, J. E., Wiens, J. A., Rodell, C. F., Anway, J. C.: A conceptual model of diet selection as an ecosystem process. J. Theoret. Biol. **60**, 93–108 (1976)

Emlen, J. M.: Ecology: An Evolutionary Approach. Reading, Mass.: Addison-Wesley, 1973, 493 pp.

Flake, L. D.: Food habits of four species of rodents on a shortgrass prairie in Colorado. J. Mammal. **54**, 636–647 (1973)

Forrester, J. W.: Industrial Dynamics. Cambridge, Mass.: MIT Press, 1961, 464 pp.

French, N. R., McBride, R., Datmer, J.: Fertility and population density in the black-tailed jack rabbit. J. Wildl. Mgmt. **29**, 14–26 (1965)

French, N. R., Maza, B. G., Aschwanden, A. P.: Life spans of *Dipodomys* and *Perognathus* in the Mojave Desert. J. Mammal. **48**, 537–548 (1967)

Gessaman, J. A.: Ecological energetics of homeotherms. Logan: Utah State Univ. Press, 1973, 155 pp.

Golley, F. B.: Energy values of ecological materials. Ecology **42**, 581–584 (1961)

Graham, N. McC.: Energy cost of feeding activities and energy expenditure of grazing sheep. Aust. J. Agr. Res. **15**, 969–973 (1964)

Grant, W. E.: Small mammal studies on the Pawnee Site during the 1971 field season. US/

IBP Grassland Biome Tech. Rep. No. 163. Fort Collins: Colorado State Univ., 1972, 51 pp.

Grodzinski, W., Gorecki, A.: Daily energy budgets of small rodents, In: Secondary Productivity of Terrestrial Ecosystems. Petrusewicz, K. (ed.). Warszawa-Krakow, Poland: Institute of Ecology, Polish Academy of Science, 1967, pp. 295–314

Gustafson, J. D., Innis, G. S.: SIMCOMP Version 3.0 user's manual. US/IBP Grassland Biome Tech. Rep. No. 218. Fort Collins: Colorado State Univ., 1973, 149 pp.

Hairston, N. G., Smith, F., Slobodkin, L. B.: Community structure, population control and competition. Am. Naturalist. **94,** 421–425 (1960)

Hansen, R. M., Flinders, J. T., Cavender, B. R.: Dietary and energy relationships of jackrabbits at the Pawnee Site. US/IBP Grassland Biome Tech. Rep. No. 14. Fort Collins: Colorado State Univ., 1969, 43 pp.

Haskell, H. S., Reynolds, H. G.: Growth, developmental food requirements, and breeding activity of the California jack rabbit. J. Mammal. **28,** 29–136 (1947)

Hodgson, J., Wilkinson, J. M.: The influence of the quantity of herbage offered and its digestibility on the amount eaten by grazing cattle. J. Brit. Grassl. Soc. **23,** 75–80 (1968)

Holling, C. S.: The functional response of predators to prey density and its role in mimicry and population regulation. Mem. Entomol. Soc. Can. No. 45, 1965

Hyder, D. N., Bement, R. E., Norris, J. J., Morris, M. J.: Evaluating herbage species by grazing cattle. Part I. Food intake, In: Proc. 10 Internat. Grassland Congr., Valtioneu-voston Kirjapaino, Helsinki, Finland, 1966, pp. 970–974

Ivlev, V. S.: Experimental Ecology of the Feeding of Fishes. New Haven: Yale Univ. Press, 1961, 302 pp.

Jameson, D. A., Dyer, M. I.: Process studies workshop report. US/IBP Grassland Biome Tech. Rep. No. 220. Fort Collins: Colorado State Univ., 1973, 444 pp.

Johnson, D. R., Groepper, K. L.: Bioenergetics of north plains rodents. Am. Midland Naturalist. **84,** 537–548 (1970)

Kemp, G. A., Keith, L. B.: Dynamics and regulation of red squirrel populations. Ecology **51,** 763–779 (1970)

Kleiber, M.: The Fire of Life. New York: John Wiley & Sons, 1961, 454 pp.

Knowlton, F. F.: Preliminary interpretations of coyote population mechanics with some management implications. J. Wildl. Mgmt. **36,** 369–382 (1972)

Kolenosky, G. B.: Wolf predation on wintering deer in east-central Ontario. J. Wildl. Mgmt. **36,** 357–369 (1972)

Lamprey, H. F.: Estimation of the large mammal densities, biomass, and energy exchange in the Tarangire Game Reserve and the Masai Steppe in Tanganyika. East Afr. Wildl. J. **2,** 1–46 (1964)

Lasiewski, R. C., Dawson, W. R.: A re-examination of the relation between standard metabolic rate and body weight in birds. Condor **60,** 13–23 (1967)

Laycock, W. A., Buchanan, H., Krueger, W. C.: Three methods of determining diet, utilization and trampling damage on sheep range. J. Range Mgmt. **25,** 352–356 (1972)

Laycock, W. A., Harniss, R. O.: Trampling damage on native forb-grass ranges grazed by sheep and cattle, In: Proc. 12th Internat. Grassland Congress, Moscow, USSR, 1974, pp. 349–354

Lechleitner, R. R.: Wild Mammals of Colorado. Boulder: Colorado: Pruett Press, 1969, 254 pp.

Liebig, J.: Chemistry and its Application to Agriculture and Physiology, 4th (Lond.) ed. New York: John Wiley & Sons, 1849, 401 pp.

MacArthur, R. H.: Population ecology of some warblers of northeastern coniferous forests. Ecology **39,** 599–619 (1958)

Main, A. R., Shield, J. W., Waring, H.: Recent studies on marsupial ecology. Monogr. Biol. **8,** 315–331 (1959)

Maynard, L., Loosli, J. K.: Animal Nutrition, 5th ed. New York: McGraw-Hill, 1962, 533 pp.

Moen, A. N.: Wildlife ecology. San Francisco: Freeman, 1973, 458 pp.

Peden, D. G.: The trophic relations of *Bison bison* to the shortgrass prairie. Ph.D. thesis, Colorado State Univ., Fort Collins, 1972, 134 pp.

Peden, D. G., Van Dyne, G. M., Rice, R. W., Hansen, R. M.: The trophic ecology of *Bison bison* L. on shortgrass plains. J. Appl. Ecol. **11**, 489–498 (1974)

Rauzi, F., Painter, L. I., Dobrenz, A. K.: Mineral and protein contents of blue grama and western wheatgrass. J. Range Mgmt. 22(1): 47–49 (1969)

Rice, R. W., Nagy, J. G., Peden, D. G.: Functional interaction of large herbivores on grasslands. In: Preliminary Analysis of Structure and Function in Grasslands. French, N. R. (ed.). Range Sci. Dep. Sci. Ser. No. 10. Fort Collins: Colorado State Univ., 1971, pp. 241–265

Rice, R. W., Morris, J. G., Maeda, B. T., Baldwin, R. L.: Simulation of ruminants on range. Federat. Proc. **33(a):** 188–195 (1974)

Rogers, J. G.: Analysis of the Coyote Population of Dona Ana County, New Mexico; MS Thesis, New Mexico State University, University Park, New Mexico, 36 pp. (1965)

Scholander, P. F., Hock, R., Walters, V., Johnson, F., Irving, L.: Heat regulation in some arctic and tropical mammals and birds. Biol. Bull. **99**, 237–258 (1950)

Smith, R. C. G., Williams, W. A.: Model development for a deferred-grazing system. J. Range Mgmt. **26**, 454–460 (1973)

Southern, H. N., Lowe, V. P. W.: The pattern of distribution of prey and predation in Tawny Owl territories. J. Anim. Ecol. **37**, 75–98 (1968)

Timin, M. E.: A multi-species consumption model. Math. Biosci. **16**, 59–66 (1973)

Van Dyne, G. M.: Measuring quantity and quality of the diet of large herbivores in a practical guide to the study of productivity of large herbivores. In: A Practical Guide to the Study of the Productivity of Large Herbivores. Golley, F. B., Buechner, H. K. (eds.). Oxford, England: Blackwell Sci. Publ., 1969, pp. 54–94

Van Dyne, G. M., Van Horn, J. L.: Distance traveled by sheep on winter range. Proc. Western Sect. Amer. Soc. Anim. Sci. **16**, 1–6 (1965)

Van Horn, D.: Grasshopper population numbers and biomass dynamics on the Pawnee Site from fall of 1968 through 1970. US/IBP Grassland Biome Tech. Rep. No. 148. Fort Collins: Colorado State Univ., 1972, 70 pp.

Vavra, M.: Diet and intake of yearling cattle on different grazing intensities of shortgrass range. Ph.D. thesis, Univ. Wyoming, Laramie, 1972, 126 pp.

Vavra, M., Rice, R. W., Bement, R. E.: Chemical composition of the diet, intake, and gain of yearling cattle on different grazing intensities. J. Anim. Sci. **36**, 2, 411–414 (1973)

Von Bertalanffy, L.; General System Theory, Foundations, Development, Applications. Revised Edition. New York; George Braziller. 1968. 295 pp.

Wagner, F. H., Stoddart, L. C.: Influence of coyote predation on black-tailed jackrabbit populations in Utah. J. Wildl. Mgmt. **36**, 329–342 (1972)

Wallace, J. C., Free, J. C., Denham, A. H.: Seasonal changes in herbage and cattle diets on sandhill grassland. J. Range Mgmt. **25**, 100–104 (1972)

Watt, K. E. F.: Ecology and Resource Management (a Quantitative Approach). New York: McGraw-Hill, 1968, 450 pp.

Zar, J. H.: Using regression techniques for prediction in homeotherm bioenergetics. In: Ecological Energetics of Homeotherms. Gessaman, J. A. (ed.) Monograph Series No. 20, Logan: Utah State Univ. Press, 1973, pp. 115–133

5. Simulation of Grasshopper Populations in a Grassland Ecosystem

CHARLES F. RODELL

Abstract

A carbon flow model of the interactions of grasshoppers with producer, decomposer, and other consumer components of a grassland ecosystem is presented. The primary driving variables are air temperature and precipitation. Grasshoppers directly effect producers through diet selection, herbivory, and damage during feeding. Interactions with the decomposers follow from death and fecal production. Grasshoppers interact with other consumers via competition for food and predation. The major objectives are to estimate grasshopper populations, energy flow through the population, and effects on the grassland system. The functionally different age groups of the population are considered where appropriate to model objectives. Simulation results of the model are presented and discussed.

5.1 Introduction

Grasshoppers are of interest to entomologists and ecologists because they often compete for food with man or his domesticated animals. Recently, grasshoppers have been studied for the role they play as consumers in ecological systems (Smalley, 1960; Wiegert, 1965; Van Hook, 1971). These studies relate grasshoppers to energy flow.

Grasshoppers are visually conspicuous primary consumers in most temperate grasslands. Their significance, however, is not obvious, except during massive population outbreaks (Shotwell, 1941; Buckell, 1945). The ecological function of grasshoppers may be viewed from three perspectives: (a) the proportion of the total energy in an ecosystem that is grasshopper biomass, (b) the amount of energy flow through grasshoppers during a given time interval, and (c) the influence the grasshopper population may exert on the dynamics of other components of the ecosystem. Extensive field and laboratory work has been done in areas (a) and (b). However, these studies offer little insight into the importance of function (c), which may constitute the most important role of grasshoppers in a grassland ecosystem.

A useful way to view the role of a species is the systems approach (Watt, 1966). This approach emphasizes the interactions between the components of the ecological system and the cause–effect pathways therein. This chapter describes a simulation model of a grasshopper population. The major objective of the model is to simulate the biomass dynamics of the grasshopper population. Additional objectives are to represent the effect of grasshoppers on the functioning of the system and to estimate energy flow via grasshoppers.

The grasshopper model reported has several features which in combination distinguish it from its predecessors (Mitchell, 1973; Randell, personal communication; Butler, personal communication; Gilbert, 1973; Gyllenberg, 1974). These features include population dynamics, energy flow, and capability of interaction with other submodels. Biologically meaningful mathematical relationships were implemented within the model (i.e., the model is mechanistic) insofar as possible. This latter feature is especially important if hypotheses involving cause–effect relationships are to be tested using the modeling approach.

Generally, the approach used in developing this model is more similar to the systems approach of Watt (1964), Holling (1959), and Coulman et al. (1972) than the approach used by MacArthur and Levins (1964) and Anderson (1970). In other words, those factors considered important with regard to the influence on various life-cycle phenomena are allowed to interact at the appropriate life-cycle stages. These influencing factors alter rates of the processes represented.

5.2 Model Overview

The linkage of the grasshopper model with other models is illustrated in Figure 5.1. The other state variables in the figure represent carbon in organic material for decomposition, in secondary consumers, and in primary production. In addition, information affecting the grasshopper model is supplied by the abiotic model.

The net change in state variable representing grasshopper or grasshopper egg carbon for a given time step in the simulation is equal to the sum of the inputs minus the outputs. For grasshoppers and their eggs the carbon dynamics are determined by the processes shown in Eqs. (5.1) and (5.2), respectively:

Change in grasshopper = forage intake + egg biomass hatching
    − grasshopper carbon oviposited − respiration − feces − mortality   (5.1)

Change in egg = Grasshopper carbon oviposited
                        − eggs carbon hatching − mortality   (5.2)

where each equation represents a time step.

Because grasshoppers are poikilothermic, air temperature is an important factor influencing forage intake, litter production, and life-cycle phenomena (hatching, development, sexual maturation, egg laying, and mortality). Other weather factors such as relative humidity and precipitation are considered where appropriate.

The grasshopper and producer models interact via the foraging activity of grasshoppers on aboveground material. The quantity and phenology of each

ELM 73 GRASSHOPPER SUBMODEL

A = Grasshopper Carbon g/m^2
B = Producer Carbon g/m^2
C = Producer Phenology
D = Temperature
E = Precipitation
F = Relative Humidity
G = Predator Carbon g/m^2

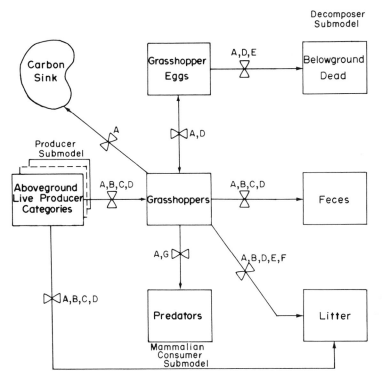

Fig. 5.1. Flow diagram of grasshopper submodel and the interactions with other submodels (solid lines indicate carbon flows; A-G represent factors influencing flows)

producer category influences the amount ingested and the damage that results from feeding activities. Damage, plant destruction without ingestion, contributes to litter production. Additional flows to the decomposition submodel are consequences of mortality and fecal production.

The interactions between the mammalian consumer submodel and the grasshopper submodel are both direct and indirect. On the Pawnee Site coyotes and grasshopper mice consume grasshoppers. Consumers also influence one another through the utilization of common foods. Consumption of plant material alters the amount of food available and thus the quantity and quality of the forage ingested.

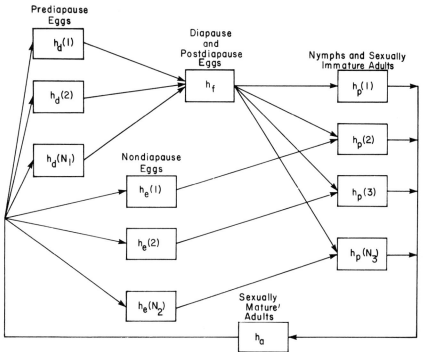

Fig. 5.2.   Flow of carbon through components of a grasshopper population

The flow of carbon through the grasshopper population is represented in Figure 5.2. Five different categories of grasshopper carbon are considered with flows between these groups representing the processes of hatching, development and sexual maturation, egg laying, and transition to diapause. The sum of the grasshopper groups, representing nymphs and sexually immature adults $[h_p(I)]$ plus sexually mature adults (h_a) in Figure 5.2, is the grasshopper state variable in Figure 5.1. In other words, all of the variables $h_p(I)$ and h_a carry on the processes associated with grasshoppers represented in Figure 5.1. Likewise, the sum of the variables, representing eggs $[h_e(I), h_d(I),$ and $h_f]$ in Figure 5.2, is the egg state variable in Figure 5.1.

The model considers grasshoppers of a give life-cycle type rather than a given species. There are two basic kinds of life cycles represented among grasshopper species in temperate grasslands, those in which: (a) the population overwinters in the later nymphal stages and metamorphose into adults in the spring and (b) eggs comprise the overwintering stage and hatching occurs in the spring. Process (a) is the typical life cycle at the Pawnee Site and is the one represented. In developing the model, the manner in which a process responds to a variable is of central interest, that is, the shape of the function relating the dependent variable (process) to the independent variable. Thus it is assumed that all species of interest respond (hatch, develop, and mature) similarly to the driving variables.

This assumption appears valid for those processes where data permit comparison. See, for example, the relationship of temperature to nymphal development rate (Shotwell, 1941; Putnum, 1963), that of temperature to reduction in egg viability (Parker, 1930; Riegert, 1967), and that of body weight to oxygen consumption (Wiegert, 1965). The representation of a life-cycle type is adequate given the objectives and the level of model resolution. The model does not simulate numbers of individuals, but C/m^2. Modeling weight density rather than numbers is sufficient for the needs of the model.

5.3 Life-cycle Dynamics

Lack (1954) showed that natural populations of animals tend to fluctuate in number within certain bounds. For most cases the upper bound was more restrictive than might be expected given their reproductive capabilities, suggesting that there were factor(s) responsible for regulation. Whether the main limiting factors were related to population density and thus density-dependent as proposed by Nicholson (1958) or unrelated to population size and therefore density-independent (Andrewartha and Birch, 1954; Thompson, 1956) was a central issue. The voluminous literature pertaining to this question of population regulation as related to grasshoppers was reviewed by Dempster (1963). The concept applied in developing this model was that for grasshopper populations, as with most animal populations, both density-independent and density-dependent factors operate simultaneously. For grasshopper populations weather seems the most important factor influencing population size. Several studies indicated correlations between grasshopper abundance and variations in weather. Edwards (1960) showed a positive correlation between numbers and mean temperature in Saskatchewan during the active adult period (July and September) for the previous 3 yr. Also for Saskatchewan, McCarthy (1956) found that temperature was an important factor along with sunshine during the egg-laying period. Others (Criddle, 1917; Wakeland, 1961) found grasshopper abundance to be negatively correlated to rainfall.

Temperature and rainfall were considered the two most important factors influencing population dynamics, the former being most influential on rates of development and egg laying and hatching. Soil water has a direct effect on embryonic development and an indirect effect through the quality and abundance of food. Precipitation and moist conditions are important influences on grasshopper mortality. Temperature has an effect through feeding rate whereby warm weather increases feeding and thus growth and fecundity, while cool weather, through reduced feeding, leads to a diminished population.

The one density-dependent factor implemented in the model was food availability (food:grasshopper ratio). This factor influences the rates of mortality and sexual maturation (Clark et al., 1967).

The diagrammatic representation of the life cycle is shown in Figure 5.2. The life-cycle processes considered in the model are nymphal development, sexual maturation, egg laying, induction of diapause, embryonic development, and hatching. For the variables representing nymphs and sexually immature adults $[h_p(I)]$, prediapause eggs $[h_d(I)]$, and nondiapause eggs $[h_e(I)]$, each box contains the value (g C/m²) attributed to that stage at a given time. For example, all of the eggs hatched on day 150 of the simulation might be assigned to variable $h_p(N)$ and those hatched on the 151st day to $h_p(N+1)$, and so on. The stage of nymphal development or sexual maturation for $h_p(N)$ is tabulated separately from the stages of the other $h_p(I)$ variables. With the completion of sexual maturation for $h_p(N)$, the contents of that box are transferred to the variable designating sexually mature adults (h_a). The same procedure is used for variables representing prediapause eggs and nondiapause eggs where embryonic development is considered.

5.3.1 Nymphal Development

After hatching nymphal development progresses as a function of daily temperature. Putnam (1963) raised nymphs of three grasshopper species (*Camnula pellucida, Melanoplus bivattatus*, and *M. bilituratus*) at six different temperature levels and averaged the results for the three species. Figure 5.3 shows Putnam's data converted to proportion of development per day plotted against temperature. The rate of development increased almost linearly as temperature increased from 23.9°C to 37.7°C. Data obtained on several grasshopper species by Shotwell (1941) were used to complete this relationship for temperatures greater than 38°C. The representation of nymphal development in this model used the straight-line segments of Figure 5.3.

Grasshoppers normally go through five or six instars during development. Modeling separate instars was not considered necessary. When differential effects on nymphs of various ages need representation, nymphal age is a fraction

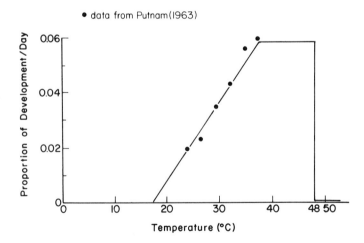

Fig. 5.3.   Effect of daily temperature on rate of nymphal development (Putnam, 1963), with functional form used in simulation model

of total development attained at that time. Two nymphal groups, 0–25 percent developed and 25–100 percent developed, are treated. The rates of processes such as mortality resulting from precipitation vary with each nymphal age group considered.

5.3.2 Sexual Maturation

In some species under favorable developmental conditions, teneral males may have sperm bundles in their follicles. However, the female gonads at the time of adult emergence were not as advanced as those of the male (Uvarov, 1966). Therefore, the model allows for a period of sexual maturation. Although several factors influence the rate of maturation, such as food quality and the presence of the opposite sex, only two factors are considered in the model.

Sexual maturation is determined by the accumulation of degree-days. It has been demonstrated that the rate of sexual maturation of nondiapausing adults increases with increasing temperatures (Brett, 1947). Thus the number of the degree-days above developmental zero, taken as 12°C (Church and Salt, 1952), is accumulated until the threshold for maturation is obtained.

Clark et al. (1967) stated that for the grasshopper *Phaulacridium vittatum,* sexual maturation occured within 2 wk given enough food. Sexual maturation is implemented in the model by calculating the available forage and compared with the weight of grasshoppers. If the forage:grasshopper ratio is above a preset value, food abundance does not inhibit maturation. If this ratio is less than the specified value, no maturation occurs and development is retarded and even regressed. This relationship is consistent with information in Clark et al. (1967). The rationale for this mechanism is that adequate food is necessary for maturation. Richards and Waloff (1954) demonstrated a weight increase concurrent with maturation, and Clark (1947) showed that noningestion of green plants delayed maturation.

5.3.3 Egg Laying

Effect of Temperature.   Parker (1930) cited field observations indicating a temperature range in which oviposition occurs. Harries's (1939) study of six diverse species of insects found identical temperature–fecundity curves for all species. As temperature increased, the activity rose exponentially to a peak and then fell off rather sharply. The shape of this function was adapted to field observations, and data from Parker (1930) were used to estimate parameters.

Fecundity is determined by assuming a 1:1 sex ratio. This assumption is consistent with the field data collected by Richards and Waloff (1954) and is reasonable given the type of sex determination in grasshoppers—a balance between female-determining X-chromosomes and male-determining autosomes with females and males having the genotypes XX and XO, respectively (White, 1973). Another assumption is that females weigh about 1.5 times as much as males. The average dry-weight ratio (male: female) of *Schistocera* with an adult age of 1–50 days was 1:1.504 (Uvarov, 1966). Live-weight ratios for *Omocertus viridulus* ranged a bit higher [1:1.5–1:2.5; Richards and Waloff (1954)].

Finally, data on five species indicated that females lose from 12.7% to 20.6% of their weight at oviposition (Richards and Waloff, 1954). These data were used as a guideline, along with model performance, to estimate the fecundity rates.

The egg production per time step (e_p in g/m^2) is expressed as in Eq. (5.3):

$$e_p = 0.6a_b r_f \, \delta t, \tag{5.3}$$

where a_b is adult carbon weight (g/m^2) and r_f is fecundity rate (g·g^{-1}·day^{-1}). The constant (0.6) implements the assumption that 60% of the adult biomass is female.

Diapause and Nondiapause Eggs.   Once the amount of egg biomass is determined, distribution between diapause and nondiapause categories is made. In many species of temperate-climate grasshoppers females lay two types of eggs, those entering diapause and those not entering, for instance, *Melanoplus bivittatus* (Church and Salt, 1952), *M. differentialis* (Slifer and King, 1961), and *M. sanguinipes* (Pickford and Randell, 1969). A single female may lay both kinds of eggs in a single egg pod. In addition, the distinction between diapause and nondiapause has a genetic component as evidenced by selection experiments (Leberre, 1953; Pickford and Randell, 1969; Slifer and King, 1961). One final note of interest regarding this phenomenon is that field observations indicated that the frequency of the nondiapause eggs increased as the season progressed (Church and Salt, 1952).

Photoperiod is the usual driving variable that operates on such phenomena in other insects (Wigglesworth, 1965). The distribution of oviposited eggs between the two categories, then, is represented as a function of time. Until a specified date, all eggs are of the diapause type. After that date the percentage of nondiapause eggs increases linearly with time from 0% to 10%. This mechanism is crude but given the paucity of data and the model objectives, is deemed adequate.

5.3.4 Embryonic Development

Diapause-type eggs develop to a certain stage, and then development ceases until the required cold treatment occurs. Reigert (1967), among others, provided data on time periods and temperatures that were effective at breaking diapause for the clear-winged grasshopper, *Camnula pellucida*. Nondiapause eggs developed as long as conditions remained above developmental zero (Pickford and Randell, 1969). Following cold conditions, a time lag is used to allow physiological processes to resume.

In addition to temperature, soil water is necessary for embryonic development (Uvarov, 1966). Modeled embryonic development proceeds under favorable temperature conditions only when sufficient soil water is present. In other words, the presence or absence of adequate soil water acts as a switch for embryonic development.

5.3.5 Hatching

Once eggs reach the diapause stage, they are stored as a single variable (h_t). The hatching process is a function of temperature history (accumulation of degree-days), as is embryonic development. Hatching, as a function of temperature history, is assumed to be normally distributed. Hatching occurs over a range of 40 degree-days above developmental zero (12°C) with the mean being the amount needed to complete embryonic development.

This process is implemented by using a cumulative frequency distribution approximating that of a normal distribution with a mean of X and a standard deviation of 10. As degree-days above developmental zero are accumulated, the proportionate increase on the ordinate of the cumulative frequency distribution in a given iteration of this simulation indicates the proportion of the initial egg mass-density that hatches at that time. Each allotment of nondiapause eggs hatches as a unit when the threshold (total developmental degree-days) is reached.

An additional requirement placed on hatching is that of a prepatory warm period. Parker (1930) observed that hatching occured during the first warm period of 3–5 days having maximum temperatures above 24°C.

5.3.6 Egg Mortality

Hatching success is a critical factor, constituting the first point where the future population of grasshoppers is determined. Unfortunately, there are few data on this factor. Four main variables appear to influence egg mortality: (a) moisture, (b) temperature, (c) parasitism, and (d) fungal attack. The influence of (a) and (b), the weather factors, is emphasized throughout the model and are again stressed here. A nominal egg mortality rate is substituted for the two latter variables.

Temperature. Data indicated the eggs of grasshoppers from cold and temperate climates to be more resistant to cold than nymphs and adults (Uvarov, 1966). However, low temperatures did detrimentally effect egg survival. Data on this process were obtained by Parker (1930) and Riegert (1967) (Fig. 5.4). These results indicated that temperatures below −20°C were increasingly detrimental, and temperatures as low as −30°C caused 100% mortality. However, temperatures above −20°C did not appear to decrease egg viability. In addition to the effect of the temperature, the length of exposure is important. As the exposure time to adverse temperatures increased, mortality also increased (Pickford, 1966b). There were no data dealing with the effect of low temperatures on nondiapause eggs; thus diapause and nondiapause eggs are treated alike. Soil

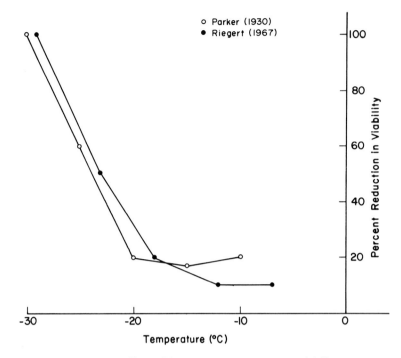

Fig. 5.4.   Effect of low temperatures on egg viability

temperatures below −20°C in the model are saved for the period during which the eggs are in the ground. These values are then weighted by time, and an estimate of low temperature induced mortality is made from a linearization of the data in Figure 5.4 for the temperature range of −20°C to −30°C.

Parker (1930) demonstrated that excessively high temperatures (≥45°C) also caused egg mortality. However, it is improbable that such temperatures would be experienced by eggs buried at least 2–5 cm in the soil. Therefore, egg mortality resulting from high temperatures is ignored.

Soil Water.   Uvarov (1966) stated, ''Data on the resistance of eggs to desiccation are neither numerous nor exact.'' However, studies do indicate that the lack of soil water is detrimental. Pickford (1966b) provided field data that show eggs of *Camnula pellucida* laid during drought conditions suffer decreased viability. If dry conditions exist, eggs would lose water during periods of embryonic development when water was normally absorbed. However, eggs in diapause are extremely resistant to desiccation (Dempster, 1963). For example, Salt (1952) exposed eggs of *Melanoplus bivittatus* to 0% relative humidity at 40°C for a period of 1 hr. His results showed that the rate of water loss was highest in the earlier stages of embryonic development.

The model allows soil-water shortage to influence egg mortality only during the prediapause period. This influence is achieved by arresting embryonic development when soil water is below a threshold level. Thus when winter sets in and these eggs have not reached the diapause stage, they die. Effects of too much water are not considered. Egg mortality in a time step for diapause eggs and nondiapause eggs is a relative rate computed as the sum of the nominal rate and the low temperature rate.

5.3.7 Grasshopper Mortality

Parker (1930) demonstrated the importance of abiotic factors in determining grasshopper survival. Mortality is determined by mechanisms available through ELM, including low and high temperatures, low humidity, precipitation, predation, and starvation. In addition to these mechanisms, a nominal mortality rate is also included to account for those factors not represented, such as age and bacterial and fungal diseases. The implementation of predation is discussed in the mammalian submodel.

The grasshopper carbon removed by death in a given time step (m_i in g C/m^2 of grasshopper age class i) is expressed by the relationship in Eq. (5.4):

$$m_i = b_i(n_i d_i + l_i + h_i + p_i + g_i)\, \delta t, \tag{5.4}$$

where b_i is grasshopper weight density of age class i(g/m^2), n_i is the nominal mortality rate (g·g^{-1}·day^{-1}), d_i is the effect of density and food availability on the nominal death rate (dimensionless), l_i is the rate of mortality due to low temperature (g·g^{-1}·day^{-1}), h_i is the rate of mortality due to high temperatures and desiccation (g·g^{-1}·day^{-1}), p_i is the rate of mortality due to precipitation (g·g^{-1}·day^{-1}), and g_i is the rate of mortality due to predation (g·g^{-1}·day^{-1}) (this variable is determined in the mammalian model).

Low Temperatures. The implementation of mortality from frost and low temperatures is not data-based, but Parker (1930) demonstrated that low temperatures could cause death of nymphs and adults. Temperatures below a specified threshold are modeled to increase mortality by way of variable l_i. As the temperature decreases below the threshold, the amount of mortality increases linearly to a maximum mortality resulting from low temperatures.

High Temperature and Low Relative Humidity. Parker (1930) showed that exposure of grasshoppers to high temperatures increased the mortality rates in both adults and nymphs, with adults being slightly more heat-tolerant. Shotwell (1941) reported an increase in nymphal mortality from 17% to 42% as temperatures increased from 95°F to 105°F in developing *Camnula pellucida,* and Pickford (1966a) reported field data indicating excessively high temperatures to be detrimental, especially to young nymphs.

The combined conditions of high temperature and low relative humidity are assumed to increase mortality by desiccation. There have been few studies on this aspect of grasshopper mortality (Uvarov, 1966), and any pertinent investigations were concerned with grasshoppers and locusts normally inhabiting arid regions. However, because desiccation does seem to be a possibility on temperate grasslands and because the input is available from the abiotic submodel, a series of days with high temperatures and low relative humidities is modeled to increase mortality.

Laboratory data on high-temperature-induced grasshopper mortality represents the response of the animal to the temperature. These data are not directly applicable in the model because grasshoppers regulate their temperature by taking advantage of microenvironmental features. For example, under hot conditions grasshoppers climb blades of grass to escape the high ground-surface temperatures. Therefore, while air temperature and ground-surface temperatures are simulated, it is difficult to estimate the environmental temperature realized by the grasshopper.

Mortality from high temperature and low humidity is represented by the fact that unless the temperatures exceed a threshold, high temperatures have no effect on mortality. Once the threshold is exceeded, the amount of mortality depends on relative humidity. As relative humidity decreases, mortality increases linearly to a maximum at zero.

Precipitation. Precipitation might be an important regulator of grasshopper populations. Moist conditions may create environments ideal for fungal and bacterial diseases. Parker (1930) suggested that moisture might be the most important natural check of grasshopper abundance. In his field studies Pickford (1966a) found that precipitation was the most important weather factor in determining mortality. This was especially true for young nymphs. Additionally, Pickford (1966a) observed drowning deaths from flooding in one of his treatments.

Given these qualitative data, mortality from precipitation is implemented to depend on precipitation and grasshopper age. Young nymphs are more sensitive to rainfall than older nymphs and adults.

Food Limitations. Weather, generally considered to operate in a density-independent manner, was the primary factor governing distribution and abundance of grasshoppers (Birch, 1957). Pianka (1970) suggested that terrestrial organisms that live less than 1 yr are primarily r-selected. For such organisms, inter- and intraspecific competition was small and variable. Mortality for such species was catastrophic and density-independent. At times of outbreak, however, food shortage could occur and starvation would function as a density related effect. No data were found relating density to mortality, presumably because density seldom plays an important role. This assumption is controversial, however (Dempster, 1963), and a density-related mortality factor is included.

The ratio of available forage (g C/m^2) to grasshoppers (g C/m^2) is calculated. If this ratio falls below a threshold level, density affects mortality. Otherwise, available forage has no effect on mortality. The density ratio (forage:grasshopper) when below this level acts inversely on the nominal mortality rate. For example, if the ratio was 1:2 or 3:4, the nominal mortality rate is multiplied by 2 or $\frac{4}{3}$ [d_i in Eq. (5.4)], respectively.

5.4 Interactions

The diagram in Figure 5.1 represents the flow of carbon to and from the grasshopper population. Death was discussed above. The implementation of functional relationships for feeding, defecation, and damage is described in this section.

5.4.1 Forage Consumption

The total amount of forage material consumed in a time step by nymph and adult grasshoppers (f_i in g C/m²) is determined by a foraging rate that varies with: (a) age (g_i in g C for forage (g C of grasshopper)$^{-1}$ day^{-1}), (b) the effect of temperature on feeding (t_f dimensionless), and (c) grasshopper density (b_i in g C of grasshoppers/m²). The subscript i indicates age class. This relationship is the following

$$f_i = g_i b_i t_f \, \delta t. \tag{5.5}$$

The sum of the f_i is the forage consumed by all grasshoppers in a time step.

Grasshopper Age. The forage rate factor for age class $i(g_i)$ is a constant that allowed nymphs to consume more forage per gram of insect than adults. Nymphs have a metabolic rate per unit weight that is two or three times that of an adult (Odum et al., 1962). Reported values for the amount of dry weight of food intake per unit dry weight of insect biomass per day for leaf-eating arthropods ranged from 0.3 to 2.5 (Mitchell, 1973). Values used for g_i in the model fall within this range (0.6, 1.3) with nymphs eating at about twice the rate of adults. The rate (g_i) is considered to be a maximum.

Temperature Influence. Field observations suggested that there was a certain range of air temperatures in which grasshoppers would forage (Parker, 1930). The model, however, is not designed to treat fractions of a day; that is, it cannot treat diurnal activity. Thus the temperature influence (t_f) is calculated by estimating the number of hours during a 24-hr period wherein the temperature exceeds a minimum for foraging. This influence is implemented by assuming the daily temperature reached its minimum at 6:00 A.M. and its maximum at 2:00 P.M. The temperature curve from 6:00 A.M. to 6:00 P.M. is assumed to be sinusoidal. From 6:00 P.M. to the next day's minimum at 6:00 A.M. the decline in temperature is assumed to be linear. If the time above the critical feeding temperature exceeds 16 h, the grasshoppers eat at the rate g_i. If, however, the time above the minimum feeding temperature falls short, forage intake is reduced proportionately.

Diet Selection. From the total forage intake (f_i) the amount of intake per forage category j, that is, warm season grass, cool season grass, and so on, by group $i(c_{ij})$ is determined by multiplying f_i by a present preference (p_j, dimensionless) for forage category j; hence

$$\Sigma p_j = 1,$$

$\sum_i c_{ij}$ is the total amount of forage category j eaten by grasshoppers, and

$$\sum_j c_{ij} = f_i.$$

The present preference p_{rj} is determined from three factors: (a) the amount of the forage category available for grasshopper consumption (f_{aj} in g C/m^2) divided by the total available forage, (b) a phenological effect of the forage category (p_{ej}, dimensionless), and (c) a preference index for the forage group (p_{bj}, dimensionless). Equation (5.6) expresses this consociation:

$$p_{rj} = \frac{f_{aj}}{\sum_j f_{aj}} p_{ej} p_{bj} \qquad (5.6)$$

The preference-index values for a forage category by the "generalized grasshopper" at the Pawnee Site were derived from data on grasshopper abundance and stomach contents (Van Horn, 1972) and the forage preferences determined in laboratory feeding trials (Mitchell, 1975). The preference-index values per forage category (p_{bj}) were calculated by weighting the preferences in the feeding trials by the relative abundance of that grasshopper species at the study site. Values of p_{bj} are for lush green material provided in abundance under laboratory conditions. Under field conditions these values are assumed to be modified by effects of phenology and relative availability. The effect of phenology is used as an indicator of desirability, with the younger, more digestible plants being more desirable. The phenological values are calculated in the producer section. Figure 5.5 shows the phenological effect on preference.

The final preference modification deals with the relative availability of a forage group. This modification is used to compensate for grasshopper behavior. It is assumed that the more available the forage, the greater the chance of contact and the likelihood of its being consumed.

5.4.2 Fecal Production

Fecal production is intake less assimilation. The proportion assimilated is a function of the weighted average phenology of the forage eaten. The rationale is that younger vegetation is higher in nutrient content and more digestible. Odum et al. (1962) stated that grasshopper assimilation efficiencies vary from 0.3 to 0.6, depending on food. Other literature values included 0.274 (Smalley, 1960), 0.31–0.48 (Husain et al., 1946), 0.16–0.20 (Hussain, 1972), 0.35–0.78 (Davey, 1954), 0.161–0.486 (Gyllenberg, 1969), and 0.226–0.322 (Mitchell, 1975). These studies varied with the grasshopper species used, the kind of food eaten, and the age of the individuals. The model uses 30% and 60% as the limits for assimilation

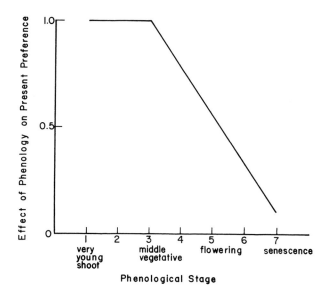

Fig. 5.5.   Relationship between phenological stage of forage category and its effect on determining diet selection

efficiency corresponding to weighted average phenologies of 7.0 and 1.0, respectively. Linear interpolation between these points determines the assimilation efficiency for other values of phenology.

5.4.3 Respiration and Production

The assimilated portion of the consumed forage is divided at constant proportions into respiration and production. Reported values for the proportion of assimilated forage that goes to production include 0.15 in a South Carolina old field (Odum et al., 1962), 0.39 and 0.35 in a Michigan old field (Wiegert, 1965), 0.45, 0.44, and 0.44 in a Finnish meadow (Gyllenberg, 1969), and 0.37 in a Georgia salt marsh (Smalley, 1960). The first two papers cited dealt with grasshopper species comparable to those found at the Pawnee Site. In addition, the temperature regime of the Michigan old field during the study was comparable with those at the Pawnee Site; thus a value of 0.37 is used.

5.4.4 Litter Production

Andrzejewska et al. (1967) stated:

> . . . if the influence of grasshoppers on the vegetation were to be determined on grounds of food intake, the result would be much below the true value of losses suffered by the plants. Grasshoppers slightly gnaw the leaf blade so that part of the leaf drops off. This way they destroy 4.8 times the amount of consumed plant biomass.

In another study Andrzejewska and Wójcik (1970) found the extent of grass-hopper destruction depends on the shape and length of the leaves bitten. When shorter leaves were involved, the amount of grass destroyed was, on the average, six times the amount eaten and was as high as ten times consumption. As plants and leaves elongated, the damage increased to an average value of destruction of 15 times the amount consumed. Andrzejewska and Wójcik (1970) stated that this ratio may go as high as 25.

Other reported values were not as high as those obtained by Andrzejewska and Wójcik. Gyllenberg (1969) estimated that leaf damage by the grasshopper *Chorthippus parallelus* was about 0.7 times the amount consumed from a meadow in Finland. Mitchell (1973), using three species of grasshoppers, *Melanoplus sanguinipes, Aulocara elliotti,* and *Melanoplus foedus* in laboratory feeding trials with three food plants, *Bouteloua gracilis, Sphaeralcea coccinea,* and *Artemisia frigida,* found, as did Andrzejewska and Wójcik (1970), that destruction depended on the plant. In addition, Mitchell concluded that destruction rates were somewhat proportional to consumption rates and the ratio of destroyed:consumed was inversely related to the plant's preference rating. His ratios of destroyed:consumed varied considerably with the species of grasshopper and plant (0.010–68.93). Generally, however, the results were in the range of 0.5–2.5.

The flow of aboveground live plant material to litter is the function of the amount consumed, the age of the grasshopper, and the plant category of Eq. (5.7):

$$l_j = \sum_i c_{ij} d_{ij} \, \delta t \tag{5.7}$$

where l_j is the amount of forage group j transferred to litter per time step as the result of grasshopper damage (g/m²); c_{ij} is the amount of forage group j eaten by grasshopper age class i(g/m²); and d_{ij} is a damage rate of forage j by grasshoppers of age i(g·g⁻¹ day⁻¹). The other factor taken into account in Eq. (5.7) is that nymphs are less destructive than adults (R. E. Pfadt, personal communication); thus d_{ij} is greater for adults. Damage rates range from 0.1 to 2.5.

5.5 Results and Discussion

5.5.1 Biomass Dynamics

A major objective of this modeling effort was to simulate the biomass dynamics of the grasshopper population. The simulation results for 1970 and 1971 are shown in Figure 5.6. Data from 1970 (Van Horn, 1972) were used to adjust the model. No data were available for 1971.

The first point of interest is that the simulated 1970 population was greater than the 1971 population. On January 1 of each year egg densities were similar—

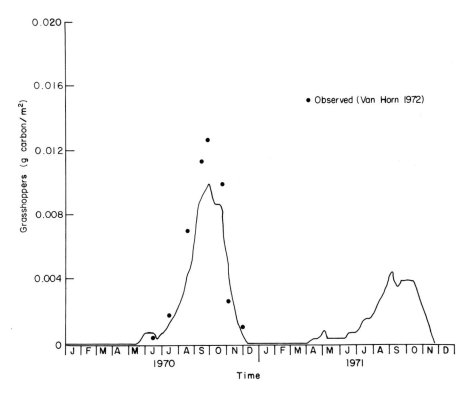

Fig. 5.6. Simulated grasshopper biomass dynamics for 1970 and 1971

0.00010 g C/m^2 in 1970 and 0.00015 g C/m^2 in 1971. If initial values for eggs was the reason for the population difference for the two years, the difference would be in the opposite direction. Also, net primary production was essentially the same for the two years. The decreased grasshopper population in 1971 was the result of a moist spring that increased nymphal mortality. This relationship is demonstrated in Figure 5.7, where rainfall events and cumulative grasshopper death are plotted. In 1971 there was about twice as much precipitation during the nymphal period as in 1970. The increased mortality early in 1971 was reflected by the greater April and May cumulative mortality values in 1971 as compared to 1970.

Net secondary production (p_s) is defined here in the usual sense (Petrusewicz and Macfadyen, 1970):

$$p_s = \int \left[\sum_i c_i(t) - f_i(t) - r_i(t) \right] \delta t \qquad (5.8)$$

where $c_i(t)$ is total consumption, $f_i(t)$ is feces production, and $r_i(t)$ is the respiration for age class i at time t.

Net grasshopper productivity amounted to 0.009% and 0.006% of the net primary production for 1970 and 1971, respectively. These values were less than

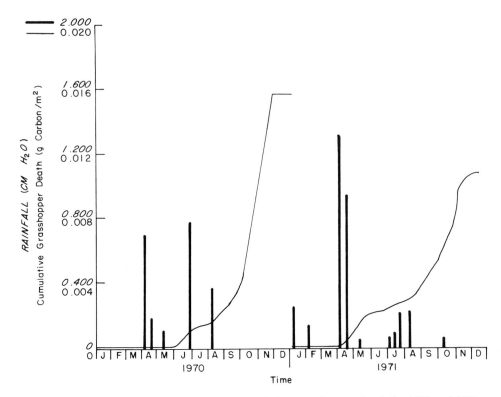

Fig. 5.7.   Simulated rainfall events and cumulative grasshopper death for 1970 and 1971

those determined in field studies of other systems (Smalley, 1960; Odum et al., 1962; Wiegert, 1965; Gyllenberg, 1969; Van Hook, 1971) and suggest that energy content of grasshoppers at the Pawnee Site was trivial for the years simulated.

The results of these simulations suggested that, for the Pawnee Site, grasshopper populations were not food limited, but were regulated primarily by weather.

5.5.2 Energy Flow

The various processes by which grasshoppers interact with other components of the grassland ecosystem are diagramed in Figure 5.1. The cumulative sums of the processes shown in Figure 5.1 by which grasshoppers interact with other model components are provided in Table 5.1 for 1970 and 1971. Grasshoppers were responsible for the removal of carbon from the producers by consumption and damage. These two processes removed 0.4% and 0.2% of the net primary production for 1970 and 1971, respectively.

The production of surface litter accounted for approximately 1.1% and 0.4% of the litter carbon decomposed for 1970 and 1971. Egg death contributed carbon to litter in the 0–4-cm soil layer. For 1970 and 1971 this input was 7×10^{-4} and 5×10^{-4} g C $\cdot$ m^{-2} $\cdot$ yr^{-1}, respectively. In each case this value was less than $1 \times 10^{-3}\%$ of the carbon decomposed in this soil layer.

Table 5.1.   Cumulative flow of carbon into and out of
grasshoppers and litter production by grasshoppers
$(g\ C \cdot m^{-2} \cdot yr^{-1})$

	Simulated year	
Process	1970	1971
Consumption	0.148	0.084
Litter production	0.699	0.399
Respiration	0.052	0.020
Grasshopper death	0.018	0.012
Feces	0.096	0.052
Egg death	0.0007	0.0005

Simulations for 1970 and 1971 indicated that grasshoppers had little effect on the cycling of carbon at the Pawnee Site. As mentioned in a previous section, the grasshopper's most important role may be that of litter production, and the ratio of litter produced to forage consumed may run as high as 25 for grassland systems. Simulations were performed that considered the effect of different levels of litter production on net primary production and the annual carbon turnover by surface-litter decomposers. The results of these simulations and their comparison to a system without grasshoppers are shown in Table 5.2. The greater grasshopper density was in 1970; yet even at high litter production rates, net primary production was reduced by less than 1%. For the same set of conditions, the surface litter carbon turnover increased 2.24%.

Even though grasshoppers did not influence energy flow significantly, they might with proper timing make an impact. One simulation experiment was performed to investigate this effect. For a 1970 simulation, grasshoppers were introduced on day 130 at a density of 0.02 g C/m². This date was one week earlier

Table 5.2.   Effect of grasshopper litter production on net primary production and surface-litter decomposition

Ratio of litter produced/ consumed for warm season and cool season grasses	Net primary production (g dry wt · m⁻² · yr⁻¹) and change (%)[a]		Total litter decomposition (g C · m⁻² · yr⁻¹) and change (%)[a]		Total transfer from live shoots to litter resulting from grasshopper foraging (g C · m⁻² · yr⁻¹)[a]	
	1970	1971	1970	1971	1970	1971
No grasshoppers present	512.94	503.67	74.69	112.14	0	0
2.5	512.04	502.98	75.01	112.16	0.70	0.40
	−0.18	−0.14	+0.43	+0.02		
15.0	510.34	502.07	75.68	112.22	2.20	1.24
	−0.51	−0.32	+1.33	+0.07		
25.0	508.74	501.30	76.36	112.27	3.66	2.04
	−0.82	−0.47	+2.24	+0.12		

[a] Compared to the simulations where no grasshoppers were present.

and at a higher density than they first appeared in earlier simulations. The result was a reduction of 14% in net primary productivity. This simulation showed that even at low grasshopper densities, if plant production is small, grasshoppers may be important in regulating plant production.

5.5.3 Exercises on the Population Dynamics

The abundance of grasshoppers on rangelands is of interest because of the forage they consume. A recent review of studies on grasshopper impact (Anderson, 1972) suggested that "grasshoppers annually consume from 6% to 12% of the available forage" on rangelands in the U.S. (Cowan, 1958, as cited by Anderson, 1972); and estimates concerned with the value of the forage set the yearly loss to grasshoppers of range and pasture grasses in the U.S. at $80 million (Parker and Connin, 1964, as cited by Anderson, 1972). These figures, along with those of grasshopper damage for years when high populations were reached (Shotwell, 1941), suggested the economic importance of identifying those factors exerting the greatest effect on population fluctuation. This section considers several environmental conditions that function as limiting factors for grasshopper populations and compares simulation results where these factors are relaxed.

Criddle (1917) was one of the first to report the coincidence of grasshopper outbreaks and periods of dry weather. Since then, it has been commonly held that grasshopper outbreaks result from periods of drought. This viewpoint was valid to the extent that excessive moisture was detrimental to the population. However, water may also be limiting, both directly during embryonic development and oviposition and indirectly via forage.

The most important characteristics of the life cycle of temperate region grasshoppers is that they generally have a single generation per year, generations are discrete, and life-cycle stages are synchronous. In other words, there are periods when the entire population consists of eggs (e.g., species that overwinter in the egg stage); and the adult and nymphal stages occur at certain times of the year, but are otherwise absent. Thus unusually favorable or unfavorable conditions can markedly affect population size. This characteristic of a single, annual, synchronous generation is unlike that of many locust species in warm desert climates.

Table 5.3 presents a number of adverse conditions that were exercised in the model. Weather, as demonstrated by several investigators (Parker, 1930; McCarthy, 1956; Edwards, 1960), was considered to be the primary factor affecting population numbers. Cole (1954) analyzed the response of r (intrinsic rate of increase) to various life-history features. In general, he argued that reduction in the age of first reproduction, an increase in fecundity, and an increase in longevity all increase r. For the model exercises in Table 5.3, condition IV affects age of first reproduction, condition V affects fecundity, and conditions I, II, and III affect longevity.

The conditions imposed in the model, simulation results, and a comparison of these results with a 2-yr control run are presented in Table 5.4. For these simulations, no conditions were changed, except for that in Table 5.4. The

Table 5.3.   Conditions affecting grasshopper life-cycle processes

Condition number	Adverse condition	Life stage involved	Life-cycle process involved	Reference
I	Inadequate soil water	Egg	Embryonic development	Salt (1949, 1952)
II	Cold soil temperatures	Egg	Egg survival	Pickford (1972)
III	Damp canopy and soil	Nymph	Nymph survival	Pickford (1966a)
IV	Cool air temperatures	Nymph; sexually immature adult	Development; sexual maturation	Parker (1930) Bhatnagar and Pfadt (1973)
V	Cool air temperatures	Adult	Oviposition	Parker (1930)

simulations ran for 2 yr and all factors except the temperature effect on oviposition were cumulative.

The most striking result was the increase (716%) in grasshopper biomass when there was no cold induced egg mortality. On the other hand, when temperatures do cause egg mortality approximately 25% of the time (this condition was imposed stochastically via Monte Carlo methods), the greatest reduction in grasshopper biomass is observed. These results indicate great sensitivity to variations in egg mortality. The role of egg mortality in determining grasshopper

Table 5.4.   Simulation results of grasshopper peak weight density from altering various weather and soil conditions

	Condition	Maximum grasshopper biomass, 1971 (g C/m²)	Deviation from control (%)[a]
I	Inadequate soil water		
	25 percent of the time[b]	0.0033	−26.7
II	Cold soil temperature		
	Never	0.0367	+716.0
	25 percent of the time[b]	0.0008	−82.2
III	Precipitation		
	50 percent of the time[b]	0.0014	−68.9
	25 percent of the time[b]	0.0027	−40.0
	Never	0.0057	+26.7
IV	Air temperature		
	$T_m - 5.0$[c]	0.0029	−35.6
	$T_m - 3.0$	0.0036	−20.0
	$T_m + 3.0$	0.0063	+40.1
	$T_m + 5.0$	0.0099	+120.1
V	Air temperature (oviposition)		
	25 percent of the time[b]	0.0041	−8.9
	Maximum rate	0.0054	+20.0
	Standard simulation	0.0045	

[a]   Compared with 1971 peak biomass.
[b]   Stochastically imposed via Monte Carlo methods.
[c] T_m is the daily maximum air temperature (°C).

population size is of biological interest. Some studies suggested that variation in egg survival did play a major role in determining population numbers (Pickford, 1972). However, investigation of factors influencing egg survival in the field has been a neglected aspect of grasshopper life-history studies. Of the factors built into this model, there was a greater need for information on the mechanisms of egg dynamics than any other part of the life cycle.

The second most sensitive process tested was that of nymphal development and sexual maturation (condition IV, Table 5.4). That a decrease in the age of first reproduction causes an increase in r was consistent with the findings of Cole (1954).

Extremely wet seasons (condition III, Table 5.4) reduced populations in the second year by 40.0% and 68.9%. Pickford's (1966a) field observations indicated that grasshopper populations were highly sensitive to properly timed rainfall. This result implied that simulated mortality due to rainfall was not as great as in nature.

Finally, the model was relatively insensitive to the influence of soil water on embryonic development and air temperature on oviposition. The conditions imposed in the model pushed these processes to reasonable extremes; for example, inadequate soil water arrested embryonic development 25% of the time, and oviposition was arrested 25% of the time in one case.

5.5.4 Grasshopper Predictability

One evaluation criterion for simulation models is their capacity for prediction, that is, the ability of the model to mimic the dynamics of a field population. Prediction capability is often measured by a validation exercise where model results are compared with data. This kind of exercise was performed on the grasshopper model.

Field Data. The Pawnee Site data from 1970 (Van Horn, 1972) were used to tune the grasshopper model. Mean sample values for these collections are illustrated in Figure 5.6. The data against which the model was tested were collected by Pfadt in 1972 (Bhatnagar and Pfadt, 1973). There were some differences between these two data sets, as they were collected by different investigators using different sampling techniques. Van Horn employed a light-weight drop cage of 2-m² area for each sample. On each collection date 30 samples were taken. Material was sorted by species, sex, and age and dry weight estimates were made from the field-collected specimens. Pfadt estimated population densities by walking a transect and counting and sexing instars and adults of each species within a 1-ft² area. At least 100 counts were taken for each sampling date. Conversion to dry weight was done with oven-dried specimens from the same area collected with a sweep net.

The collection locations for the two data sets were not identical but in close proximity to each other, and the plant species composition was essentially the same. Only data involving grasshopper species overwintering in the egg stage

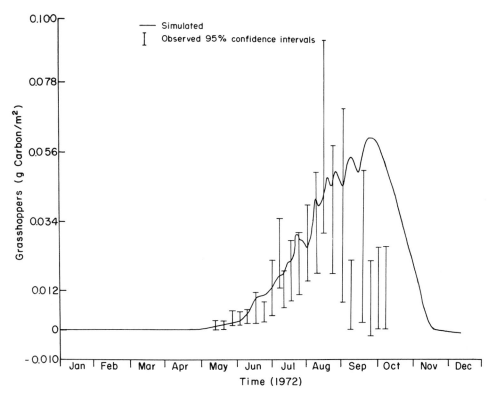

Fig. 5.8. Test run simulation of grasshopper biomass dynamics compared to field observations for 1972

were used. This data selection was necessary to be consistent with the assumptions in the model.

Test Run. The initial model values for the 1972 test run were final values obtained from a 2-yr simulation for 1970 and 1971. Thus the tuning data and the test data were separated by a year for which no data were available.

The results of the test simulation are shown in Figure 5.8, along with the 95% confidence intervals on Pfadt's data. Of the 22 sample dates, the simulated population dynamics fall within the confidence intervals 16 times or 73% of the time. Until September 24 the simulated dynamics fall outside the confidence intervals only once. The subsequent diversion was not surprising in that simulated values at time t are dependent on those at time $t - 1$. Likewise, field data collected at time t are dependent on the population at time $t - 1$. Thus if the simulated value at time t is not in the field-data confidence interval, the probability is lowered that the computed value will be in the confidence interval at time $t + 1$.

It appeared that the simulated population did not die as rapidly in the fall as did the field population, suggesting the model implementation of adult mortality to be inadequate. The model digression from field data could result from one or more of the five mortality mechanisms in the model or some other factor(s) not considered, such as migration. This test simulation illustrates the advantage to having the modeler also serve as the experimenter. From this dual role the modeler can obtain a qualitative understanding of the population. Such an ideal creates other problems and was not realized in this effort.

Field data indicated the grasshopper population at the Pawnee Site to be five times larger in 1972 than in 1970 (Figs. 5.6 and 5.8). The model successfully predicted this increase. Thus although the predicted time of peak biomass was different from the observed time, the model did predict year-to-year changes in population size.

5.5.5 Population Modeling

Populations are complicated entities. They are so complicated, in fact, that treatment of populations in even more complicated structures like ecosystems is difficult. Populations influence and are influenced by other populations and other trophic levels of the ecosystem. The modeling of populations, as part of an effort to understand their behavior, has been attempted at several levels of complexity. It has been said that an understanding of the principles governing populations and ecosystems is achieved only through the development of simple and general mathematical models (Anderson, 1970; Maynard Smith, 1974). Earlier attempts at generalizations regarding factors affecting population regulation have been controversial and largely unsuccessful (Nicholson, 1958; Thompson, 1956). Additional knowledge of populations and the components of their biotic and abiotic environment are needed before general principles can be developed. The construction of mechanistic simulation models provides an opportunity for such investigations.

Acknowledgments. I want to express my gratitude to Dr. G. M. Van Dyne and the Natural Resource Ecology Laboratory at Colorado State University whose interest and support made this modeling effort possible.

My special gratitude is due to Dr. G. S. Innis for introducing me to simulation modeling and providing several helpful suggestions throughout the course of this exercise. Invaluable stimulation and discussions were provided by my colleagues Drs. J. C. Anway, G. C. Cole, H. W. Hunt, W. J. Parton, R. H. Sauer, and R. G. Woodmansee.

Personal help and information have been provided by Mr. J. D. Gustafson, Dr. J. E. Ellis, and Dr. R. Andrews of the Natural Resource Ecology Laboratory, Colorado State University.

This chapter reports on work supported in part by National Science Foundation Grants GB-31862X, GB-31862X2, GB-41233X, and BMS73-02027 AO2 to the Grassland Biome, U.S. International Biological Program, for "Analysis of Structure, Function, and Utilization of Grassland Ecosystems."

References

Anderson, F. S.: Simple elementary models in population dynamics, In: Proceedings for Advanced Study, Institute of Dynamics Numbers Populations. Oosterbeek, The Netherlands, den Boer, P. J., Gradwell, G. R. (eds.), 1970, pp. 358–365

Anderson, N. L.: The assessment of range losses caused by grasshoppers, p. 173–179. In: Proceedings of the International Study, Conference of Current and Future Problems of Acridology. Hemming, C. F., Taylor, T. H. C. (eds.), London, England, 1972, pp. 173–179

Andrewartha, H. G., Birch, L. C.: The Distribution and Abundance of Animals. Chicago: Univ. Chicago Press, 1954

Andrzejewska, L., Breymeyer, A., Kajak, A., Wójcik, Z.: Experimental studies on trophic relationships of terrestrial invertebrates, p. 477–495. In: Secondary Productivity of Terrestrial Ecosystems (Principles and Methods), Vol. 2. Petrusewicz, K. (ed.). Warsaw, Poland: Panstwowe Wydawnictwo Naukowe, 1967, pp. 477–495

Andrzejewska, L., Wójcik, Z.: The influence of Acridoidea on the primary production of a meadow (field experiment). Ekologia Polska 18, 89–109 (1970)

Bhatnagar, K. N., Pfadt, R. E.: Growth, density, and biomass of grasshoppers in the shortgrass and mixed-grass associations. US/IBP Grassland Biome Tech. Rep. No. 225. Fort Collins: Colorado State Univ., 1973, 120 pp.

Birch, L. C.: The role of weather in determining the distribution and abundance of animals. Cold Spring Harpor Symp. Quant. Biol. 22, 203–218 (1957)

Brett, C. H.: Interrelated effect of food, temperature, and humidity on the development of the lesser migratory grasshopper: *Melanoplus mexicanus mexicanus* (Sauss.) (Orthoptera). Stillwater: Oklahoma Agr. Exp. Sta. Tech. Bull. No. T-26, 1947, 50 pp.

Buckell, E. R.: The grasshopper outbreak of 1944 in British Columbia. Can. Entomol. 77, 115–116 (1945)

Church, N. S., Salt, R. W.: Some effects of temperature on development and diapause in eggs of *Melanoplus bivittatus* (Say.) (Orthoptera:Acrididae). Can. J. Zool. 30, 173–184 (1952)

Clark, L. R.: An ecological study of the Australian plague locust (*Chortoicetes terminifera* Walk.) in the Bogan–Macquarre outbreak area, N. S. W. Council Sci. Ind. Res. Aust. Bull. No. 226, 1947, 71 pp.

Clark, L. R., Geier, P. W., Hughes, R. D., Morris, R. F.: The Ecology of Insect Populations in Theory and Practice. London: Metheun, 1967

Cole, L. C. The population consequences of life history phenomena. Quart. Rev. Biol. 29, 103–137 (1954)

Coulman, G. A., Reice, S. R., Tummala, R. L.: Population modeling: A systems approach. Science 175, 518–521 (1972)

Cowan, F. T.: Trends in grasshopper in the United States. 10th Internat. Congr. Entomol. 3, 55–58 (1958)

Criddle, N: Precipitation in relation to insect prevalence and distribution. Can. Entomol. 49, 77–80 (1917)

Davey, P. M.: Quantities of food eaten by the desert locust, *Schistocerca gregaria* (Forsk.), in relation to growth. Bull. Entomol. Res. 45, 539–551 (1954)

Demptster, J. P.: The population dynamics of grasshoppers and locusts. Biol. Rev. 38, 490–529 (1963)

Edwards, R. L.: Relationship between grasshopper abundance and weather conditions in Saskatchewan, 1930–1958. Can. Entomol. 92, 619–623 (1960)

Gilbert, B. J.: Flow of forage to herbivores. M.S. thesis, Colorado State Univ., Fort Collins, 1973

Gyllenberg, G.: The energy flow through a *Chorthippus parallelus* (Zett.) (Orthoptera) population on a meadow in Tvärminne, Finland. Acta Zool. Fennica 123, 4–74 (1969)

Gyllenberg, G.: A simulation model for testing the dynamics of a grasshopper population. Ecology 55, 645–650 (1974)

Harries, F. H.: Some temperature coefficients for insect oviposition. Ann. Entomol. Soc. Am. **32,** 758–776 (1939)

Holling, C. S.: The components of predation as revealed by a study of small mammal predation of the European pine sawfly. Can. Entomol. **91,** 293–320 (1959)

Husain, M. A., Mathur, C. B., Roonwal, M. L.: Studies on *Schistocerca gregaria* (Forskal). XII. Food and feeding habits of the desert locust. Ind. J. Entomol. **8,** 141–163 (1946)

Hussain, N.: Consumption and utilization of crested wheatgrass and western wheatgrass by the big-head grasshopper, *Aulocara elliotti* (Thomas). Ph.D. thesis, Univ. Wyoming, Laramie, 1972, 142 pp.

Lack, D.: The Natural Regulation of Animal Numbers. London: Oxford Univ. Press, London, 1954.

Leberre, J. R.: Contribution a l'étude biologique du criquet migratoria des landes (*Locusta migratoria gallica* (Remaudiere)). Bull. Biol. France Belg. **87,** 227–273 (1953)

MacArthur, R. H., Levins, R.: Competition, habitat selection, and character displacement in a patchy enviroment. Proc. Nat. Acad. Sci. **51,** 1207–1210 (1964)

Maynard Smith, J: Models in Ecology. London: Cambridge Univ. Press, 1974, 146 pp.

McCarthy, H. R.: A ten-year study of the climatology of *Melanoplus mexicanus mexicanus* (Sauss.) (Orthoptera:Acrididae) in Saskatchewan. Can. J. Agr. Sci. **36,** 445–462 (1956)

Mitchell, J.: A model of food consumption by three grasshopper species as determined by differential feeding trials. Ph.D. thesis, Colorado State Univ., Fort Collins, 1973, 165 pp.

Mitchell, J.: Variation in food preferences of three grasshopper species (Acrididae:Orthoptera) as a function of food availability. Am. Midlands Naturalist **94,** 267–283 (1975)

Nicholson, A. J.: Dynamics of insect populations. Annu. Rev. Entomol. **3,** 107–136 (1958)

Odum, E. P., Connell, C. E., Davenport, L. R.: Population energy flow and three primary consumer components of old-field ecosystems. Ecology **43,** 88–96 (1962)

Parker, J. R.: Some effects of temperature and moisture upon *Melanoplus mexicanus mexicanus* Saussure and *Camnula pellucida* Scudder (Orthoptera). Montana Agr. Exp. Sta. Bull. 223, 1930, 132 pp.

Parker, J. R., Connin, R. V.: Grasshoppers: Their habits and damage. Agr. Information Bull. No. 287, 1964, 28 pp.

Petrusewicz, K., Macfadyen, A.: Productivity of terrestrial animals: Principles and methods. IBP Handbook No. 13. Philadelphia: F. A. Davis, 1970, 190 pp.

Pianka, E. R.: On *r*- and *K*-selection. Am. Naturalist **104,** 592–597 (1970)

Pickford, R.: Development, survival, and reproduction of *Camnula pellucida* (Scudder) (Orthoptera:Acrididae). Can. Entomol. **101,** 894–896 (1966a)

Pickford, R.: The influence of the date of oviposition and climatic conditions on hatching of *Camnula pellucida* (Scudder) (Orthoptera:Acrididae). Can. Entomol. **98,** 1145–1159 (1966b)

Pickford, R.: The effects of climatic factors on egg survival and fecundity in grasshoppers, In: Proc. Internat. Study Conf Current and Future Problems of Acridology, London. Hemming, C. F., Taylor, T. H. C. (eds.), 1972, pp. 257–260

Pickford, R., Randell, R. L.: A non-diapause strain of the migratory grasshopper, *Melanoplus sanguinipes* (Orthoptera:Acrididae). Can. Entomol. **101,** 894–896 (1969)

Putnam, L. G.: The progress of nymphal development in pest grasshoppers (Acrididae) of western Canada. Can. Entomol. **95,** 1210–1216 (1963)

Richards, O. W., Waloff, N.: Studies of the biology and population dynamics of British grasshoppers. Anti-locust Bull. **17,** 1–182 (1954)

Riegert, P. W.: Some observations on the biology and behavior of *Camnula pellucida* (Orthoptera:Acrididae). Can. Entomol. **99,** 952–971 (1967)

Salt, R. W.: Water uptake in eggs of *Melanoplus bivattatus* (Say.). Can. J. Res. **27(D):** 236–242 (1949)

Salt, R. W.: Some aspects of moisture absorption and loss in eggs of *Melanoplus bivittatus* (Say.). Can. J. Zool. **30,** 55–82 (1952)

Shotwell, R. L.: Life histories and habits of some grasshoppers of economic importance on the Great Plains. USDA Tech. Bull. No. 774, 1941, 47 pp.

Slifer, E. H., King, R. L.: The inheritance of diapause in grasshopper eggs. J. Hered. **52,** 39–44 (1961)

Smalley, A. E.: Energy flow of a salt marsh grasshopper population. Ecology **41,** 672–677 (1960)

Thompson, W. R.: The fundamental theory of natural and biological controls. Annu. Rev. Entomol. **1,** 379–402 (1956)

Uvarov, B.: Grasshoppers and locusts. London: Cambridge Univ. Press, 1966, 481 pp.

Van Hook, Jr., R. L.: Energy and nutrient dynamics of spider and Orthopteran populations in a grassland ecosystem. Ecol. Monogr. **41,** 1–26 (1971)

Van Horn, D. H.: Grasshopper population numbers and biomass dynamics on the Pawnee Site from fall 1968 through 1970. US/IBP Grassland Biome Tech. Rep. No. 148. Fort Collins: Colorado State Univ., 1972, 70 pp.

Wakeland, C.: The replacement of one grasshopper species by another. USDA Products Res. Rep. No. 42, 1961, 9 pp.

Watt, K. E. F.: The use of mathematics and computers to determine optimal strategy and tactics for a given insect pest control problem. Can. Entomol. **96,** 202–220 (1964)

Watt, K. E. F.: The nature of systems analysis, In: Systems Analysis in Ecology. Watt, K. E. F. (ed.). New York: Academic Press, 1966, pp. 1–14

White, M. J. D.: Animal cytology and evolution. London: Cambridge Univ. Press, 1973

Wiegert, R. G.: Energy dynamics of the grasshopper populations in old field and alfalfa field ecosystems. Oikos **16,** 161–176 (1965)

Wigglesworth, V. B.: The principles of insect physiology. London: Methuen, 1965

6.   A Simulation Model for Decomposition in Grasslands

H. WILLIAM HUNT

Abstract

A model has been developed to simulate the dynamics of decomposers and substrates in grasslands. Substrates represented are humic material, feces, and dead plant and animal remains. Except for humic material, substrates are further divided into a rapidly and a slowly decomposing fraction. The proportion of rapidly decomposing material in a substrate is predicted from its initial nitrogen content. The belowground portion of the system is divided into layers because temperature and soil water, the most important driving variables for the model, vary with depth. Decomposition rates are predicted from temperature, water tension, and inorganic nitrogen concentration.

Taxonomic groups of decomposers are not distinguished, but a distinction is made between active and inactive states, which differ in both respiration and death rates and in that only active decomposers assimilate substrate.

The model's predictions compare favorably to data on carbon-dioxide evolution and to litter-bag experiments, but not to ATP estimates of active microbial bioass. The model indicates a profound influence of soil depth on decomposition rates and on decomposer biomass dynamics, growth yield, and secondary productivity.

6.1 Introduction

The objectives of the ELM decomposition submodel are related to those of the whole model, namely, to simulate the biomass dynamics of grassland ecosystems and the responses of grasslands to grazing, irrigation, and fertilization and to provide a tool for investigating management problems.

Table 6.1 lists some simulation models for decomposition in grassland and tundra. The number of state variables for substrates and for decomposers provide an indication of the objectives. Patten's (1972) model includes the greatest variety of microbial and invertebrate decomposers. His objectives, in part, were "to underscore the ecological significance of the detritus subsystem and draw attention to the paucity of data and gaps in knowledge concerning decomposition processes." In contrast, the model of Bunnell and Dowding (1973) has no state variables for decomposers, since its objectives are to facilitate an intersite comparison among decomposition rates and factors affecting the rates, and not to examine differences in the decomposers themselves. The objectives of the models of Bledsoe et al. (1971), Timin et al. (1973), and Bunnell

Table 6.1.   Decomposition submodels of some published ecosystem simulation models

Citation	Biome	Number of substrate state variables	Number of decomposer state variables	Decomposition output presented?
Bunnell (1973)	Tundra	6	3	No
Bunnell and Dowding (1973)	Tundra	7	0	No
Timin et al. (1973)	Tundra	5	0	Yes
Randell et al. (1972)	Grassland	4	5	Yes
Randell et al. (1972)	Grassland	5	4	Yes
Bledsoe et al. (1971)	Grassland	9	6	No, described as invariable
Patten (1972)	Grassland	5	23	Yes
Anway et al. (1972)	Grassland	10	4	Yes
This chapter	Grassland	13	8	Yes

(1973) are to test hypotheses, to stimulate more precise, critical statements of hypotheses, and to provide a framework for organizing and analyzing data. The ELM model differs from other published ecosystem simulation models in its treatment of substrate heterogeneity and the activity of decomposers. Complex substrates are divided into either rapidly or slowly decomposing components, and decomposers are considered to be either active or inactive, a distinction more important for our objectives than their taxonomic identity.

Figure 6.1 is the compartment diagram for the decomposition model. Substrates and decomposers are divided into four groups according to their location, because temperature and water tension vary significantly with depth. Driving variables for each level are provided by the heat flow, water flow, and nitrogen submodels of ELM. Material inputs to the substrate compartments are from the producer and consumer submodels. Decomposition is represented by flows from the substrate compartments to active decomposers, and respiration by flows from both active and inactive decomposers to the CO_2 sink. Death is represented by flows from decomposers to the substrate compartments. Mechanical processes move substrate from the surface to the upper belowground compartments, and leaching moves the labile component from each layer to the layer below.

6.2 The Problem of Heterogeneous Substrates

A pure substance decomposing at constant temperature and water tension might lose a constant proportion of the amount present per unit of time. That is,

$$\frac{dX}{dt} = -kX,$$
$$X_t = X_0 e^{-k(t)},$$

where X_t is the amount of material remaining at time t; X_0 is the amount present initially; and k is a rate constant. Decay of a pure substance would depart from the exponential model if the rate of decomposition changes as decomposition proceeds, as might occur if the size or composition of the microbial population or the size of the substrate particles change with time. The rate of decomposition increases with time in Clark's (1970) data on the decomposition of cellulose filter paper buried in soil at constant temperature and moisture. Presumably, it takes time for the buildup of a microbial population able to use a pure cellulose substrate. Clark observed that when nitrogen was added to the filter paper the decomposition rate did not change dramatically with time.

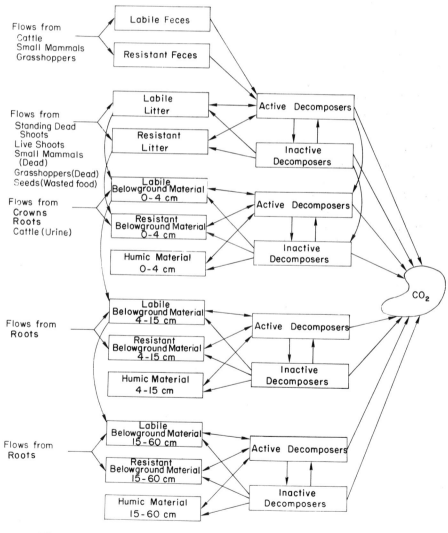

Fig. 6.1.   Compartment diagram for decomposition submodel of ELM

Heterogeneous substrates decomposing under constant conditions do not, in general, conform to the exponential decay model. Observed rates are greater than those predicted by the exponential in early stages of decomposition when the more easily decomposed components are present and less than those predicted in later stages when more resistant components predominate (Pinck et al., 1950; Floate, 1970; Martel and Paul, 1970).

Minderman (1968) suggested that the individual components of a heterogeneous substrate might show exponential decay. This assumption was used by Grill and Richards (1964) to explain losses of nitrogen and phosphorus from decomposing diatoms and by Otsuki and Hanya (1972) for carbon loss from decomposing green algae. Modeling each chemical component in a complex substrate by an exponential decay curve is unsatisfactory because: (a) pure substances might not conform to exponential decay, (b) some chemical components are formed from others during the decomposition process, so that the generation of substances must be considered, and (c) the presence of some substances can inhibit the decomposition of others (Gaudy et al., 1963; Lewis and Starkey, 1968). Therefore, each substance can hardly be treated separately. Nevertheless, some plant materials decompose faster than others and the rate of decomposition decreases as decomposition proceeds.

A useful compromise between treating a substrate as homogeneous and treating each of its components separately is to consider heterogenous materials to consist of two fractions. This approach has been employed to describe changes in the biological oxygen demand during sewage treatment (Eckenfelder, 1970). The labile or rapidly decomposing fraction (e.g., sugars, starches, proteins) and the resistant or slowly decomposing fraction (e.g., cellulose, lignins, fats, tannins, waxes) are assumed to decay exponentially under constant conditions. Thus A_t, the proportion of material remaining on day t, is given by

$$A_t = Se^{-k(t)} + (1 - S)e^{-h(t)} \qquad (6.1)$$

where S is the initial proportion of labile material; $(1 - S)$ is the initial proportion resistant; k is the rate constant for the labile component; and h is the rate constant for the resistant component.

The utility of Eq. (6.1) was tested with data (Pinck et al., 1950) on the decomposition of various plant materials in soil at constant temperature and moisture. The carbon/nitrogen ratio (C/N) and lignin content of the materials were given along with the amount present after 1.5, 4, and 12 months. Table 6.2 gives least-squares estimates of the parameters S, k, and h. Regression analysis was used to establish the relationship between the model's parameters and the nitrogen and lignin content. The estimates of S, but not of k or h, were significantly related to C/N and to lignin content. Inclusion of C/N and lignin content simultaneously in the regression did not explain significantly more variability than either variable alone. The C/N ratio was used for further development of the model, because the nitrogen content of materials is more commonly and reliably estimated than lignin.

Table 6.2. Estimates of parameters of decomposition model [Eq. (6.1)] fitted to data[a] of Pinck et al. (1950)

Material added	Simultaneous estimates			
	S	k	$h \cdot 10^4$	S_0[b]
Casein	0.816	0.0545	5.24	0.829
Bluegrass	0.580	0.0463	5.65	0.573
Young millet	0.559	0.0494	8.03	0.573
Oats	0.602	0.0437	5.21	0.589
Soybeans	0.597	0.0482	1.93	0.566
Oak leaves	0.223	0.0409	9.71	0.254
Corn stover	0.270	0.0449	11.07	0.317
Wheat straw	0.324	0.0236	7.35	0.293
Cellulose	0.135[c]	[c]	3.87[c]	0.098
Lespedeza	0.454[c]	[c]	8.87[c]	0.498
Intermediate millet	0.478	0.0521	7.62	0.492

[a] "Organic carbon retention in unlimed Cecil clay loam following addition of 1 per cent carbon . . . without supplemental nitrogen."
[b] Estimated, assuming $h = 0.00071$ and $k = 0.044$ (see text).
[c] The iterative procedure used did not converge, so the least-squares estimates of S and h were found by assuming a value of 0.048 for k.

Table 6.2 shows a similarity of k (and h) values for various materials. If the k (and h) values are identical for all materials, the labile (or resistant) components from all sources will have the same decomposition dynamics, and only the total labile and total resistant material need be simulated without regard to origin. This simplification was tested by using the average values of k and h in Table 6.2 and estimating only S in Eq. (6.11). The resulting estimates, S_0 (Table 6.2), agree well with the values obtained by estimating all three parameters simultaneously. The relationship between S_0 and C/N is nonlinear, but transformation of C/N to $\sqrt[3]{N/C}$ produces an approximately linear relation (Fig. 6.2). The equation for predicting S_0 from N/C is

$$S_0 = 0.070 + 1.11 \sqrt[3]{N/C},$$
$$R^2 = 0.98, P < 0.001. \tag{6.2}$$

Using values for S estimated from nitrogen content [Eq. (6.2)] and the average values of h and k gives a good fit to the data ($R^2 = 0.90$, $P < 0.001$). Thus differences among materials can be accounted for by the proportion labile, and the decay rates h and k may be assumed constant among fresh substrates.

Feces, a material having already undergone some decomposition, might not be expected to conform to Eq. (6.2). Floate (1970) measured the rates of decomposition of four plant materials and of the feces of sheep fed the materials. Equation (6.1) was applied to his data to obtain estimates of S, h, and k. For the

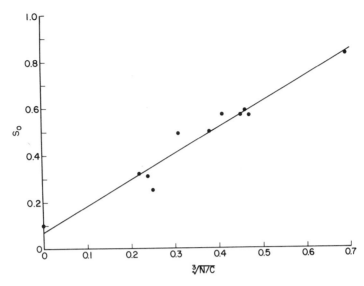

Fig. 6.2.   Relationship between the proportion of labile constituents in a substrate and its nitrogen content

plant materials, the relationships among the three parameters and nitrogen content were the same as for the data of Pinck et al. (1950); that is, only S was significantly related to $\sqrt[3]{N/C}$. For feces, however, none of the three parameters were significantly related to $\sqrt[3]{N/C}$ or to each other. The estimates of S were less variable for feces (0.06–0.12) than for the plant materials (0.13–0.46), so a constant value of 0.08 was assumed for ruminant feces. The decay rates h and k were half as great for feces as for plant materials, so feces are treated separately.

Humic material is included in the model because it is such a large compartment. However, the turnover time for humus is great, as much as 1000 yr (Campbell et al., 1967). In the model the decomposition rate of humic material is set to 0.0014 that of the resistant component of plant material, a rate allowing decomposition of about 0.001 of the humic material per year.

Theories for the formation of humic material are discussed by Felbeck (1971). The theory adopted here is that humic materials are produced intracellularly by microbes and released on their death. Jensen (1932) found that microbial tissue decomposes more rapidly than mature plant material but leaves a residue of about 0.3% humus-like material very resistant to decomposition. Therefore, in the model, microbial material on death is split between humic material (0.003), resistant substrate (0.5), and labile substrate (0.497).

Although h and k are relatively constant for a given set of conditions, their values depend on temperature, soil water, soil aeration, and perhaps soil fertility and pH. Table 6.3 gives mean values of h and k for laboratory experiments conducted under different conditions. Predictably, there is a strong correlation in the degree to which conditions favor the decomposition of the labile and resistant

Table 6.3.   Average values of h and k in data on decomposition of herbage

Source of data	Number of pairs of estimates	k	$h \cdot 10^4$
Millar et al. (1936)	12	0.198	67.2
Floate (1970)	3	0.0939	31.6
Broadfoot and Pierre (1939)	19	0.0285	11.6
Tenney and Waksman (1929)	6	0.0381	11.2
Pinck et al. (1950), Allison et al. (1949)	35	0.0475	6.4

components. The intercept of the regression line between average values of h and k does not differ significantly from zero, and it seems reasonable that conditions completely halting decomposition of one component would also halt decomposition of the other; hence k is assumed proportional to h:

$$k = 29.81\ h$$
$$R^2 = 0.98,\ P < 0.001. \tag{6.3}$$

The right-hand side of Eq. (6.3) is substituted for k in Eq. (6.1), and one parameter is eliminated from the model.

The data of Pendleton (1972) on the decomposition of cattle manure and several grasses in soil at constant temperature and moisture was used to test the model. Values of S for the grasses were predicted from their nitrogen content according to Eq. (6.2). The values of h and k that satisfy Eq. (6.3) and give the best least-squares fit to the data for the grasses are 0.00722 and 0.219 day^{-1}, respectively. Values for feces are assumed to be half as great (i.e., 0.00361 and 0.109 day^{-1}). Figure 6.3 gives predictions of the model along with the observed values. One parameter was estimated from the data because we could not predict the effect of soil-specific factors.

To test the applicability of simple exponential decay to Pendleton's data, an exponential curve was fitted to each of the five materials. Even though five parameters were estimated, the overall fit was poorer than in Figure 6.3, for which only one parameter was estimated from the data.

6.3 Decomposition under Variable Conditions

In field experiments the decomposition rate is greatly affected by water and temperature (Jenny et al., 1949). Changing environmental conditions should cause departures from exponential decay. Indeed, the data of Jenny et al. (1949), Koelling and Kucera (1965), Minderman (1968), and Clark (1970) show departures from the exponential. In contrast, the data of Witkamp (1966) and part of

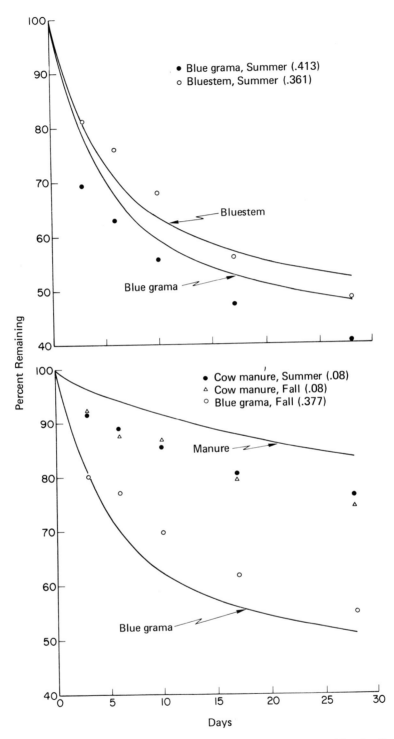

Fig. 6.3.   Predicted (line) and observed (points) course of decomposition for feces and hay [data from Pendleton (1972); *S* values given in parentheses]

the data of Witkamp and Olson (1963) conform to an exponential model. In both latter cases the experiments were initiated in the winter, as appropriate for a deciduous forest. According to Witkamp (1966), "The effect of stage of decay is less significarnt possibly because it is composed of two counteracting influences, viz. chemical impoverishment and improving physical conditions." That is, heterogeneity of the substrate would have led to decreasing relative losses if physical conditions had not continued to improve with decomposition.

6.3.1 The Effect of Water

The effect of water tension on the decomposition rate is taken from the data of Bhaumik and Clark (1947) on peak rates of CO_2 evolution from five soils amended with corn stover (Fig. 6.4). Except for the most sandy soil the curves

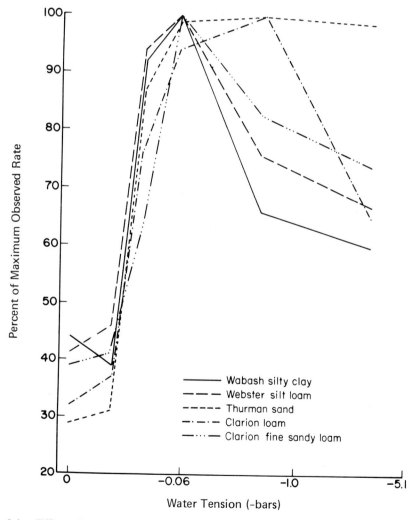

Fig. 6.4.   Effect of water tension (log scale) on rate of decomposition [data of Bhaumik and Clark (1947)]

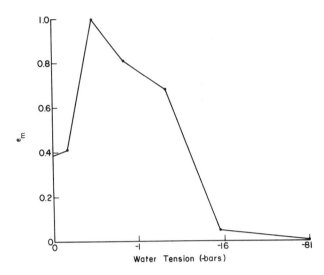

Fig. 6.5.   Relationship between water tension (log scale) and its effect (e_m) on decomposition rate

are quite similar. By averaging the values of Figure 6.4 for the four less sandy soils and adjusting to give a peak value of 1.0, a curve for e_m (Fig. 6.5), the effect of water, is obtained that should hold for many soils. To formulate Figure 6.5 it was assumed that decomposition is nearly 0 at -15 bars. The extension of the curve to zero at -81 bars allows for slight activity in dry soil. Water tension in each soil layer is supplied by the water-flow submodel (Chapter 2).

The abiotic submodel does not compute water tension in the litter, but predicts dew formation and the interception of rainfall. The decomposition rate of litter is assumed to increase proportionately to litter water up to 0.67 g H_2O per g dry weight of litter. In addition, water in the top 1.5 cm of soil is assumed available to the litter, and when the effect of tension in this layer is greater than that determined by dew and rainfall, the higher rate is used.

6.3.2 The Effect of Temperature

The effect of temperature (e_t) on the rate of decomposition is given in the model by

$$e_t = \exp(-5.66 + 0.240\ T - 0.00239\ T^2) \tag{6.4}$$

where T is temperature in °C ($0 < T < 38$). This curve fits the data of Drobnik (1962) on initial rates of CO_2 evolution from unamended soil at constant water content and at temperatures of 8–38°C. The rate of CO_2 evolution is small near 0°C (Douglas and Tedrow, 1959; Van Cleve and Sprague, 1971) and is set to zero in the model. A linear decline to zero is assumed at 38–45°C. The heat flow submodel (Chapter 2) provides estimates of temperature at the surface and at 15-cm increments in the soil.

6.3.3 The Effect of Nitrogen

In model runs without fertilization, the concentration of nitrate and ammonium N in the soil varies from 0.02 to 0.43 g N/(m² · cm). Over this range the effect of inorganic nitrogen on the rate of decomposition (e_N) increases linearly from 0.7 to 1.2 (Fig. 6.6). The form of this function is arbitrary since experiments showing that fertilization increases the decomposition rate (Allison et al., 1949; Pinck et al., 1950; Clark, 1970) do not show how fertilization affects inorganic N concentration. Although the effect of inorganic nitrogen on the decomposition rate probably depends on the nitrogen content of the substrate, the classification of a substance as labile or resistant does not necessarily indicate its N content. Thus inorganic N is assumed to affect the rate of decomposition of labile and resistant fractions alike. The concentration of inorganic nitrogen in the soil is simulated in the nitrogen submodel (Chap. 7).

6.3.4 Calculating the Decomposition Rate

In the model, the rate of decomposition in soil is a function of the effect of moisture (e_m), the effect of temperature (e_t), the effect of inorganic nitrogen (e_N), and a rate constant, the value of which depends on the soil and the nature of the decomposing material, that is, whether it is a labile or resistant component of plant material or feces. Decomposition of litter and feces is assumed to be

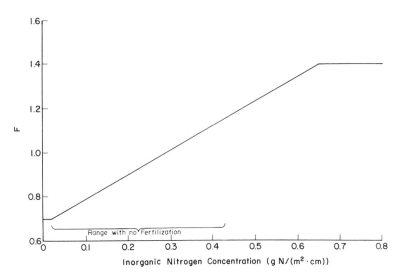

Fig. 6.6.   Relationship between concentration of inorganic nitrogen and its effect (e_N) on decomposition rate

independent of soil inorganic nitrogen, since they are in less intimate contact with soil than belowground material. Decomposition of humic material also is related only to temperature and moisture.

The model is operated as a difference equation model, and proportional losses per time step dt are calculated as

$$p_k = k' \cdot dt,$$
$$p_h = h' \cdot dt, \tag{6.5}$$

where

$$k' = r_k \cdot e_m \cdot e_t \cdot e_N,$$
$$h' = r_h \cdot e_m \cdot e_t \cdot e_N. \tag{6.6}$$

p_k and p_h are the proportional losses per time step from the labile and resistant components, respectively, and k' and h', the loss rates, are products of the effects of temperature, moisture, inorganic nitrogen, and the maximum decomposition rates r_h and r_k. Values for r_h and r_k were chosen by adjusting the model to fit data.

Calculating the joint effect of temperature, moisture, and inorganic nitrogen as the product of independently formulated functions is undoubtedly simplistic, but data are not available to describe interactions among the three factors. The possibility of substrate depletion—the exhaustion of readily available substrates under favorable conditions—is usually overlooked. Bartholomew and Norman (1946) measured CO_2 evolution from oat straw at several combinations of temperature and moisture. In their data the apparent interaction between effects of temperature and moisture changes between the first and third days of incubation, and this change seems in part attributable to substrate depletion. In field studies, substrate quality or quantity could confound the effects of temperature and moisture. Changes observed in the respiratory quotient of soil cores (Klein, 1977) establish the existence of seasonal changes in the quality of substrates.

In cases where flows occur only out of and not into a compartment, such as in a litter-bag experiment, the relationships between k' and h' of Eq. (6.5) and the instantaneous rates k and h of Eq. (6.1) are

$$h' = [1 - \exp(-h \cdot dt)]/dt,$$
$$k' = [1 - \exp(-k \cdot dt)]/dt$$

Therefore, the relationship between h' and k' corresponding to Eq. (6.3) is

$$k' = [1 - (1 - h' \cdot dt)^{29.81}]/dt \tag{6.7}$$

Equation (6.7) is used to relate k' and h' for litter-bag experiments; otherwise, Eq. (6.3) is used.

6.4 Physical Transfers

6.4.1 Leaching

Leaching may be an important mechanism for weight loss from decomposing materials. Broadfoot and Pierre (1939) give the water-soluble component of 31 samples of tree leaves and the weight of each sample after 2, 4, and 6 mo of decomposition. Equation (6.1) was fitted to their data to obtain estimates of the labile proportion (S). There is a significant correlation ($r = 0.60$, $P < 0.01$) between S and the proportion water-soluble. Brown and Frederick (1968) show directly that the water-soluble part of Sudan grass decomposes more rapidly than the water-insoluble part.

Leaching is represented in the model by the transfer of material between the labile components of substrates. Koelling and Kucera (1965), Brown and Frederick (1968), and Pendleton (1972) give the nitrogen content and water-soluble fraction of several grasses, from which it may be calculated that about 45–65% of the labile component may be leached in as short a time as 4 h. Nykvist (1959a,b, 1961a,b) shows the amount of material leached to increase with temperature. Rate of leaching (g · m^{-2} · day^{-1}) is calculated as the product of an effect of temperature, an effect of the rate of water movement and the amount of labile substance. The effect of temperature varies linearly from 0.3 to 1.0 as the temperature increases from 0° to 42°C, and the effect of water movement increases linearly from 0 to 0.5 as water movement rises from 0 to 1 cm/day. For example, at 25°C a maximum of 36% per day of the labile component can be transferred by leaching. Leaching of litter is driven by infiltration and that of belowground substrates, by percolation of water through the soil. The heat- and water-flow submodels (Chap. 2) provide the driving variables for leaching.

6.4.2 Mechanical Transfer

Apparently there is no information on the rate at which particulate material is incorporated into the soil by freezing–thawing, animal activity, or rainfall. In the model a constant proportion (0.00014) of litter and litter decomposers is transferred to the soil daily. Thus about 5% of the litter material is transferred yearly to the upper soil layer.

6.5 Decomposer Biomass

The model treats decomposers as if they consisted entirely of microbes. The justification is that decomosition in the shortgrass prairie is mediated largely by microbes; at least 95% of the energy flow through the detritus food chain is through microbes (Andrews et al., 1974). The possibility that the soil fauna has an effect out of proportion to its share of the energy flow is refuted by the finding that the exclusion of fauna from litter bags has little effect on the decomposition rate in grasslands, in contrast to forests (Curry, 1969).

The model does not divide microbes taxonomically because of the very uneven knowledge of the groups. Although generalizations can be made about the physiology and ecology of bacteria, fungi, and actinomycetes, distinguishing these groups in the model would create more problems than it would solve. For example, it is difficult to specify even the general effects of freezing and drying on the survival and activity of microbes in soil, let alone the difference between groups in these effects. Unfortunately, results from particular species often must be used for generalizations about all microbes.

Perhaps the most satisfying approach to the problem of modeling microbes would be to calculate the energy requirement of the population from its size and activity and predict the decomposition rate from that energy requirement. However, even the most promising methods for measuring the biomass of active microbes in the soil, such as ATP assay (Sparrow and Doxtader, 1973), are still in the developmental stage. Plate counts are not generally useful (Brock, 1971; Schmidt, 1973) because inactive stages may give rise to colonies, and active stages may fail to grow in a nonsoil environment (Alexander, 1971). It is even possible that direct microscopic counts, generally considered to overestimate the number of live microbes (Alexander, 1961), detect a small part of those revealed by electron microscopy (Webley and Jones, 1971). Ignorance of the biomass of active microbes in soil impairs attempts at modeling biomass directly.

The model proceeds from the assumption that the rate of decomposition in grassland ecosystems is independent of the size of microbial populations. This assumption depends on the observation (Clark and Paul, 1970; Babiuk and Paul, 1970; Gray and Williams, 1971) that the input of energy in grasslands is insufficient to maintain the activity of populations of the size observed. Microbes must be inactive much of the time. Presumably when substrate is added, a microbial population sufficient to sustain the maximum decomposition rate is already present. The population response, seen when a great amount of easily assimilable substrate is added to soil (Shields et al., 1973), may be unusual in the natural ecosystem. The problem becomes one of predicting the biomass of microbes, given the rate of disappearance of substrate.

6.5.1 Maintenance-energy Requirement

The model distinguishes metabolically active, but not necessarily growing, microbes from spores and other "resting stages" having reduced metabolism (Park, 1965; Gray and Williams, 1971). The maintenance-energy requirement (E) of active microbes is predicted from temperature according to the Arrhenius equation (Giese, 1968). *Escherichia coli* has a temperature characteristic of about 20 kcal $\cdot$ mol^{-1} (Marr et al., 1963). Therefore, E is related to absolute temperature T_a by

$$E = c \cdot \exp(-10,000/T_a),$$

where c is a constant. Schulze and Lipe (1964) give the E of *E. coli* at 30°C as 1.32 g glucose per g cell weight per day, from which c may be found: $2.82 \cdot 10^{14}$. Thus

$$E = 2.82 \cdot \exp(32.24 - 10,000/T_a). \qquad (6.8)$$

Equation (6.8) is used to predict the grams of substrate carbon per gram cellular carbon per day required to support active microbes. The maintenance-energy requirement E takes values of 0.13, 0.76, and 2.8 at 10°, 25°, and 37°C, respectively. It might be argued that *E. coli*, native to a warmer and wetter environment than the soil, would have a higher maintenance requirement than soil microbes, but similar values are reported for yeast (Wase and Hough, 1966) and two species of *Aerobacter* (Pirt, 1965).

In early versions of the model Eq. (6.8) was applied to the whole microbial biomass, but the supply of substrate was insufficient to maintain the observed populations of 50–100 g dry weight/m² (Reuss, 1971). This confirms the conclusion that soil microbes must usually be in an inactive state.

Dormant stages of microbes continue respiration at a reduced rate (Lamanna and Mallette, 1965). Ensign (1970) found that the rate of loss of cell weight from starving *Arthrobacter* dropped in 3 days to about 1.7% of the initial rate and remained at this low level for at least 31 days. Respiration of specialized resting structures, such as spores, is probably much slower, so the model sets the respiration of inactive microbes to 0.07% of the maintenance requirement of active microbes. This number was chosen in part to prevent net yearly changes in predicted microbial biomass.

6.5.2 Growth and Activity

When the growth rate of a microbial population is limited by its carbon source, the growth yield decreases at low growth rates (Herbert, 1958; Pirt, 1965; Postgate, 1973). The model sets microbial respiration (r_m) to the following function of the maintenance-energy requirement (E), the active biomass (b), the maximum growth yield (Y), and the amount of material decomposed (d):

$$r_m = E \cdot b + (1 - Y)(d - E \cdot b). \qquad (6.9)$$

When the growth rate is high ($d \gg E \cdot b$), a proportion (almost $1 - Y$) of d is respired and the rest goes for growth. The maximum growth yield Y is taken as 0.60 (Lamanna and Mallette, 1965; Payne, 1970). When growth is slow ($d \approx E \cdot b$), most of the assimilated substrate is respired, and the growth yield is low.

Starving microbes may lose weight through the utilization of stored reserves (Park, 1965; Boylen and Ensign, 1970), but this condition probably does not persist, since microbes may change from an active to an inactive state. Starving *Arthrobacter* cultures may lose about 10% of their weight on the first day and decreasing amounts on subsequent days (Ensign, 1970). In the model, $d < E \cdot b$

implies starvation. Since respiration of starving microbes does not decrease with time in the model, the maximum loss per day must be reduced from the observed 10% to 1%. Values much above 1% lead to great population declines when the soil dries slowly, because an active population will remain active in spite of losses to respiration. Therefore, if the weight loss ($Eb - d$) is no more than 1% per day of the active biomass, all microbes remain active but if greater, a proportion becomes inactive or dies. The proportion remaining active is chosen so that its maintenance requirement exceeds d by 1% of the biomass; that is, starving microbes lose no more than 1% of their weight per day.

Survival of starving bacteria depends on the original condition of the culture (Harrison, 1960; Ensign, 1970). Under favorable conditions laboratory cultures may show high survival for long periods, but I know of no data pertaining to the survival of starving active microbes in the soil. The model assumes that 85% of microbes in excess of those which can be supported by the available substrate die and 15% remain viable in an inactive state. Factors breaking dormancy in microbes include temperature fluctuations, drying, and wetting (Sussman, 1965), but the effects of a change from a dormant to an active state would be felt only under conditions favorable for growth. Therefore, transition from the inactive to the active state proceeds at a constant rate when $d > Eb$. Thus the transition depends on temperature, water, inorganic nitrogen concentration, the amount of substrate present, and the biomass of active microbes.

6.5.3 Effects of Freezing and Drying

Most studies of the effects of freezing and drying on soil pertain to nitrogen and carbon mineralization. The effects on microbial populations are difficult to assess in studies following population changes by plate counting, since the aggregation and fragmentation of propagules could be affected by these treatments. Nevertheless, freezing and drying probably kill many microbes, particularly the vegetative stages (Jager, 1967). In the model freezing kills 14% of active microbes, but once frozen no more die unless they are thawed and refrozen. Drying is assumed to kill 35% of active microbes per day as long as soil-water potential is less than -26 bars.

Sneath (1962) plated soil from the roots of old herbarium specimens and found that viable counts decreased approximately exponentially with time for the first 150 y, with a daily mortality rate of about 0.00027. The model sets the constant daily mortality rate of inactive microbes to half this value, 0.00013, since respiration also leads to a decrease in the inactive biomass.

6.6 Tuning the Model

Final values for some of the parameters in the decomposer submodel were chosen by trial and error in order to achieve a balance in the state variables over a 2-year (1970–1971) period. Parameters with the least support in theory and data were chosen for tuning. Data used in tuning were 1970–1971 litter-bag experiments and estimates of litter. Initial values for belowground substrates were obtained from total root data by assuming that 40% of the roots were dead, and those for decomposers were chosen to agree with direct count estimates of total microbial biomass (Doxtader, 1969).

The tuning process is difficult to document because: (a) it proceeded simultaneously with changes in the structure of the model, (b) the appropriate values for parameters depend on the performance of the other submodels of ELM, and (c) the simultaneous processes of model development and tuning have required at least 500 model runs.

6.7 Model Performance

None of the reports in Table 6.1 compare model output directly to data. This chapter presents two types of comparison. First, output is compared to data that were used in tuning. While it is important that the model can be tuned to fit a set of data, the resulting fit is not a test of the predictive power of the model, which is tested only by the second kind of comparison: that with data not used in model development.

6.7.1 Litter-bag Experiments

Equation (6.5) was tested with data on the decomposition of litter in nylon-mesh bags buried at 5 cm (Clark, 1970; F. E. Clark, personal communication) at the Pawnee Site. Figure 6.7 shows data for 1970–71, when the litter material used was bluestem hay with a N content of 0.75%. The initial proportion of labile constituents was 37 percent, according to Eq. (6.2). Equation (6.6) was used to calculate h', and k' was computed from Eq. (6.7). A value for r_h (0.04) was chosen to achieve a good fit to the data. Thus Figure 6.7 shows data used to "tune" the model, and Figure 6.8 compares the model's predictions to 1972 data (D.C. Coleman, personal communication) not used in tuning the model. The hay used in 1972 had a slightly smaller labile component (0.33) than that used in 1970–1971. The model predicts the amount of material remaining after most of the labile component has been lost better than it forecasts the timing of the early more rapid loss. Departures of the model from the data probably can be accounted for by the following:

1. The experiments were carried out at a point over a mile from the place where precipitation was recorded and important differences could occur.

2. Hay from the same collection was used in 1970 and 1971, but its resistance to decay may have changed during the year of storage.

3. The model was driven by average water tension at 0–4 cm, which might vary at times in its relationship to the tension in the litter bags (buried at ≈5 cm).

4. There is error in predicting the proportion of labile constituents from the N content.

A more stringent test of the model would be provided by experiments that follow decomposition for longer periods or employ a variety of different materials decomposing under a greater range of environmental conditions.

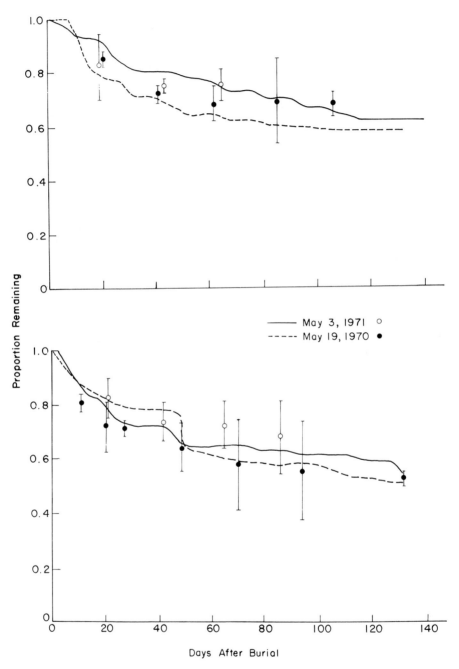

Fig. 6.7.   Comparison of model (lines) and data (means and 95% confidence intervals) for decomposition of bluestem hay buried on the indicated dates

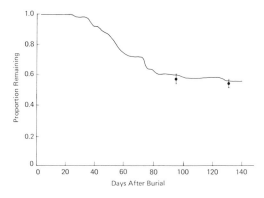

Fig. 6.8.   Validation test for litter-bag model (line). Means and 95 percent confidence intervals are for "litter" (collected from standing dead of 1971) buried on March 27, 1972

6.7.2 Decomposer Biomass

Figure 6.9 shows model output for active and inactive decomposers in the litter and the three soil layers. Some interesting differences are predicted for the behavior of decomposers at the various levels. Warm periods in winter may lead to short bursts of activity near the surface, but the deep-soil temperature remains too low. The biomass of both active and inactive decomposers is more variable nearer the surface, undoubtedly because temperature and soil water in the model show this pattern. Also, decomposers in the litter attain population peaks later in the year than do those in the soil, a pattern also seen in the 1970 and 1971 model runs. This results from different temporal patterns of input from the producer submodel, with more than 80% of the input to litter occurring between June and August, while root death is distributed more uniformly throughout the year. Decomposers in the litter and top soil layer show a net increase, while those in the bottom two soil layers show a net decrease in 1972. The net yearly change in a layer varies depending on the availability of soil water for that layer.

The only time series data available for decomposer biomass at the Pawnee Site in 1970–1972 consist of ATP measurements taken in 1971 (Sparrow and Doxtader, 1973). Adenosine triphosphate is presumed to reflect the biomass of active microbes, and Figure 6.10 compares predictions of active biomass in the 0–15-cm layer to the estimates based on ATP levels. The model was not tuned to these data, obviously. The significance of the departures of the model from the data is difficult to judge since no measure of variability was given for the data. A real discrepancy might reflect an inadequate formulation of the model or a varying ATP:biomass ratio. Since that ratio depends on the microfloral composition, the stage of growth, oxygen tension, nutrient condition, and solute concentration (Sparrow and Doxtader, 1973; Ausmus, 1973), it probably changes with time. Measuring or predicting the biomass of active microbes in soil remains a problem.

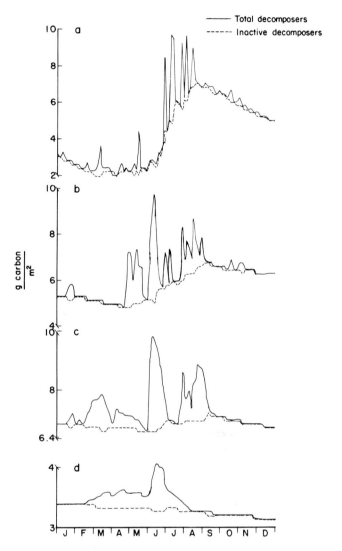

Fig. 6.9.   Model output of the biomass of inactive decomposers and total decomposers in: (a) litter, (b) 0–4-cm soil layer, (c) 4–15-cm layer, and (d) 15–60-cm layer; simulation is for 1972 with no cattle

6.7.3 Carbon-dioxide Evolution

Figures 6.11 and 6.12 compare the observed rates of CO_2 evolution on the ungrazed treatments at the Pawnee Site in 1971–1972 (D.C. Coleman, personal communication) to daily soil CO_2 output in the model (93% from decomposers and 7% from live roots and crowns). The model was not tuned to either year's data, not even to adjust respiration to the general levels observed in the data.

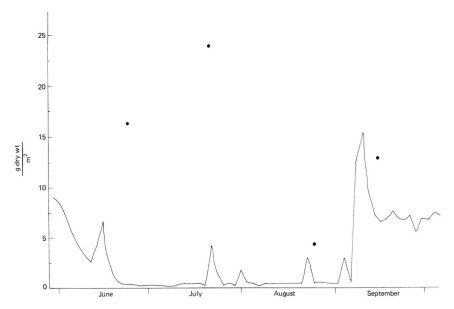

Fig. 6.10. Model predictions of active decomposer biomass (line) and estimates based on ATP (points) in the 0–15-cm soil layer at Pawnee Site in 1971

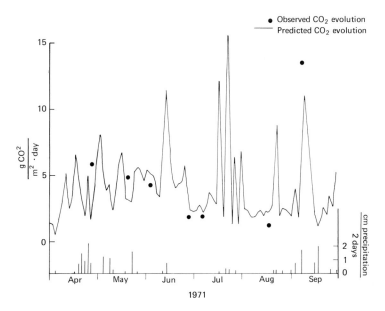

Fig. 6.11. Precipitation and CO_2 evolution in 1971

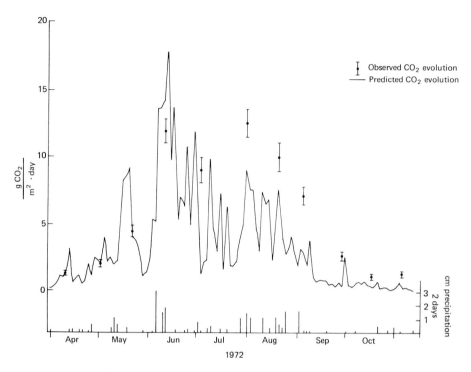

Fig. 6.12.   Precipitation and CO_2 evolution in 1972 (95% confidence intervals given for the data)

This fit is considered satisfactory despite the fact that the model usually does not come within the 95% confidence intervals. The square root of the mean-squared difference between model and data is 2.8 for 1971 and 3.5 for 1972. The fact that the model does about as well in two different years suggests that the most important factors affecting the rate of CO_2 evolution are reasonably represented and that confidence can be placed in the model's predictions.

6.7.4 Substrate Dynamics

The dynamics of the substrate compartments are determined jointly by inputs from the producer and consumer submodels and by outputs to decomposers; hence most of these results are presented elsewhere. Figure 6.13 shows the general pattern, with decreases in total substrate during the summer months and increases in the winter. The labile component increases relative to the resistant component during periods when litter is accumulating and decreases when litter is decreasing.

With the rather long turnover time (1000 yr) assumed for humus, the model predicts for 1972 that humus decomposition amounts to only 3.46 g C · m^{-2} · yr^{-1}. The assumption that humus derives entirely from dead decomposers leads to a humus production of only 0.12 g C · m^{-2} · yr^{-1}. This imbalance, which

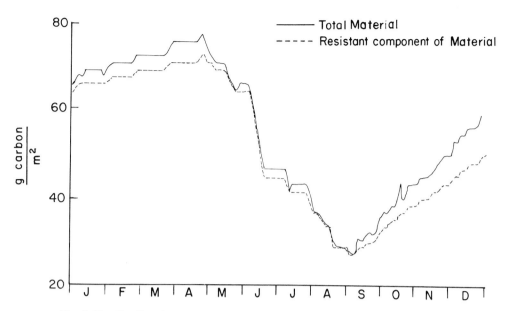

Fig. 6.13.   Predicted amount of the resistant component and of total substrate in the 0–4-cm soil layer in 1972

applies to 1970–1971 as well as 1972, is of little consequence to the performance of the model, since the humus compartment is so large (~5600 g/m²). Several possible explanations for the imbalance are that: (a) the turnover time of humus is 28 times (3.46 ÷ 0.12) longer than assumed, (b) the model underestimates the death of decomposers by a factor of 28, (c) the humic component of decomposers is 8.4% instead of 0.3%, or that decomposers are not the sole source of humus. The last explanation seems by far the most likely, and of course there is evidence that humus originates from plant material as well as from microbes (Martin and Haider, 1971). This model result suggests that most humus originates from plant materials.

6.7.5 Carbon-flow Budget

It is interesting to compare the magnitudes of transfers operating in the model, since such information is practically impossible to obtain in the ecosystem. Figure 6.14 presents the model output as a carbon budget for decomposers. Secondary productivity (death plus net change in biomass) drops from 60 g C in the surface litter to 22, 10, and 2 g in the three soil layers, for a total of 94 g C · m⁻² · yr⁻¹. A related trend is the drop in the net yield (production/substrate decomposed) from 0.52 in the litter to 0.23, 0.13, and 0.08 in the three soil layers. The net yield in the litter is nearer to the maximum possible yield (0.60) than is that in the soil because the yield is high only when growth is rapid, which occurs more often near the surface. Because of high death rates in the litter there is a greater recycling of material within the decomposer compartment; that is, a large proportion of material decomposed in the litter consists of dead decomposers. The ratio of productivity to the initial value of biomass, a quantitative measure of turnover, decreases from 19 yr^{-1} in the litter to 3.2, 1.4, and 0.5 in the soil layers. The overall value in soil is 2.0 yr^{-1}. Death from both freezing and drying and from starvation are relatively more important near the surface, where the environment is more hostile. Surprisingly, starvation is more important than freezing and drying, even in the litter. The model suggests that many decomposers die of starvation as the environment becomes colder or dryer, so that by the time conditions deteriorate sufficiently to kill active stages, much of the biomass is already in an inactive, resistant state.

6.8 Discussion

Many benefits of modeling accrue during the period of model development. Difficulties encountered in tuning the model draw attention to areas of ignorance and help direct an experimental effort. Several examples follow.

In an early version of the model, it was assumed that differences between the decomposition rates in litter and at various depths in the soil could be accounted for by differences in the temperature and moisture regimes, but this assumption led to a net decrease in substrate in the deeper layers. Incorporation of the effect of inorganic nitrogen concentration did little to correct the imbalance. The problem was finally handled by allowing the maximum decomposition rates [r_k

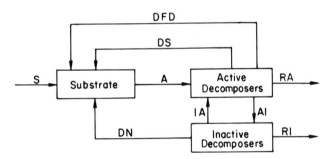

and r_h, Eq. (6.6)] to vary in the ratio of 7.5:8.0:4.5:1.0 from the litter to the upper, middle, and lower belowground layers, respectively. The decompositon rate might decrease with depth because of decreasing oxygen concentration, decreasing organic matter content of the soil, or less favorable physical structure of the soil. Such a decrease in the intrinsic rate of decomposition with depth has not been reported, possibly because studies that have compared decomposition at different depths (Weaver, 1947) have not accounted for the effects of temperature and moisture. Other possible explanations are that the driving variables for decompositon (temperature, soil water, and nitrogen) or the quantity or quality of inputs to belowground litter from producers do not change appropriately with depth in the model. It seems unlikely, however, that these could account for the observed eightfold difference between the top and bottom layers. The question could be resolved by laboratory experiments at constant temperature and water

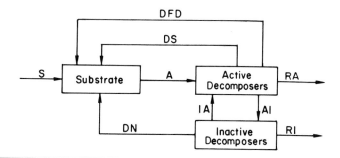

Budget Item[a]	Layer			
	Litter	Soil 0–4 cm	Soil 4–15 cm	Soil 15–60 cm
A = substrate decomposed	114	93	75	22
RA = respiration of active decomposers	54	71	65	20
RI = respiration of inactive decomposers	0.3	0.5	0.4	0.2
AI = change from active to inactive state	8.9	2.3	1.4	0.011
IA = change from inactive to active state	6.8	1.4	0.9	0.005
DFD = death from freezing and drying	6.8	7.5	1.3	0.6
DS = death from starvation	51	13	8.1	0.8
DN = nominal death	0.2	0.4	0.3	0.2
S = substrate addition from producers and consumers	28	47	54	33
Initial value for active decomposers	0.03	0.07	0.07	0.03
Final value for active decomposers	0	0	0	0
Initial value for inactive decomposers	3.2	6.9	6.9	3.4
Final value for inactive decomposers	4.9	7.0	6.7	3.2
Initial value for substrate (excluding humic material)	44	68	65	116
Final value for substrate (excluding humic material)	16	44	54	129

[a] Flows are in g carbon · m⁻² · year⁻¹; state variables are g carbon · m⁻².

Fig. 6.14. Carbon budget for decomposers on the ungrazed treatment in 1972

pronounced depth-dependence of the physical attributes of the soil leads to different responses, even of genetically identical organisms. The parameter chosen to vary with depth in order to achieve a balance in each layer is the rate of transformation from the inactive to active state, which decreases from 0.0078 per day in the litter to 0.00077, 0.00057, and 0.0000065 in the three belowground layers. The rationale for this choice is that when temperature and moisture conditions become favorable for a change to the active state, a greater proportion of inactive decomposers near the surface will find themselves near suitable substrate, because substrate is more abundant relative to the inorganic soil matrix near the surface. Also, cold and heat shock, factors breaking dormancy, are more pronounced near the surface.

Whereas the structure of the decomposition model can be described without considering the other submodels, its performance depends on predicted temperature, soil water, inorganic nitrogen, and dead plant and animal material. The decomposition model in turn affects the performance of the phosphorus submodel through decomposer dynamics and affects the water flow submodel through the interception of precipitation by litter and through the effect of belowground litter on field capacity.

Most of the insights derived from this modeling exercise apply at a low level of biological resolution, because the model is of low resolution. The level of resolution could be increased by incorporating a variety of groups of decomposers, an explicit treatment of interactions among the various groups, a more intimate association with the nutrient submodels, and a classification of substrates based on chemical composition. The danger of formulating a model too advanced beyond its information base (Bunnell, 1973) must be kept in mind, but the model described here is probably not yet at that point. There remain a number of interesting ideas about decomposers and decomposition to be tested using ELM.

Acknowledgments. This work was greatly influenced by US/IBP Grassland Biome scientists. D. C. Coleman provided continuous support and advice, and F. E. Clark generously shared his data. Some of the basic ideas in the model were discussed in the Grassland Biome Process Study Workshops held in 1972. Participants included F. E. Clark, D. C. Coleman, D. Davidson, K. G. Doxtader, R. D. Giffin, J. D. Gustafson, D. A. Klein, P. A. Mayeux, R. M. Pengra, S. Shushan, and the author. Helpful comments were provided by G. W. Cole, F. L. Bunnell, A. B. Clymer, and R. G. Woodmansee. Finally, the authors of the other submodels of ELM, particularly C. F. Rodell, R. H. Sauer, W. J. Parton, J. C. Anway, and our director, G. S. Innis, provided stimulation and assistance.

This chapter reports on work supported in part by National Science Foundation Grants GB-31862X, GB-31862X2, GB-41233X, BMS73-02027 AO2, and DEB73-02027 AO3 to the Grassland Biome, U.S. International Biological Program, for "Analysis of Structure, Function, and Utilization of Grassland Ecosystems."

References

Alexander, M.: Introduction to Soil Microbiology. New York: John Wiley and Sons, 1961, 472 pp.

Alexander, M.: Microbial Ecology. New York: John Wiley and Sons, 1971, 511 pp.

Allison, F. E., Sherman, M. S., Pinck, L. A.: Maintenance of soil organic matter: I. Inorganic soil colloid as a factor in retention of carbon during formation of humus. Soil Sci. **68**, 463–478 (1949)

Andrews, R., Coleman, D. C., Ellis, J. E., Singh, J. S.: Energy flow relationships in a shortgrass prairie ecosystem. In: Proceedings of the First International Congress of Ecology. Centre for Agr. Publ. and Doc., Wageningen, The Netherlands, 1974, pp. 22–28

Anway, J. C., Brittain, E. G., Hunt, H. W., Innis, G. S., Parton, N. J., Rodell, C. F., Sauer, R. H.: Elm: Version I. O. US/IBP Grassland Biome Tech. Rep. No. 156. Fort Collins: Colorado State Univ., 1972, 285 pp.

Ausmus, B. S.: The use of the ATP assay in terrestrial decomposition studies. Bull. Ecol. Res. Commun. (Stockholm) **17**, 223–234 (1973)

Babiuk, L. A., Paul, E. A.: The use of fluorescein isothiocyanate in the determination of the bacterial biomass of grassland soil. Can. J. Microbiol. **16,** 57–62 (1970)

Bartholomew, W. V., Norman, A. G.: The threshold moisture content for active decomposition of some mature plant materials. Soil Sci. Soc. Am., Proc. **11,** 270–279 (1946)

Bhaumik, H. D., Clark, F. E.: Soil moisture tension and microbiological activity. Soil Sci. Soc. Am., Proc. **12,** 234–238 (1947)

Bledsoe, L. J., Francis, R. C., Swartzman, G. L., Gustafson, J. D.: PWNEE: A grassland ecosystem model. US/IBP Grassland Biome Tech. Rep. No. 64. Fort Collins: Colorado State Univ., 1971, 179 pp.

Boylen, C. W., Ensign, J. C.: Intracellular substrates for endogenous metabolism during long-term starvation of rod and spherical cells of *Arthrobacter crystallopoietes.* J. Bacteriol. **103,** 578–587 (1970)

Broadfoot, W. M., Pierre, W. H.: Forest soil studies: I. Relation of rate of decomposition of tree leaves to their acid–base balance and other chemical properties. Soil Sci. **48,** 329–348 (1939)

Brock, T. D.: Microbial growth rates in nature. Bacteriol. Rev. **35,** 39–58 (1971)

Brown, J. R., Frederick, L. R.: Decomposition of the water-soluble fraction of sudangrass residue in soil. Plant Soil **28,** 467–470 (1968)

Bunnell, F. L.: Decomposition: Models and the real world. Bull. Ecol. Res. Commun. (Stockholm) **17,** 407–415 (1973)

Bunnell, F. L., Dowding, P.: ABISKO—A generalized decomposition model for comparisons between sites. US/IBP Tundra Biome Rep. 73-6. College, Alaska: Univ. Alaska, 1973

Campbell, C. A., Paul, E. A., Rennie, D. A., McCallum, K. J.: Applicability of the carbon-dating method of analysis to soil humus studies. Soil Sci. **104,** 217–224 (1967)

Clark, F. E.: Decomposition of organic materials in grassland soil. US/IBP Grassland Biome Tech. Rep. No. 61. Fort Collins: Colorado State Univ., 1970, 23 pp.

Clark, F. E., Paul, E. A.: The microflora of grasslands. Adv. Agron. **22,** 375–435 (1970)

Curry, J. P.: The decomposition of organic matter in soil. Part I. The role of the fauna in decaying grassland herbage. Soil Biol. Biochem. **1,** 235–258 (1969)

Douglas, L. A., Tedrow, J. C. F.: Organic matter decomposition rates in Arctic soils. Soil Sci. **88,** 305–312 (1959)

Doxtader, K. G.: Microbial biomass measurements at the Pawnee Site: Preliminary methodology and results. US/IBP Grassland Biome Tech. Rep. No. 21. Fort Collins: Colorado State Univ., 1969, 16 pp.

Drobnik, J.: The effect of temperature on soil respiration. Folia Microbiol. **7,** 132–140 (1962)

Eckenfelder, W. W., Jr.: Water quality engineering for practicing engineers. New York: Barnes & Noble, 1970, 328 pp.

Ensign, J. C.: Long term starvation survival of rod and spherical cells of *Arthrobacter crystallopoites.* J. Bacteriol. **103,** 569–577 (1970)

Felbeck, G. T., Jr.: Chemical and biological characterization of humic matter. In: Soil Biochemistry, Vol. 2. McLaren, A. D., Skujins, J. (eds.) New York: Marcel Dekker, 1971, pp. 36–59

Floate, M. J. S.: Decomposition of organic materials from hill soils and pastures. II. Comparative studies on the mineralization of carbon, nitrogen and phosphorus from plant materials and sheep faeces. Soil Biol. Biochem. **2,** 173–185 (1970)

Gaudy, A. F., Jr., Komolrit, K., Bhatla, M. N.: Sequential substrate removal in heterogeneous populations. J. Water Pollut. Control Fed. **35,** 903–922 (1963)

Giese, A. C.: Cell Physiology, 3rd ed. Philadelphia: Saunders, 1968, 671 pp.

Gray, T. R. G., Williams, S. T.: Microbial productivity in soil. p. 255–285. In: Microbes and Biological Productivity. 21st Symp. Soc. Gen. Microbiology. Hughes, D. E., Rose, A. H. (eds.). Oxford and New York: Cambridge Univ. Press, (1971)

Grill, E. V., Richards, F. A.: Nutrient regeneration from phytoplankton decomposing in seawater. J. Marine Res. **22,** 51–69 (1964)

Harrison, A. P., Jr.: The response of *Bacterium lactis aerogenes* when held at growth temperature in the absence of nutriment: An analysis of the survival curves. Roy. Soc. (Lond.), Proc. B. **152**, 418–428 (1960)

Herbert, D.: Some principles of continuous culture, In: Recent Progress in Microbiology. VII International Congress of Microbiology. Tunevall, G. (ed.). Stockholm: Almqvist and Wiksell, 1958, pp. 381–396

Jager, G.: Changes in the activity of soil microorganisms influenced by physical factors (drying–remoistening, freezing–thawing). In: Graff, O., Satchell, J. E. (eds.) Progress in Soil Biology. Amsterdam: North Holland, 1967, pp. 178–191

Jenny, H., Gessel, S. P., Bingham, F. T. Comparative study of decomposition rates of organic matter in temperate and tropical regions. Soil Sci. **68**, 419–432 (1949)

Jensen, H. L.: The microbiology of farmyard manure decomposition in soil. III. Decomposition of the cells of micro-organisms. J. Agr. Sci. **22**, 1–25 (1932)

Klein, D. A.: Seasonal carbon flow and decomposer parameter relationships in a semiarid grassland soil. Ecology **58**, 184–190 (1977).

Koelling, M. R., Kucera, C. L.: Dry matter losses and mineral leaching in bluestem standing crop and litter. Ecology **46**, 529–532 (1965)

Lamanna, C., Mallette, M. F.: Basic Bacteriology, 3rd ed. Baltimore: Williams & Wilkins, 1965, 1001 pp.

Lewis, J. A., Starkey, R. L.: Vegetable tannins, their decomposition and effects on decomposition of some organic compounds. Soil Sci. **106**, 241–247 (1968)

Marr, A. G., Nilson, E. H., Clark, D. J.: The maintenance requirement of *Escherichia coli*. Ann. New York Acad. Sci. **102**, 536–548 (1963)

Martel, Y., Paul, E. A.: An example of a process model: The carbon turnover for use in ecosystem studies in grasslands. In: Grassland Ecosystems: Reviews of Research. Range Sci. Dep. Sci. Ser. No. 7. Coupland, R. T., Van Dyne, G. M. (eds.). Fort Collins: Colorado State Univ., 1970, pp. 179–189

Martin, J. P., Haider, K.: Microbial activity in relation to soil humus formation. Soil Sci. **111**, 54–63 (1971)

Millar, H. C., Smith, F. B., Brown, P. E.: The rate of decomposition of various plant materials in soils. J. Am. Soc. Agron. **28**, 914–923 (1936)

Minderman, G.: Addition, decomposition and accumulation of organic matter in forests. J. Ecol. **56**, 355–362 (1968)

Nykvist, N.: Leaching and decomposition of litter. I. Experiments on leaf litter of *Fraxinus excelsior*. Oikos **10**, 190–211 (1959a)

Nykvist, N.: Leaching and decomposition of litter. II. Experiments on needle litter of *Pinus silvestris*. Oikos **10**, 212–224 (1959b)

Nykvist, N.: Leaching and decomposition of litter. III. Experiments on leaf litter of *Betula verrucosa*. Oikos **12**, 249–263 (1961a)

Nykvist, N.: Leaching and decomposition of litter. IV. Experiments on needle litter of *Picea abies*. Oikos **12**, 264–279 (1961b)

Otsuki, A., Hanya, T.: Production of dissolved organic matter from dead green algal cells. I. Aerobic microbial decomposition. Limnol. Oceanogr. **17**, 248–257 (1972)

Park, D.: Survival of microorganisms in soil. In: Ecology of Soil-borne Plant Pathogens. Baker, K. F., Synder, W. C. (eds.). Berkeley: Univ. California Press, 1965, pp. 82–98

Patten, B. C.: A simulation of the shortgrass prairie ecosystem. Simulation **19**, 177–186 (1972)

Payne, W. J.: Energy yields and growth of heterotrophs. Annu. Rev. Microbiol. **24**, 17–52 (1970)

Pendleton, D. F.: Degradation of grassland plants. M.S. thesis, Colorado State Univ., Fort Collins, 1972, 37 p.

Pinck, L. A., Allison, F. E., Sherman, M. S.: Maintenance of soil organic matter II. Losses of carbon and nitrogen from young and mature plant materials during decomposition in soil. Soil Sci. **69**, 391–401 (1950)

Pirt, S. J.: The maintenance energy of bacteria in growing cultures. Roy. Soc. (London) Proc. B. **163,** 224–231 (1965)

Postgate, J. R.: The viability of very slow-growing populations: A model for the natural ecosystem. Bull. Ecol. Res. Commun. (Stockholm) **17,** 287–292 (1973)

Randell, R. L., Gyllenberg, G. G., Kae, S. L., Jones, D. C., Kowal, H.: Data analysis and modelling. In: Matador Project Fifth Annual Report. Coupland, R. T. (ed.). Saskatoon, Saskatchewan: Canadian IBP Grassland Zone Programme, 1972, pp. 74–169

Reuss, J. O.: Decomposer and nitrogen cycling investigations in the Grassland Biome. In: Preliminary Analysis of Structure and Function in Grasslands. Range Sci. Dep. Sci. Ser. No. 10. French, N. R. (ed.). Fort Collins: Colorado State Univ., 1971, pp. 133–146

Schmidt, E. L.: Chairman's summary of panel discussion. 1. The traditional plate count technique among modern methods. Bull. Ecol. Res. Commun. (Stockholm) **17,** 453–454 (1973)

Schulze, K. L., Lipe, R. S.: Relationship between substrate concentration, growth rate and respiration rate of *Escherichia coli* in continuous culture. Archiv für Mikrobiologie **48,** 1–20 (1964)

Shields, J. S., Paul, E. A., Lowe, W. E., Parkinson, D.: Turnover of microbial tissue in soil under field conditions. Soil Biol. Biochem. **5,** 753–764 (1973)

Sneath, P. H. A.: Longevity of micro-organisms. Nature **195,** 643–646 (1962)

Sparrow, E. B., Doxtader, K. G.: Adenosine triphosphate (ATP) in grassland soil: Its relationship to microbial biomass and activity. US/IBP Grassland Biome Tech. Rep. No. 224. Fort Collins: Colorado State Univ., 1973, 161 pp.

Sussman, A. S.: Dormancy of soil microorganisms in relation to survival. In: Ecology of Soil-borne Plant Pathogens. Baker, K. F., Snyder, W. C. (eds.). Berkeley: Univ. California Press, 1965, pp. 99–110

Tenney, F. G., Waksman, S. A.: Composition of natural organic materials and their decomposition in the soil: IV. The nature and rapidity of decomposition of the various organic complexes in different plant materials, under aerobic conditions. Soil Sci. **28,** 55–84 (1929)

Timin, M. E., Collier, B. D., Zich, J., Walters, D.: A computer simulation of the Arctic tundra ecosystem near Barrow, Alaska. US/IBP Tundra Biome Rep. 73-1. College, Alaska: Univ. Alaska, 1973

Van Cleve, K., Sprague, D.: Respiration rates in the forest floor of birch and aspen stands in interior Alaska. Arctic Alpine Res. **3,** 17–26 (1971)

Wase, D. A. J., Hough, J. S.: Continuous culture of yeast on phenol. J. Gen. Microbiol. **42,** 13–23 (1966)

Weaver, J. E.: Rate of decomposition of roots and rhizomes of certain range grasses in undisturbed prairie soil. Ecology **28,** 221–240 (1947)

Webley, D. M., Jones, D.: Biological transformations of microbial residues in soil, In: Soil Biochemistry, Vol. 2. McLaren, A. D., Skujins, J. (eds.) New York: Marcel Dekker, 1971, pp. 446–485

Witkamp, M.: Decomposition of leaf litter in relation to environment, microflora, and microbial respiration. Ecology **47,** 194–201 (1966)

Witkamp, M., Olson, J. S.: Breakdown of confined and nonconfined oak litter. Oikos **14,** 138–147 (1963)

7. A Grassland Nitrogen-flow Simulation Model[1]

J. O. Reuss and G. S. Innis

Abstract

A dynamic simulation model of nitrogen flow in a grassland ecosystem has been developed. State variables in the model include the nitrogen in the following major components: soil ammonium, soil nitrate, live roots, dead roots, soil organic matter, live tops, and litter. Belowground components are subdivided into four depth layers. The model includes a simple producer–decomposer submodel, but the nitrogen sections have also been incorporated into the ELM model.

Temperature and soil-water driving variables are supplied from external sources (data or other models). Nitrogen flows, plant growth, and plant decomposition are controlled by temperature, soil water, phenological stage of plant development, and the nitrogen status of the various components of the system. The interaction of the nitrogen components with production and decomposition rates allow nitrogen status to act as a control variable in ecosystem processes, an important feature in a grassland model.

Tests indicate that the model performs in a realistic manner. Simulation of fertilizer effects has been satisfactory, and the model has been useful in identifying critical processes where further research is necessary to advance understanding of the system.

7.1. Introduction

Nitrogen is a key element in the grassland ecosystem. Soil nitrate and ammonium levels are normally low because of their rapid utilization by grasses and decomposer organisms. The productivity of grasslands is almost universally limited by the supply of plant-available nitrogen, except under conditions of severe temperature or moisture stress. Nitrogen concentrations are an important factor controlling rates of decomposition of plant material as well as diet selection and nutritional status of consumer organisms. Thus development of a nitrogen flow model was essential to the ELM modeling effort.

A dynamic simulation model describing nitrogen flow in a grassland ecosystem has been developed, and its essential elements are included in this paper. The model has gone through several stages of development, and an earlier version was described by Reuss and Cole (1973). The present version is satisfactory for fairly extensive testing and allows us to evaluate the suitability of the

[1] Supported in part by the Colorado State University Experiment Station and published as Scientific Series Paper No. 2047.

mechanisms and constants utilized. Thus at present we have attained a plateau of development, such that the model in this form can be used as a module in a grasslands ecosystem model. In many cases it was necessary to use empirical relationships that provide for the system to respond in a manner consistent with experience. With a more complete understanding of the nitrogen flows in this system, these can be replaced with more biologically meaningful relationships.

The model described in this chapter contains simple production and decomposition submodels, and only soil-water and temperature data are required from external sources. The discussion here focuses on the nitrogen-flow sections of the model and the mechanisms of interaction with production and decomposition rates since the present producer and decomposer submodels only drive the nitrogen model for testing purposes.

The nitrogen-flow sections of this model and the interactions with production and decomposition rates have been incorporated into the ELM model with only those modifications necessary to accommodate the structure of the larger model.

7.2. Model Structure

7.2.1. State Variables

Nitrogen State Variables.   This is a compartmental model incorporating eight major components with 23 state variables in the nitrogen-flow section as follows (see Fig. 7.1): (a) soil nitrate nitrogen at four depths, (b) soil ammonium nitrogen

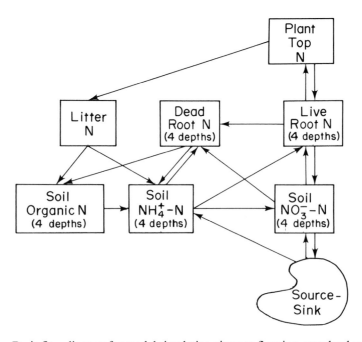

Fig. 7.1.   Basic flow diagram for model simulating nitrogen flow in a grassland ecosystem

at four depths, (c) live-root nitrogen at four depths, (d) dead-root nitrogen at four depths, (e) soil organic nitrogen at four depths, (f) live-plant top nitrogen, (g) nitrogen in aboveground dead plant parts and surface litter, and (h) source or sink.

The five belowground compartments are divided into four subcompartments representing four depth layers. The depths represented by these subcompartments are flexible and may be varied according to the characteristics of the particular site. For the simulations shown in this chapter the first layer was taken to be 0.025 m thick, and the three lower-depth layers were 0.05 m each. The units of the nitrogen state variables are g N/m². In the case of the belowground components the units are g N/m² in each individual depth layer.

Producer–Decomposer State Variables.   The producer–decomposer submodel includes five major components and 11 state variables representing above- and belowground biomass as follows: (a) live-plant tops, (b) live roots at four depths, (c) dead roots at four depths, (d) litter and dead tops, and (e) source or sink.

Units of the biomass state variables are g dry matter/m². Again in the case of the belowground compartments the units are g dry matter/m² in a depth layer.

7.2.2. Temperature and Soil-water Effects

Many of the processes described by the flow functions in both the nitrogen section and the producer–decomposer submodel are limited by temperature and by soil water availability. All flow rates are first determined at 25°C and 0.3 bars soil-water suction. These rates are then modified by temperature and soil-water coefficients calculated for each time step. Temperature coefficients for all processes in the model are taken from the relationship shown in Figure 7.2. This curve is based on a Q_{10} of 2.0 in the range of 20–30°C. It is similar to that given by Sabey et al. (1969) for the nitrification process and consistent with the work of Stanford et al. (1973). Effects of temperatures above 35°C are probably different for the various processes. However, such high temperatures are rarely encountered in temperate soils except where soil-water suctions are sufficiently high to severely limit most biological processes.

Two different relationships are used to describe the effect of soil water on the various processes. The curve relating soil-water suction to soil-water coefficients for nitrogen uptake by plants is shown in Figure 7.3. The same relationship is used for all other water-dependent processes except for nitrification, where the rate drops to zero near saturation due to low oxygen tensions. While some refinement of the soil-water coefficients for the various processes would be desirable, we would expect the model to be relatively insensitive to moderate changes in this relationship. This occurs because in most soils the bulk of the available water is held at soil-water suctions of a few bars or less, the range within which soil-water coefficients for nitrogen uptake are near 1.0. Soil temperature and soil-water data must be furnished to the model as driving variables. The temperature and soil water effects on rate processes are assumed to be multiplicative in all cases.

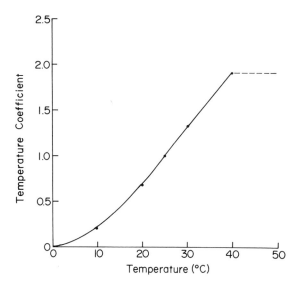

Fig. 7.2. Temperature coefficients for temperature-dependent rate processes in the grassland nitrogen model

7.2.3 Nitrogen Flows

All nitrogen flows shown in Figure 7.1 are described as a function of one or more of the following: (a) time, (b) soil temperature, (c) soil water, (d) daily growth, (e) death or decomposition, and (f) nitrogen content. Nitrogen concentrations for each biomass compartment are internally calculated and in many cases are used as control variables. Rates are calculated on a daily basis, and 1-day time steps are used in the simulation.

The methods of calculation for those flows included in the present version of the model are given below. In all cases precise mathematical formulations are used. In general, this does not imply a theoretical basis or derivation, but rather that the general form of the relationship is known or can be reasonably assumed and that an arbitrary function has been fitted for convenient calculation. Rela-

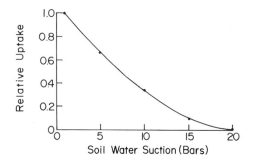

Fig. 7.3. Soil-water dependence assumed for nitrogen uptake by grass roots

tionships such as those shown in Figures 7.2 and 7.3 are calculated by means of linear interpolation.

Nitrogen Inputs or Losses.   Nitrogen inputs in grassland ecosystems include nitrogen found in precipitation, fixation by free-living organisms, and symbiotic fixation. The grassland systems studied as part of the Grassland Biome Program have very low rates of fixation, generally below 0.1 g N $\cdot$ m^{-2} $\cdot$ yr^{-1} (Copley and Reuss, 1972). Precipitation inputs are somewhat larger and have been estimated for the Pawnee Site at 0.3 g N $\cdot$ m^{-2} $\cdot$ yr^{-1} (Reuss and Copley, 1971). A total input of 0.5 g N $\cdot$ m^{-2} $\cdot$ yr^{-1} from these two sources, evenly divided between ammonium and nitrate forms and uniformly distributed in the rainfall, is included in the model. Symbiotic fixation is site-specific and must be determined for each site considering the species and density of legumes. The simulations shown here assume negligible symbiotic fixation.

In nature losses to the sink in Figure 7.1 could occur from leaching, microbial denitrification, and volatilization in various forms. Leaching or denitrification would only be significant when substantial amounts of nitrate are present in wet soils. This combination of conditions rarely occurs in arid or semiarid grasslands, so these processes are not included, although they may be necessary for simulating more mesic systems, particularly when fertilized. Little is known about the rate and magnitude of nitrogen volatilization from these systems although, our data suggest the ammonia loss from animal excreta could amount to as much as one-half the rainfall input. There is no provision for this flow in the model, but this omission should not significantly affect short-term simulations.

Mineralization of Soil-Organic Nitrogen.   The mineralization of soil organic nitrogen to ammonium is a key process in most natural systems. In the model this rate is dependent on soil water, temperature, and the amount of organic N present. A first-order rate constant of 6.7 $\cdot$ 10^{-4}/day at 25°C and 0.3 bars soil-water suction is assumed. This rate is based on an average annual turnover of 2% of the soil organic nitrogen, a value widely used in soil fertility evaluations and confirmed by Geist et al. (1970). A consideration of the seasonal temperature and soil-water regimes commonly encountered in arid and semiarid grasslands led to the assumption that this occurs in the equivalent of 30 days at 25°C and 0.3 bars soil-water suction. No reverse flow is provided as microbial immobilization is assumed to occur through the dead roots rather than directly into soil organic matter.

Oxidation of Ammonium to Nitrate.   The oxidation of ammonium to nitrate is a two-step process in which ammonium is oxidized to nitrite by organisms of the genus *Nitrosomonas* and the nitrite oxidized to nitrate by *Nitrobacter*. It is only necessary to represent the rate-limiting step in the model that is normally the oxidation of ammonium to nitrite.

The rate of oxidation of ammonium to nitrate is initially assumed to be a function of ammonium concentration of the form:

$$\frac{-dS}{dt} = \frac{AS}{K + S},$$

where S is the substrate concentration, A is the maximum rate, and K is the concentration at which half the maximum rate is achieved, as suggested by McLaren (1973). Values used are $A = 40$ g $\cdot$ m^{-3} $\cdot$ day^{-1} and $K = 90$ g/m^3, arrived at by fitting published oxidation rates determined in calcareous soils (Justice and Smith, 1962).

At low ammonium concentrations (i.e., 5 g/m^3) this initial rate provides for the oxidation of about 40% of the amount present per day, declining to 20% at a soil ammonium concentration of 100 g/m^3. This oxidation is a biological process, and the rate is also dependent on the microbial populations present. These populations respond to increases in substrate, but rates of growth are dependent on microbial generation times. For the oxidation of ammonium to nitrite by organisms of the genus *Nitrosomonas,* Alexander (1965) gives generation times in the range of 8–40 h. In the model we limit the increase in rates above 2 g $\cdot$ m^{-3} $\cdot$ day^{-1} to a maximum of 1.25 times the previous day's rate, equivalent to a generation time of 20.5 h.

Uptake by Live Roots.   The rate of uptake of nitrate nitrogen per unit of live root is determined by the nitrate concentration. It is considered to be the sum of two processes, each of which can be described by maximum rate M, a half-saturation constant K, and the substrate (nitrate) concentration S:

$$U = \frac{M_A S}{K_A + S} + \frac{M_B S}{K_B + S}$$

The total uptake U is expressed in g N $\cdot$ g root^{-1} $\cdot$ day^{-1} while the nitrate concentration is in g N/m^3.

Values for the constants in the present version of the model were arrived at empirically and are: (a) $M_A = 2.00 \cdot 10^{-3}$ g N $\cdot$ g root^{-1} $\cdot$ day^{-1}, (b) $M_B = 0.40 \cdot 10^{-3}$ g N $\cdot$ g root^{-1} $\cdot$ day^{-1}, (c) $K_A = 84$ g N/m^3, and (d) $K_B = 4.8$ g N/m^3. At low nitrate levels uptake rises rapidly as nitrate concentration increases. At high soil-nitrate levels the uptake rate levels off, approaching $(M_A + M_B)$ as the soil nitrate concentration becomes very high. The constants chosen are such that the K_B and M_B largely control uptake rates at low soil nitrate concentration. Above approximately 20 g N/m^3 this process has nearly reached its maximum, and further concentration increases are reflected through the process described by K_A and M_A. Total uptake of $0.7 \cdot 10^{-3}$ g N $\cdot$ g root^{-1} $\cdot$ day^{-1} is predicted at a nitrate concentration of 20 g/m^3. These parameter values were selected to provide rates consistent with solution uptake rates calculated from data by Rumburg and Sneva (1970) and field rates calculated from data shown by Power (1967).

The above processes could result in roots with very high N concentrations if uptake rates exceed rates of translocation to the tops. In nature this buildup is apparently prevented by physiological controls that effectively decrease uptake as root-nitrogen concentrations increase. This control mechanism is represented in the model by a reverse flow that decreases net uptake at high root-nitrogen concentrations. While the actual biological control mechanisms may or may not include a reverse flow, this is a convenient way of modeling the control. The reverse rate factor B is a fraction of the uptake U, and net uptake U_n in g N $\cdot$ g root^{-1} $\cdot$ day^{-1} is calculated by:

$$U_n = U(1 - B), 0 \leqslant B \leqslant 1.$$

Thus when $B = 0$, no reverse flow occurs and net uptake U_n is equal to uptake; when $B = 1$, reverse flow is equal to forward flow and no net uptake occurs. The value of B increases as root nitrogen concentrations R_c increase and decreases with increasing values of soil nitrate S:

$$B = [\exp(R_c - 0.006)Z] - 1,$$

where
$$Z = 40 + 35e^{-0.15S}.$$

If this calculation results in $B < 0$, B is set to 0; if it results in $B > 1$, B is set to 1.

The constants were chosen so that reverse flow is zero and net uptake U_n is equal to uptake U when root nitrogen concentrations are below 0.6%. The root nitrogen concentration above which net uptake is zero, that is, where reverse flow is equal to uptake, varies with soil nitrate. At soil nitrate concentrations approaching zero this equilibrium is established at 1.5% N in the roots, while at nitrate concentrations of 20 g/m^3 or greater, net uptake is zero if the roots contain more than 2.5% N. Plant nutrient uptake processes are a function of temperature and soil water, and appropriate corrections for these factors are applied to the net uptake. Finally, total flow is calculated from net uptake per gram of root and total live-root biomass.

Relative uptake rates of ammonium and nitrate are difficult to assess and are undoubtedly a function of plant species. Solution concentrations of ammonium are much lower than those for nitrate at similar soil concentrations, due to adsorption of ammonium on the exchange complex. This relationship varies with the mineralogical properties of the soil. For the present we are using an uptake function for ammonium identical to that for nitrate except for the application of a coefficient to reduce effective ammonium concentration due to adsorption. This coefficient is set at 0.1, as suggested by Frere et al. (1970). Additional data relevant to grass species would be very helpful.

Translocation.   The translocation functions devised for this model are empirical, but seem to provide adequately for the observed seasonal changes in nitrogen distribution within the plant (Whitehead, 1970; Rumburg et al., 1964). The assumptions are that: (a) the plant maintains a ratio between the concentration of nitrogen in the tops and the concentration in the roots and (b) this ratio varies between 3.0 and 1.0 as a function of stage of growth.

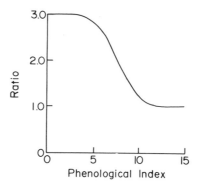

Fig. 7.4. Ratio maintained between live-top and live-root nitrogen concentrations as a function of phenological index of plant development

The stage of growth is identified by a phenological index funished by the producer–decomposer submodel. This index provides 14 growth stages (French and Sauer, 1974) and is essentially an earlier version of that described by Sauer (1973). The ratios maintained between live-top and live-root nitrogen concentration as a function of this phenological index is shown in Figure 7.4.

The producer–decomposer submodel imposes growth restrictions on tops and roots based on nitrogen concentrations, thus effectively preventing excessive dilution of nitrogen in either live-root or top systems. Where nitrogen in the aboveground live tissue is relatively high in late season, this mechanism results in some transfer back to live roots, a phenomenon consistent with field observations. No specific temperature and soil-water limitations are included, as the effects of these parameters are implicit through their role in controlling growth rates.

Organ-death Transfers. The transfers of nitrogen from live-top to litter compartments and from live-root to dead-root compartments are determined by death rates of tops and roots from the producer–decomposer submodel. The nitrogen flows are calculated by multiplying the biomass flow by the mean nitrogen concentration in the donor compartment. This transfer is, of course, a simplification as it assumes uniform nitrogen concentration within each compartment when, in fact, the senescent material may be lower in nitrogen than the remaining live tissue. Inclusion of this effect must await development of a somewhat more sophisticated producer–decomposer submodel.

Decomposition-related Flows. The transfers of nitrogen resulting from the decomposition of dead roots and litter comprise a complex interacting subset. Decomposition rates of both materials are calculated in the producer–decomposer submodel. These rates are not independent of the nitrogen system, but are a function of the nitrogen concentration as described below. The nitrogen released by decomposition is then partitioned between soil organic matter and ammonium. The amounts transferred to each of these forms are a function of the nitrogen concentration (C/N ratio) of the decomposing material.

Transfers from Decomposing Roots.   The biomass of dead roots decomposed daily is furnished by the producer–decomposer submodel. The amount of nitrogen transferred daily from dead roots to soil organic matter and ammonium is then calculated by multiplying the mean dead-root nitrogen concentration by the amount of dead roots decomposed that day. The fraction of this daily total transferred to soil organic matter is determined by the nitrogen concentration as shown in Figure 7.5, while the remainder is transferred to soil ammonium. The assumptions involved in the relationship are that: (a) when high-C-concentration root material is decomposed (C/N ratio > 50), virtually all nitrogen in this material is converted to the stable organic matter fraction and (b) in high-N-concentration materials when the requirement of formation of the stable organic matter is satisfied, most of the nitrogen in excess of this requirement is mineralized, (i.e., transferred to ammonium). Specifically, we assume that all N is incorporated into organic matter when the nitrogen concentration is less than 1%. When the concentration is 1.7%, we assume 85% of the total to be converted to organic matter. As the nitrogen concentration increases a greater fraction of the N in the decomposing roots is mineralized. Figure 7.5 is derived by assuming that the N concentration of the decomposing material is uniformly distributed from 0 to twice the modeled concentration. The fraction of the material y at a concentration c less than 1.7% (working only with concentrations in excess of 1%) is $0.017y/2c$. We want 15% of that N at average concentrations less than 1.7% (but > 1%) mineralized and 85% of that above 1.7% mineralized. Thus the fraction to soil organic matter, Z, in Figure 7.5 is computed as:

$$Z = 0.15(1 - 0.017/c) + 0.85\,(0.017/c).$$

The 1.7% N value was chosen as being typical of the nitrogen concentration below which net mineralization is generally considered to be very small (Harmsen and Kolenbrander, 1965).

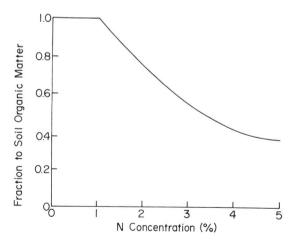

Fig. 7.5.   Fraction of the daily nitrogen transfer out of the dead-root compartment that becomes part of the stable organic matter complex as a function of dead-root nitrogen concentration

Transfers from Decomposing Litter.   The transfer of litter nitrogen to soil organic matter is very similar to the flow from dead roots to organic matter. The litter decomposition rate returned from the producer–decomposer submodel is a function of nitrogen concentration in the litter. The amount of nitrogen removed from the litter is calculated by multiplying the daily decomposition by the nitrogen concentration. This amount of nitrogen is then apportioned between the organic and ammonium fractions of the uppermost soil layer. A suitable data base is absent, but it is assumed that a significant fraction is returned as soluble amines and amino acids that are removed from the litter and enter the ammonium pool. Thus the fraction entering soil organic matter is lower than for dead roots. The present partitioning is constructed from the following assumptions: (a) at nitrogen concentrations less than 0.6%, half of the nitrogen in the decomposing litter enters the stable soil organic matter, (b) at a concentration of 1.8 percent N, 20 percent of this goes to soil organic matter, and (c) all nitrogen in the litter in excess of the 1.8% is returned to ammonium.

Based on these assumptions a curve relating the fraction of the nitrogen in the daily litter decomposition (F_L) to the nitrogen concentration of the litter was developed. Transfer to organic matter is the total decomposition nitrogen times F_L; the remaining fraction is transferred to ammonium.

Immobilization by Decomposing Roots.   It is well recognized that where plant materials with a high C/N ratio are decomposing in soils containing substantial amounts of mineral nitrogen, the soil ammonium and nitrate are utilized by the decomposer organisms. The model provides for this immobilization phenomenon in root decomposition, but it is assumed that soil ammonium and nitrate are positionally unavailable for use in the decomposition of aboveground plant materials. While immobilized nitrogen actually flows to microorganisms, these organisms are largely engaged in decomposition of root residues. The model does not contain a separate compartment for microbial nitrogen, so the decomposing organisms are included in the dead-root compartment. Therefore, immobilization is represented as a flow to dead roots, from which the nitrogen will subsequently be either transferred to stable organic matter or remineralized. Our attempt at quantification of the immobilization process is described below.

Estimation of immobilization rates is difficult. There is a general relationship with plant nitrogen concentration, and we would not expect immobilization to occur with nitrogen contents above 2.5%. Power (1968) found a small net mineralization at 30°C using grass roots with 1.4% N, but in this case immobilization probably occurred early, followed by mineralization.

He also found that grass roots containing 0.84% N added to soil containing 55 μg nitrate nitrogen per g soil at a rate of 0.0025 g root/g soil immobilized about 8.4 μg N/g soil over a 3-wk period at 30°C. Assuming a bulk density of 1.2, we can calculate that $3.5 \cdot 10^{-6}$ g nitrate N was immobilized per g root per day for each g/m^3 of nitrate N in the soil. Converting to 25°C (Fig. 7.2), we have a value of $2.6 \cdot 10^{-6}$ for roots with 0.84% N. Assuming zero immobilization in roots with 2.5% N, we can linearly interpolate/extrapolate to estimate an immobilization rate for roots of any nitrogen concentration, in terms of g N per g root per day per g/m^3 of soil nitrate nitrogen. Daily immobilization is then calculated by multiplying this rate by root biomass and nitrate concentrations.

The immobilization of ammonium nitrogen during decomposition of root material is calculated in the same manner as for nitrate. While the solution concentrations of soil ammonium are lowered by adsorption on the exchange complex, ammonium nitrogen is also known to be preferentially utilized by the decomposing organisms (Jansson et al., 1955; Broadbent and Tyler, 1962). On the assumption that these effects will balance, the same rate calculations are used for both sources. Provision is made for altering the relative immobilization rates of the two sources through an "adsorption coefficient," presently set to 1.0. All basic immobilization flows are adjusted for temperature and moisture prior to final flow calculation.

It should be recognized that for ammonium immobilization this is a reverse flow of the transfer from dead root to soil ammonium described above. Net transfer is the difference between these flows and may result in net gain or loss of either dead-root nitrogen or soil ammonium.

7.2.4 Producer–Decomposer Submodel

This submodel is very simple and in the present form is intended only to drive the nitrogen model for testing purposes prior to incorporation into ELM. Relative growth rates appropriate to the system being simulated are set externally. These growth rates vary for the various phenological stages of the plant as reflected in a phenological index. At present the phenological index in this submodel is determined only by time and not by the more sophisticated methods involving temperature and soil-water history of the system (Chap. 3; Sauer, 1973; French and Sauer, 1974). Growth rates are then modified by temperature, soil water, and the nitrogen concentration of the live-top biomass. The effect of nitrogen concentration on relative rates is shown in Figure 7.6. Note that the nitrogen concentration resulting in a given decrement of relative growth rate decreases as the season progresses. With a phenological index that varies from 0

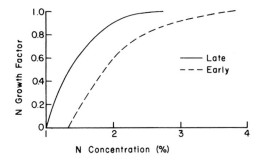

Fig. 7.6.   Effect of live top nitrogen concentration on growth rate; basic relative growth rates are modified by this factor prior to calculation of daily growth

at the dormant stage before growth starts to 14 at full maturity, the "early" curve on Figure 7.6 is used for all phenological stages below 5.0 and the "late" curve, for stages later than 10.0. For phenological stages between 5.0 and 10.0 a linear interpolation between the two curves is performed.

Root growth is based on relative growth rates that are modified by top biomass. The assumption is that root growth requires photosynthate, which in turn requires top biomass; thus relative root growth rates in this submodel are reduced when top biomass is very low. Root death is assumed as a constant proportion of root growth plus a pulse or major dieback of live roots occurring during the dormant season. Live roots increase during the growing season, and dead roots decrease as decomposition proceeds. Root growth rates are also modified by nitrogen concentration of the roots, increasing as concentration increases. Water and temperature limitations are imposed on root growth. Top death rates are controlled by the phenological stage and soil water. As water becomes limiting, the death rate increases. Initial daily rates for both root and litter decomposition are included as a function of nitrogen concentration:

$$D = 0.013 + 0.0027\,(C_N)$$

where D is the decomposition per day per unit of material present and C_N is the nitrogen concentration expressed as a percentage of the total biomass in the compartment. Temperature and water corrections are then applied to determine the daily decomposition.

7.3 Model Output

The simulations that have been run with the simplified producer–decomposer model represent hypothetical systems, although growth rates and many of the initial conditions are similar to those of the Pawnee Site. Temperature and soil-water driving variables were selected to test certain reactions of the model rather than to simulate any particular site or year. The model output selected for presentation and discussion represents three runs using the same temperature and soil-water patterns. Soil temperatures and soil-water suctions programmed for these runs are shown in Figure 7.7. The soil-water pattern selected represents

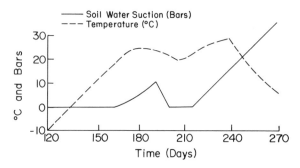

Fig. 7.7. Temperature and soil water regimes imposed in simulation shown

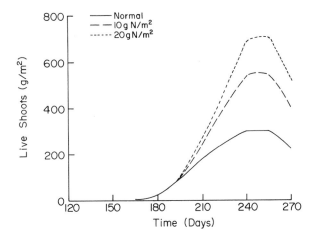

Fig. 7.8.   Live-top biomass production in simulated normal and fertilized systems

a condition of low stress through most of the season with severe drought imposed in the late stages.

The three runs represent different initial levels of soil ammonium and nitrate. In the first or "normal" run initial conditions were set to provide a total of 2.3 g N/m² in these forms. In the additional runs soil inorganic nitrogen was increased by 10 and 20 g N/m² at the start of the growing season to simulate fertilized systems. This amount was equally divided between the ammonium and nitrate forms and distributed by depth in the same proportion as the live-root biomass. The output available, even from these few selected runs, is voluminous, and presentation of the complete set is not feasible. Hopefully, those sections selected for presentation will provide some insight into performance of the model.

The simulated live top biomass production is shown in Figure 7.8. Nitrogen deficiency was severely limiting growth in the normal system where peak live-top biomass reaches 300 g/m². This increased to 540 and 692 g/m² for the 10-g and 20-g N/m² addition treatments, respectively. The additional nitrogen resulted in 24 g and 19 g biomass per g N added for the two treatments, values typical of nitrogen-fertilizer responses on grasses where soil water does not severely limit production (Whitehead, 1970). Thus the top biomass production seems to be responding in a satisfactory manner, except that small treatment differences might have been expected earlier in the season. The rapid late-season decline reflects increased death rates as a result of soil-water stress.

The general pattern of nitrogen concentration with time as shown in Figure 7.9 seems realistic and compares well with published results (Whitehead, 1970; Rumburg, et al.. 1964) and with data collected from the grassland sites. Late-season differences may be somewhat smaller than expected, but examination of other model variables show that late-season growth restrictions due to plant nitrogen levels occurred even on the 20-g N/m² run in which case we would expect concentration differences to be small. The total nitrogen accumulation in the live tops tends to lead biomass accumulation by a few days. This is consistent

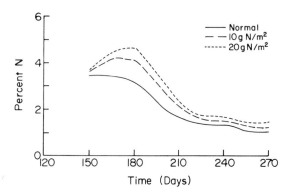

Fig. 7.9. Simulated live-top nitrogen concentration in normal and fertilized systems

with published results such as shown by Rumburg et al. (1964), but the effect was much less extreme than that found by Green, as quoted by Whitehead (1970).

Root-biomass changes, as a result of nitrogen additions, were much less marked than top biomass changes. At day 240 (late August) live-root biomass was about 20% higher in both nitrogen-supplemented simulations, and dead-root biomass averaged about 8% higher. Total root biomass increased about 11%, agreeing satisfactorily with results shown by Lauenroth (1973) and Lauenroth and Sims (1976). It must be recognized that dead-root biomass is dynamic. Thus the 8% increases in biomass resulted from a 30% increase in dead-root production, partially offset by a 5% increase in decomposition.

The simulated live-root nitrogen patterns shown in Figure 7.10 are intriguing. Total nitrogen in this compartment increases early in the season as the nitrate and ammonium in the soil are taken up. During the period of maximum biomass

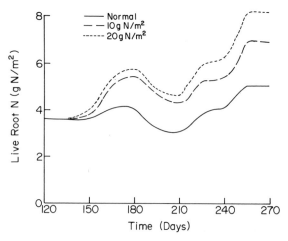

Fig. 7.10. Total nitrogen contained in live roots in simulated normal and fertilized grassland systems

accumulation translocation to the tops exceeds uptake, and live-root nitrogen declines. This decline reflects a decreased concentration as live-root biomass is increasing slightly during this period. Later in the season live-root nitrogen increases. The high values at end of season reflect both increased root biomass due to seasonal growth and a late-season increase in concentration. The increased nitrogen comes from uptake from the soil mineral pool and from translocation from live tops. The dormant season or winter dieback, set at 50% for these simulations, will result in a drop in total live-root nitrogen from these high values prior to end of season. We are unaware of data either substantiating or refuting the drop in live-root nitrogen during the period of maximum growth. Accurate measurements of this type are difficult to obtain for perennial grasses, due to the masking effects of large amounts of nonliving tissue and the difficulty in distinguishing live roots from dead. However, a consideration of the transfers involved indicate that such a pattern probably does occur. If it did not, either mineralization of organic forms of nitrogen would have to be very rapid at the time of maximum growth, or ammonium and nitrate nitrogen concentrations would have to build up substantially prior to this time. Neither of these effects is generally observed in the field and, in fact, this seems unlikely. We regard this distribution as an example of a useful hypothesis for experimental testing resulting from use of the model.

The simulated dead-root nitrogen concentrations are shown in Figure 7.11. Dead-root biomass decreases during the season due to decomposition exceeding root death, but in the normal system there was very little change in root nitrogen concentration. In the fertilized system nitrogen concentration in the dead-root compartment increases rapidly as soon as temperatures exceed 10°C. This increase reflects the rapid immobilization of ammonium and nitrate forms that occurs with high concentrations of these forms. Concentration remains relatively stable later in the season.

The percentage distribution of the two 10-g/m² increments at approximately

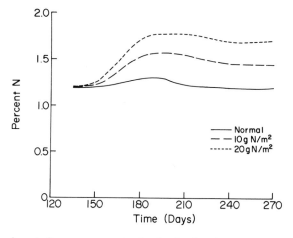

Fig. 7.11.   Dead-root nitrogen concentration in simulated normal and fertilized systems

Table 7.1. Percentage distribution at peak live
top biomass (day 240 or late August) of two
successive 10 g N/m² increments of added
nitrogen in simulated fertilized systems

| | Increment (%) | |
Compartment	1	2
Ammonium	0.7	1.5
Nitrate	7.2	32.9
Live roots	12.0	8.6
Dead roots	23.4	12.3
Soil organic	11.9	5.4
Live tops	38.6	34.8
Litter	6.0	4.5

the date of peak live top biomass is shown in Table 7.1. The values shown as increment 1 represent the difference between the amount in a given compartment in the 10-g N/m² simulation run and the amount in that compartment in the normal run expressed as a percentage of the 10-g total. Increment 2 represents the difference between the 20-g and 10-g N/m² runs similarly expressed. This would be analogous to the usual method of expressing percentage recovery in the various parts of a system when conducting a fertilizer experiment. Note that about 39% of the first increment was recovered in the live tops. This is lower than most recoveries quoted by Whitehead (1970), but he deals largely with systems having lower root:shoot ratios and multiple harvest systems, both of which tend to increase recovery in the tops. Power (1967) recovered 40% in the tops over a 2-yr period during which 50 g N/m² were applied to *Bromus inermis* L. He recovered 25–30% in the roots and 5–10% in the mineral form. These are generally consistent with our simulation results, except for our higher soil nitrate level resulting from the second increment. Total simulated recovery in the roots was about 25% in the first increment and 21% in the second.

The pool in this system with the longest turnover time is in the soil organic matter. Only 12% and 5% of the first and second increments, respectively, were transferred to this pool by day 240 (late August). Thus the bulk of the applied nitrogen still remains active in the system in forms that can be expected to influence future production and/or decomposition rates.

7.4 Discussion

We constructed a dynamic simulation model of nitrogen flow in a grassland ecosystem. We consider the model to be at a plateau in development. Although many of the flow functions may be simplistic and must be regarded as preliminary, the model is internally consistent. Model output agrees with experience and published results, and responses to the perturbations imposed reflect many characteristics of the real system.

The biological accuracy and experimental validity of many of the flow functions are open to question. In several cases revisions are being prepared as a result of further experience in modeling these systems and in light of additional relevant data found in the literature or collected by grassland biome researchers. In some cases these are simply coefficient changes that can easily be incorporated. In other cases increased realism will require more complex mechanisms that amount to virtual submodels with their own internal problems such as stability and generality. A specific example would be the transfer from dead roots to stable organic matter and ammonium. A more realistic section would provide for the narrowing of the C/N ratio as decomposition proceeds and perhaps for an initial division of soluble or labile nitrogen within the root from the more resistant fraction, as suggested by F. E. Clark (personal communication). Another example would be a more sophisticated treatment of mineralization of soil organic matter, perhaps utilizing the concepts of Stanford and Smith (1972), involving a readily decomposable fraction. When developed, such subsections can be assembled, inserted in the structure, and tested as part of the whole. Without a structure such as the present form, the assembly of such submodels into a coherent unit could be a well-nigh hopeless task. Further improvement may also be possible by adding sections to provide for additional flows that are important in some systems or under certain conditions. Leaching, microbial denitrification, and other gaseous losses are important examples. Such additions would add generality to the model.

Recognizing such limitations, the present model is very useful. The usual field experiments provide information on the levels of various state variables at certain points in time, but often provide an unsatisfactory basis for an understanding of the system dynamics. In contrast, the model response can be easily examined in detail. Such examination allows us to forumulate hypotheses of system response that are experimentally testable such as the predicted pattern of live-root nitrogen concentration. Alternatively, if we wish to examine the plausibility of a hypothesis of system dynamics, it can be inserted in the model and tested for compatibility with the rest of the system.

Simple and plausible explanations for apparently complex or even contradictory phenomena can often be formulated in terms of system dynamics. Porter (1969) has pointed out instances from the literature where both increased and decreased biomass of grass roots have been observed as a consequence of nitrogen-fertilizer application. When we consider the dynamics of the model in which both root decomposition and production rates are a function of nitrogen concentration and in which immobilization and uptake processes compete for a common pool, moderate differences in live- and dead-root distributions at the time of high soil N concentration could result in such apparent contradictions.

The soil–plant nitrogen system has been extensively studied. Traditionally such studies have involved either detailed laboratory examination of processes or field experiments that measure one or more state variables a few times per season. As processes are understood in greater detail and field measurements are proliferated, it has become increasingly difficult to relate new results to the total system, except perhaps in a qualitative manner. A model such as this provides a mechanism whereby the effects of individual rates and processes can be quantitatively evaluated in terms of overall system performance.

Acknowledgments.   The authors gratefully acknowledge the contributions of Dr. C. V. Cole of the USDA/ARS and Drs. Robert Woodmansee and George Cole of the Natural Resource Ecology Laboratory for their contributions to various aspects of the work reported here.

This chapter reports on work supported in part by National Science Foundation Grants GB-31862X, GB-31862X2, GB-41233X, and BMS73-02027 A02 to the Grassland Biome, U.S. International Biological Program, for "Analysis of Structure, Function, and Utilization of Grassland Ecosystems."

References

Alexander, M.: Nitrification. In: Soil Nitrogen. Am. Soc. Agron. Monogr. 10. Bartholomew, W. V., Clark, F. E. (eds.). Madison, Wisc., 1965, pp. 309–343

Broadbent, F. E., Tyler, K. B.: Laboratory and greenhouse investigation of nitrogen immobilization. Soil Sci. Soc. Am., Proc. **26**, 459–462 (1962)

Copley, P. W., Reuss, J. O.: Evaluation of biological N_2 fixation in a grassland ecosystem. US/IBP Grassland Biome Tech. Rep. No. 152. Fort Collins: Colorado State Univ., 1972, 90 pp.

French, N., Sauer, R. H.: Phenological studies and modeling in grasslands. In: Phenology and Seasonality Modeling. Lieth, H. (ed.). New York: Springer-Verlag, 1974, pp. 227–236

Frere, M. H., Jensen, M. E., Carter, J. N.: Modeling water and nitrogen behavior in the soil–plant system. In: Proc. 1970 Summer Sim. Conf., Vol. 2. La Jolla, Calif.: Simulation Councils, 1970, pp. 746–749

Geist, J. M., Reuss, J. O., Johnson, D. D.: Prediction of nitrogen requirements of field crops. II. Application of theoretical models to malting barley. Agron. J. **62**, 385–389 (1970)

Harmsen, G. W., Kolenbrander, G. J.: Soil inorganic nitrogen, In: Soil Nitrogen. Amer. Soc. Agron. Monogr. 10. Bartholomew, W. V., Clark, F. E. (eds.). Madison, Wisc., 1965, pp. 43–93

Jansson, S. L., Hallam, M. J., Bartholomew, W. V.: Preferential utilization of ammonium over nitrate by microorganisms in the decomposition of oat straw. Plant Soil **6**, 382–390 (1955)

Justice, J. K., Smith, R. L.: Nitrification of ammonium sulfate in a calcareous soil as influenced by combinations of moisture, temperature, and levels of added nitrogen. Soil Sci. Soc. Am., Proc. **26**, 246–249 (1962)

Lauenroth, W. K.: Effects of water and nitrogen stresses on a shortgrass prairie ecosystem. Ph.D. thesis, Colorado State Univ., Fort Collins, 1973, 115 pp.

Lauenroth, W. K., Sims, P. L.: Evapotranspiration from a shortgrass prairie subjected to water and nitrogen treatments. Water Resources Res. **12**, 3, 437–442 (1976)

McLaren, A. D.: A need for counting microorganisms in mineral cycles. Enciron. Lett. **5**, 143–154 (1973)

Porter, L. K.: Nitrogen in grassland ecosystems. In: The Grassland Ecosystem: A Preliminary Synthesis. Range Sci. Dep. Sci. Ser. No. 2. Dix, R. L., Beidleman, R. G. (eds.). Fort Collins: Colorado State Univ., 1969, pp. 377–402

Power, J. F.: The effect of moisture on fertilizer nitrogen immobilization in grasslands. Soil Sci. Soc. Am., Proc. **31**, 223–226 (1967)

Power, J. F.: Mineralization of nitrogen in grass roots. Soil Sci. Soc. Am., Proc. **32**, 671–673 (1968)

Reuss, J. O., Cole, C. V.: Simulation of nitrogen flow in a grassland ecosystem. In: Proc. 1973 Summer Computer Sim. Conf., Vol. 2, La Jolla, Calif.: Simulation Councils, 1973, pp. 762–768

Reuss, J. O., Copley, P. W.: Soil nitrogen investigation on the Pawnee Site, 1970. US/IBP Grassland Biome Tech. Rep. No. 106. Fort Collins: Colorado State Univ., 1971, 44 pp.

Rumburg, C. B., Sneva, F. A.: Accumulation and loss of nitrogen during growth and maturation of cereal rye. Agron. J. **63**, 311–313 (1970)

Rumburg, C. B., Wallace, J. D., Raleigh, R. J.: Influence of nitrogen on seasonal production of dry matter and nitrogen accumulation from meadows. Agron. J. **56**, 283–286 (1964)

Sabey, B. R., Frederick, L. R., Bartholomew, W. V.: The formation of nitrate from ammonium nitrogen in soils. IV. Use of the delay and maximum rate phases for making quantitative predictions. Soil Sci. Soc. Am., Proc. **33**, 276–278 (1969)

Sauer, R. H.: PHEN: A phenological simulation model. In: Proc. 1973 Summer Computer Sim. Conf., Vol. 2. La Jolla, Calif.: Simulation Councils, 1973, pp. 830–834

Stanford, G., Smith, S. J.: Nitrogen mineralization potential in soils. Soil Sci. Soc. Am., Proc. **36**, 465–472 (1972)

Stanford, G., Frere, M. H., Schwaninger, D. H.: Temperature coefficient of soil nitrogen mineralization. Soil Sci. **115**, 321–323 (1973)

Whitehead, D. C.: The role of nitrogen in grassland productivity. Commonwealth Bureau of Pasture and Field Crops Bull. No. 48. Hurley, Berkshire, England: 1970, pp. 5, 71, 73, 125.

8. Simulation of Phosphorus Cycling in Semiarid Grasslands[1]

C. Vernon Cole, George S. Innis, and J. W. B. Stewart

Abstract

A simulation model of the phorphorus cycle in semiarid grasslands was developed and tested. When used with appropriate data sets for biotic and abiotic driving variables at the Pawnee (Colorado) and Matador (Saskatchewan) Sites, this model predicted plant and decomposer uptakes and turnover rates of the principal phosphorus compartments. Daily phosphorus requirements for plant and decomposer uptake are taken from pools of labile inorganic phosphorus in each soil layer. Mineralization of labile organic phosphorus and leaching of water-soluble forms from standing dead biomass and litter are the main sources of replenishment of the labile inorganic pools. Phosphorus solubility, soil-water content, and rates of diffusion of phosphorus through soil are the primary controls on rates of uptake by the active fraction of the live-root biomass.

Model dynamics are more sensitive to soil parameters than to plant parameters. Simulation results indicate rates of decomposer uptake four to five times greater than plant uptake in semiarid grasslands. Simulated phosphorus concentrations in live plant tops are highly responsive to the pattern of seasonal rainfall, which agrees well with published data. The most critical informational needs revealed by model development and operation are in the areas of activity and morphology of roots and the rates of mineralization of organic phosphorus as affected by soil depth.

8.1 Introduction

Phosphorus performs an essential role in all forms of life on earth. Mammalian respiration, photosynthesis in algae or green leaves, or microbial turnover in decomposing litter require adequate levels of P in specialized biochemical forms. Thus the distribution and flow of P forms provide a valuable index to the levels and kinds of biological activity in an ecosystem. Phosphorus is second only to nitrogen as a nutrient limiting primary productivity in semiarid grasslands.

Although there is considerable information concerning chemical and physiological transformations of P, little attempt has been made to incorporate this information into comprehensive simulation models of P cycling. Difficulties in modeling arise from gaps in the information available. For example, there are

---

[1] Supported in part by the Colorado State University Experiment Station and published as Scientific Series Paper No. 2051.

excellent data on P flux across root membranes, but not on root distribution and activity.

A major difficulty in assessing the interaction of the many processes involved in P cycling in the soil–plant system is the lack of sufficient time to observe effects—particularly when the system is subject to periodic disturbances, such as tillage, crop production, and fertilization. Undisturbed ecosystems, for example, native grasslands, offer a valuable opportunity because the system is observed at "equilibrium," particularly for P cycling since there are no substantial gains or losses from the system on the time scale considered. The levels and distribution of labile inorganic and organic forms of P in the profile remain essentially unchanged over the long term, although they display seasonal dynamics. Both the seasonal dynamics and the steady-state distribution reveal the interaction of abiotic factors with producers, consumers, and decomposers. As the P cycle in grasslands in controlled by metabolic processes of biota as well as by abiotic factors, it is an example of the homeostatic condition postulated by Pomeroy (1970) in describing strategies of nutrient cycling.

The objective of this study was to construct a simulation model describing the processes of P flow in a grassland, indicating constraints and variables that control these processes. Certain processes are more clearly defined than others; assumptions were made where adequate information was not available. These assumptions were chosen to produce consistent flows throughout the system.

Early versions of the P model (Anway et al., 1972; Reuss, et al., 1973) stressed inorganic P solubility and plant uptake processes. These versions provided reasonable simulations of P concentrations in aboveground plant and root biomass. However, the overall cycle was not balanced because of inadequate representation of decomposer function and organic phosphate turnover.

Detailed studies of P budgets at the Matador Site (latitude 50°42′N, longitude 107°43′W) in Saskatchewan (Halm, 1972; Halm et al., 1972) indicated that organic phosphate turnover rate was limiting P cycling in this grassland. In August 1973 Stewart participated in a month of intensive model development in Fort Collins, where concepts of organic phosphate turnover were incorporated into the P model. Discussions with Francis Clark (ARS, USDA Nitrogen Laboratory) were useful in developing hypotheses for microbial uptake and turnover.

Data from both the Pawnee Site (Table 8.1) and Matador Site were used in the development of the P model, and the model was tested through simulation of conditions at both sites.

8.2 Review of Literature

Phosphorus cycling studies at the Matador Site in Saskatchewan have provided the most complete data available for evaluating compartments and flows in semiarid grasslands. These studies, reported in papers by Halm (1972) and Halm et al. (1972), included evaluation of soil inorganic and organic P, rooting activity of grasses as a function of soil depth, and estimates of microbial turnover rates. These investigators concluded that labile organic P plays a more important role in the availability of P in grasslands than has been generally recognized. These

Table 8.1. Distribution of soil phosphorus fractions in the Ascalon soil at the Pawnee Site

Depth (cm)	Total P[a]	Resin extractable	HCO$_3$ extractable P		Total organic P
			Inorganic	Organic	
0.0–2.5	301 ± 13	26.3 ± 0.9	22.5 ± 0.9	8.5 ± 0.4	88 ± 5
2.5–7.5	286 ± 4	15.7 ± 1.3	16.4 ± 0.6	8.0 ± 0.3	78 ± 8
7.5–12.5	265 ± 7	9.5 ± 1.6	10.9 ± 1.3	8.7 ± 0.6	102 ± 15
12.5–20.0	232 ± 5	4.1 ± 1.0	4.8 ± 0.9	9.3 ± 0.3	110 ± 4

[a] All values μg P/g soil. All values are means ± SD. For conversion to volume basis, bulk density = 1.22.

concepts were incorporated into the Matador version of the P model (Stewart et al., 1973).

A mechanistic simulation model of P and potassium flux in a tallgrass prairie was developed by Sheedy et al. (1973). This model, although at a low degree of resolution in terms of the number of processes represented, predicted time series of values of aboveground plant P that agreed well with observed values in northeastern Oklahoma. They viewed the model as a series of hypotheses concerning ecosystem structure and function.

Greenwood et al. (1971) incorporated a generalized theory of fertilizer response of vegetable crops into a mathematical model that related crop yield to nutrient requirements, the levels of the nutrients in the soil, and the ability of the soil to absorb them. They recognized their model as a steady-state conceptualization of a system that is actually constantly changing. However, by carefully fitting parameters for each crop, they obtained excellent agreement between predicted and observed yields.

Scaife and Smith (1973) described a dynamic model of P uptake and growth of lettuce. Phosphorus uptake was based on the soil solution P concentrations.

Mathematical studies of ionic diffusion through soil water films verified by thorough experimentation contributed to better understanding of the movement of nutrients to plant roots (Olsen and Watanabe, 1963, 1966; Olsen et al., 1965; Olsen and Kemper, 1968). A concept arising from this treatment is the close relationship between P concentration in the soil-solution phase and levels of labile inorganic P in the solid phase.

The role of the plant in P uptake was evaluated by examination of the anatomical features of roots and of the distances from the roots to nutrients. Using this approach, Bar-Yosef et al. (1972a) developed a high resolution simulation model of plant P uptake. In a second paper (Bar-Yosef, et al. 1972b) the predictions of their model were compared with the experimental results obtained with a new procedure for evaluating nutrient uptake in the field. In addition to indicating the kinds of soil and plant parameters required for simulations at this level of resolution, they determined the sensitivity of the system to the various parameters and concluded that computed P uptake is more sensitive to variations in soil parameters than plant parameters.

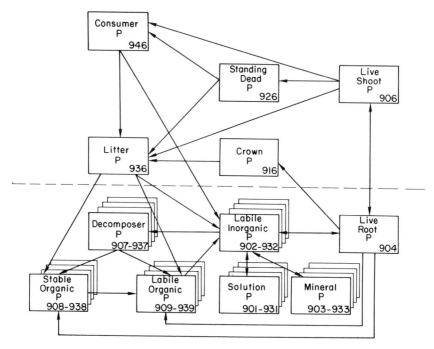

Fig. 8.1. Forrester diagram of the phosphorus submodel: dashed line separates the belowground from the aboveground compartments; in the state-variable designation the first digit, (9), identifies the P submodel, the second digit in the belowground variables refers to the soil layer, and the third refers to the P form or location [i.e., the figures 902, 912, 922, and 932 designate the amounts of labile inorganic P (last digit of 2) in the first to fourth soil layers (second digits of 0, 1, 2, 3, respectively)]

8.3 Model Structure and Development

The organization of the P submodel is illustrated in Figure 8.1. There are five aboveground state variables and a compartment containing live root P for all four soil layers. An additional six belowground state variables are replicated for each of four soil layers giving a total of 30 state variables. The system is about 7% connected, that is, the number of flow paths divided by the number of possible paths is 0.07. Units for the aboveground and root P variables are g P/m² and for the remaining 24 belowground state variables are g P/volume of soil in a layer that is 1 m · 1 m · h, where h is layer thickness in meters.

Phosphorus uptake by plant roots is simulated to move P from each soil layer into the common pool of live root P that has uniform P concentrations with depth. Reported root P concentration differences with depth were attributed to differences in the fractions of live versus dead, relative age, and differentiation in juvenile, nonsuberized, and suberized forms.

A major assumption in the present model is that there is no vertical P movement in the soil except within plant root systems. Downward movement of inorganic P in percolating waters is limited except in sandy soils. Organic phosphates are much more mobile, however, and may move downward appreciably (Hannapel et al., 1964a,b). Downward movement is assumed to have little effect on model results in simulations of seasonal dynamics for periods up to 5 yr.

Daily root uptake from the solution pool during periods of high root activity and near ideal temperature and water status may be 50 times the amount of P in the solution pool. The solution pool is replenished from the labile pool at a potential rate of 250 times daily. Since the model is intended to operate on 1 day or longer time steps, P uptake is modeled to occur directly from the pools of labile inorganic P. Flow rates per unit weight of live root in a given soil layer are controlled by the values of P concentration in solution, labile inorganic P, and soil temperature and water content.

8.4 Soil Phosphorus Solubility and Plant Uptake

A basic assumption for this model is that P supply to the plant is explained by the diffusion of water-soluble forms and that auxiliary mechanisms such as contact exchange (Olsen and Kemper, 1968) are unnecessary. As the P concentration at the root surface is reduced by plant uptake, a concentration gradient is established in the adjacent soil-water films, and phosphate ions flow toward the root. Replenishment of soluble P in a depleted zone will depend on the supply of P from soluble phosphate minerals, phosphate adsorbing surfaces, and organic P mineralization.

8.4.1 Phosphorus Solubility

Chemical forms and solubility relationships of P in soil have been thoroughly reviewed by Larsen (1967) and Olsen and Flowerday (1971). Phosphorus solubility is complicated by common ion, ion association, and pH effects (Cole and Olsen, 1959a). Phosphorus solubility increases with the amounts of phosphorus adsorbed in the monolayer region on the surfaces of soil minerals. Differences in P solubility between soils of varying texture are explained by variations in surface area and adsorption capacity of the soils (Cole and Olsen, 1959b).

Solution P equilibrates rapidly with the labile fraction of the adsorbed P. Values for the rate constant, c_{21}, for P flow from labile to solution range from 50 to 250 per day, depending on phosphorus status and soil texture (Olsen and Watanabe, 1966). These values were determined by anion-resin equilibration techniques that lower the solution P concentration to very low values, essentially stopping the reverse reaction.

Labile inorganic P is defined as that fraction of the soil P that can enter solution by isoionic exchange within a given time span. Methods for determination include anion-resin equilibration (Amer et al., 1955) and isotope-exchange techniques (Olsen and Watanabe, 1963; Sadler, 1973). The capacity factor, b, is the ratio of soil labile inorganic P concentration to the concentration of P in solution as shown in the following equation:

$$x_{902}(x_{901}^{-1} \; \theta) = c_{12}c_{21}^{-1} = b$$

where x_i is the amount (g P/m^2) in the state variable i; θ is the volumetric water content; c_{21} is the rate constant for P transfer into solution; and c_{12} is the rate constant for precipitation of soluble P. The P solubility of the Ascalon soil at the Pawnee Site, for example, is correlated with labile P ($r^2 = 0.71$) over a 10-fold range. This soil has a relatively high P solubility with $b = 17.9$, whereas the Sceptre clay at the Matador Site has a very low P solubility with $b = 1500$.

8.4.2 Phosphorus Diffusion

A theoretical background for the evaluation of P-supplying capacity and soil-textural effects on P diffusion is given by Olsen and Watanabe (1963). The porous system diffusion constant, D_p, is calculated from the bulk solution diffusion constant by the following equation:

$$D_p = D_0 \, \gamma \, (L/L_e)^2 \, \theta,$$

where D_0 is the diffusion coefficient of P in bulk solution in cm^2/s; γ is a dimensionless coefficient for diffusivity reduction due to ionic interaction and increased viscosity of water near mineral surfaces; $(L/L_e)^2$ is the tortuosity factor; and θ is the volumetric water content. The porous diffusion coefficient is experimentally evaluated by either transient or steady-state diffusion experiments (Olsen et al., 1965) and ranges from 0.4 to 15 $\times$ 10^{-7} cm^2/s. Diffusion coefficients are estimated from the clay content by the relationship found by Olsen and Watanabe (1963) for those soils in which they have not been determined directly. Their equation relating P uptake by plant roots to the diffusion coefficient and the phosphate capacity, b, showed that the P uptake rate is proportional to $(bD_p)^{1/2}$ when all other factors are constant and the absorption period is 1 day or less. The soil properties affecting P diffusion are expressed in a parameter, c_k, used in calculations of root uptake.

8.4.3 Effects of Soil Water

The effects of volumetric soil-water content on P uptake by plant roots are calculated separately from the diffusion effects for simplicity. Olsen et al. (1961) found that on each of four soils P uptake rate by corn roots was a linear function of θ between $\frac{1}{3}$ bar and 15 bars suction. No appreciable P uptake was found at suctions greater than 15 bars. This is because $(L/L_e)^2$ extrapolates to zero at finite values of θ, indicating that the water films break up.

To compute the effect of soil water on relative P-uptake rates, a linear function (e_w) is used, varying from 0 to 1 as suction varies from 15 to $\frac{1}{3}$ bar.

8.4.4 Root Morphology

Information on root distribution, activity, and morphology are essential for reliable simulations of P-uptake processes. Dahlman and Kucera (1965) in studies using ^{14}C estimated a 25% annual turnover of roots in tallgrass systems. Power (1968) observed similar rates of turnover for native grasslands near Mandan, North Dakota. One difficulty in evaluating root activity is distinguishing between live and dead tissues. Halm (1972) determined distribution of the active root system of the four major plant species at Matador by ^{32}P-labeling techniques and compared the results with those obtained with root-washing techniques. Ninety percent of the active roots were found in the top 30 cm of soil.

Singh and Coleman (1973, 1977) evaluated functional root biomass and translocation of photoassimilated ^{14}C in natural *Bouteloua gracilis*-dominated plots at the Pawnee Site. They found an average of 62% of the total root biomass to 60 cm was metabolically functional. These estimates were further confirmed by Ares (1976), who differentiated juvenile, nonsuberized, and suberized fractions in root samples from *Bouteloua gracilis*-dominated plots that were hand-separated into friable (dead) and nonfriable (live) forms. Ares also provided estimates of root density and surface area. The mean diameter of juvenile roots of *Bouteloua gracilis* is less than 0.05 mm. Diameters of nonsuberized roots range from 0.115 to 0.158 mm and suberized roots from 0.148 to 0.202 mm. Estimates of root density of nonfriable suberized roots in a May sampling were 23.1 (0 to 5 cm) and 32.4 (5 to 15 cm) cm/cm^3 soil.

Ares's criteria for distinguishing the nonsuberized roots included size, color, and rates of water absorption. The juvenile and nonsuberized forms, probably the only ones active in nutrient uptake, account for approximately one-third of the total root biomass depending on season, soil depth, and soil water conditions. His estimate agrees with the 36.2% value determined by Clark (1977) in ^{15}N studies on a blue grama prairie. In this model one-third of the total root biomass was considered active in P uptake.

8.4.5 Root Uptake

The principal driving force for P uptake by plant roots is the transfer across membranes at the root surface. Metabolic energy is expended to move P across the plasmalemma into the cytoplasm where the P concentration may be 50–100 times higher than that in the surrounding soil solution. This flow establishes a concentration gradient in the soil solution which drives P diffusion and desorption processes. Root uptake rates in a well-stirred nutrient solution vary widely with the physiological activity and nutrient status of the plant. Uptake-rate dependence on P concentration is analogous to two simultaneously operating carrier mechanisms (Hagen et al., 1957; Hagen and Hopkins, 1955). An equation of the Michaelis–Menten form for two sites of P uptake has been useful in predicting the P uptake rates over a wide range of P concentrations in solution (Carter and Lathwell, 1967):

$$\text{Uptake rate, } c_{24} = \frac{u_a}{1 + \dfrac{k_a}{P_c}} + \frac{u_b}{1 + \dfrac{k_b}{P_c}},$$

where P_c is the P concentration at the root surface. Values of c_{24} derived from studies in which the solution is kept well stirred are higher than those observed in soil studies.

Calculations of P uptake in the field indicate a range of actual uptake rates from 1.5×10^{-5} to 6.2×10^{-5} g P $\cdot$ g root$^{-1} \cdot$ day^{-1} (Quirk, 1967; Sayre, 1948). Growth analysis studies by Loneragan and Asher (1967) and Loneragan (1968) confirmed that uptake rates of 3.1 and 6.2×10^{-5} g P $\cdot$ g root$^{-1} \cdot$ day^{-1} will maintain adequate P concentration in plants growing at rates of 8% and 10% per day, respectively.

A limit on P concentration in the roots is achieved in the model by a return flow to accord with numerous observations that beyond certain P concentrations in the root cytoplasm the plasmalemma becomes leaky. This return flow is calculated as an exponential function of root P concentration ranging from 0 to 1 between the values of 0.10% and 0.50% P (dry wt).

The studies of Carter and Lathwell (1967) on temperature-dependence of P uptake by corn roots support the position that chemical processes are rate-limiting for both high- and low-concentration mechanisms. The temperature function e_t, as used in the nitrogen model by Reuss and Innis (1977; also Fig. 7.2 of this volume), was used to calculate temperature effects on P uptake. This function adequately described the responses of *Bouteloua gracilis*, a warm-season grass, while a function with a lower temperature optimum was necessary to simulate responses of *Agropyron dasystachium*, a cool-season grass. Wide differences in adaptation of P uptake to temperature were found among various sedge species (Chapin, 1974).

The effects of mycorrhizal fungi on P uptake have been documented for a variety of plant species (Mosse, 1973). Increases in P uptake by plants infected with vesicular–arbuscular mycorrhizae are facilitated by the greater root–soil contact afforded by extensive development of fungal hyphae (Sanders and Tinker, 1971). Uptake by the fungi is metabolically controlled as it is for root uptake but extends to lower concentrations of P in the soil solution (Bowen et al., 1977). These properties of mycorrhizal uptake indicate that P nutrition of plants benefits mainly when diffusion rates through soil to roots are too slow or at very low P concentrations in the soil solution. Mycorrhizal effects have not been included in this model, but are considered a promising avenue for future development (Bowen and Rovira, 1973).

Major determinants of plant P uptake are the demands made by growth, functioning of plant parts, and the external P supply. Other nutrients influence P uptake through their effects on plant growth and metabolism. Nitrogen–phosphorus interactions provide a good example as effects of N on P uptake have been observed in many plant species under a wide range of cultural conditions.

Nitrogen stimulation of rates of P uptake is a consequence of metabolic changes in the plant as demonstrated by Cole et al. (1963). In these studies P uptake by corn roots was highly correlated with root N content. In this model P-uptake rates were adjusted for simulated root N content. The function, e_N, ranging from 0 to 2, is calculated by linear interpolation for root N values between 0.8% and 1.6%, assuming a base level of root uptake at 1.2% N and a doubling of this rate at 1.6% N, and is used to influence root uptake as shown below. Cessation of uptake below 0.8% is consistent with the observation that grass roots grow little, if at all, at such low N content. Further testing with other plant species at several grassland sites will be necessary to validate these limits.

The root uptake calculation requires values of active root biomass (b_r) for each soil layer, the uptake rate constant c_{24}, the parameter relating to capacity and diffusion properties c_k, and the effects of soil water, temperature, and root N content (e_w, e_t, and e_N, respectively), as well as a function, f, allowing for return flow to the soil.

$$\text{Root uptake from each soil layer} = b_r c_{24} c_k e_w e_t e_N (1 - f)$$

The multiplication of these factors was deemed the most suitable means of expressing the combined effect of these determinants of P uptake.

The calculation of P-uptake days, obtained by summing the daily products of e_t and e_w, is a useful diagnostic for evaluating climatic effects on P availability for each soil layer.

8.5 Plant Translocation and Utilization of Phosphorus

The dynamics of P transport within plants are controlled by the activities of meristematic tissues within newly developing plant organs. This results in a pattern of nutrient flow into developing leaves in early growth stages and export at later stages. Active recycling of P has been demonstrated using radioactive P (Koontz and Biddulph, 1957; Biddulph et al., 1958; Tanaka, 1961; Sosebee and Wiebe, 1973). Tanaka's studies showed the patterns of translocation of carbohydrates and major nutrients throughout a developing rice plant.

Phosphorus-transport processes among developing organs are simplified in this model. Plant translocation is depicted as an interaction of the developing compartments of aboveground biomass and the P supply in the roots (push–pull).

An operational constraint on P translocation is that a minimum P level must be maintained within the exporting tissue. Thus no translocation of P from roots can be expected for grasses below a level of approximately 0.05% P. Furthermore, P translocation from roots is considered proportional to root P concentration between this lower limit and an upper limit (~0.5% P) above which increases in root P concentration do not further enhance translocation.

The relative effect of root P concentration (P_r) on translocation to live tops (t_r) varies exponentially from 0 to 1 as concentration varies from 0.05% to 0.5% P as expressed by:

$$g(P_r, a, b) = \exp[(P_r - a)b] - 1$$

where $a = 0.0005$ and $b = 154$.

The buildup of excess P in live tops, for example, during a period of adequate P supply in the roots but limited top growth, is prevented by incorporation of a return flow term, which is calculated as

$$g \ (P_t, \ 0.0025, \ 92.4),$$

where P_t is top P concentration.

The phenological index of a given plant species, as supplied by the producer model, is the best guide available to indicate growth responses to abiotic factors. This index regulates P translocation, in part allowing rapid flows during juvenile stages and little flow when there is no growth. The species-specific function p_h ranges from 0 to 1 over the range of the phenological index (Fig. 8.2). This function adjusts for the different periods of peak translocation in cool- and warm-season grasses.

Net daily translocation flows are calculated as the product of the maximum daily rate, live top biomass, the effects of root P concentration, phenological stage, and one minus the return flow. Values for the maximum daily transloca-tion rate are 1×10^{-3} and 2.2×10^{-3} g P $\cdot$ g live-top biomass$^{-1} \cdot$ day^{-1} for the Matador and Pawnee Sites, respectively. Phosphorus translocation to crowns is calculated similarly, except that a constant designating the active fraction of crowns is included and lower limits on the return flows are provided.

During senescence of plant parts, P is conserved by translocation to more active organs, as described earlier. Thus in computing the flow of P from live to dead tops only a fraction of P in the live tops is transferred. Depending on the level of P concentration in the live tops, there is minimum conservation at low levels of P ($\sim$0.1% P) and up to 50% conservation at high levels (0.5% P). The flow from live to dead tops is calculated using the top death rate from the producer submodel multiplied by live-top P concentration and the fraction conserved. Transfer of P from crowns to litter is calculated similarly. The flow from shoots to litter represents a direct transfer when plant parts are severed by insect feeding, animal trampling, or hail damage, and all of the P contained in these tops is transferred to litter.

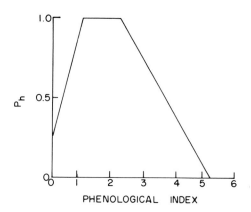

Fig. 8.2.   Effect of plant phenological stage (SPHEN) on translocation of P from roots to live tops (PHENP)

8.6 Decomposer Uptake and Turnover of Phosphorus

Large amounts of P are required by decomposers during plant residue decomposition (Chang, 1940). The ratio of cellulose decomposition to organic P formed varied from 64 to 115 g C/g organic P depending upon conditions of incubation. Hannapel et al. (1964a,b) followed the synthesis of organic P by [32]P- and ^{14}C-labeling experiments and found that a significant proportion of the P redistributed through the soil was microbially synthesized. They also found that for every part of organic P of plant origin there were three parts organic P derived from microbial mobilization of soil inorganic P.

The P uptake by microbial and other decomposers of plant material is the largest annual P flow in this submodel. Unfortunately, data to document this flow are scarce. Flow rates are derived from estimated seasonal populations of decomposers, turnover rates, and scattered data on P composition of various decomposer species.

The soil fungal and bacterial populations at the Matador Site were closely monitored in 1968 and at intermittent stages in other years (Clark and Paul, 1970). Estimates of P composition of fungi and bacteria from Alexander (1961) and Porter (1947) were later confirmed by analysis of fungal and bacterial populations in agar-containing solutions from the Sceptre soil.

The effect of low soil temperature on the viability of microorganisms has been reported by Biederbeck and Campbell (1971), who observed freeze–thaw cycle to reduce the microbial population to a much greater extent than did a freeze alone. Studies of microbial populations at the Matador Site (Babiuk and Paul, 1970; Clark and Paul, 1970; Shields and Paul, 1973; McGill et al., 1973) suggested that turnover of the microbial population is not rapid, averaging once or twice per year. The activity of a ^{14}C-labeled microbial population did not change appreciably over the growing season, but decreased markedly during winter.

Microbial biomass at the Pawnee Site was estimated by direct microscopic techniques (Doxtader, 1969) and from ATP measurements (Sparrow, 1973). From these data it was estimated that annual P turnover through decomposers was 3–4 g/m^2.

The relationship between decomposer P uptake and P concentration in the soil solution is modeled by the same form of relationship used for root uptake. The values for maximum uptake for the two mechanisms were selected to provide annual turnover rates that agreed with estimates from field studies. Parameter values that provide consistent dynamics indicate that decomposers are 20 times more efficient in P uptake per unit biomass than are roots. Values of P concentrations at which the respective sites are half saturated were selected at one-half the corresponding values for the root uptake functons, reflecting the competitive nature of microbial uptake. Thus when P supplies become limiting, microbial uptake continues at the expense of root uptake.

Temperature and moisture effects are combined multiplicatively as in root uptake processes, except that the moisture function from the decomposer model (see Fig. 6.5) is used instead of e_w, which is based on root function. The values for microbial biomass are supplied from the decomposer submodel. The control on decomposer P content is effected by using a return flow. Values of this function range exponentially from 0 to 1 between decomposer P values P_d of 0.5% and 1.0%: g (P_d, 0.005, 138.6). Thus net decomposer uptake is calculated as the product of a rate constant, microbial biomass, temperature and moisture parameters, and one minus the return flow.

8.7 Mineralization of Organic Phosphorus

The transformation of organic phosphates from plant, animal, and decomposer residues into inorganic phosphate is essential to complete the cycle of P utilization in natural systems. Much of the P in these residues is in the form of P esters in many chemical forms. The major chemical groups containing P are nucleic acids, phospholipids, inositol phosphates, sugar phosphates, and nucleoproteins. The stability of these compounds to enzymatic hydrolysis varies widely; the resistant compounds, especially the inositol phosphates, have long residence time in soils. Large amounts of P are contained in organic residues with residence times ranging from 350 to 2000 yr (Halm, 1972).

From the latest review of soil organic P (Halstead and McKercher, 1975) transformations of organic P associated with plant nutrition remain uncertainly defined; P present as nucleic acids and phospholipids may be readily hydrolyzable and hence presumably readily available to plants, whereas P present as inositol P is not.

Thompson et al. (1954) studied mineralization of C, N, and P in a group of soils developed over a wide range of vegetative and climatic conditions. Mineralization under field conditions was estimated from losses due to cultivation at paired cultivated and virgin sites for each soil. These results were compared with those of laboratory incubation over a 3-mo period. The net losses of organic C, N, and P from cultivated soils averaged 33%, 32%, and 24%, respectively. Moreover, relative rates of P mineralization under laboratory incubation exceeded those of C and N mineralization. Organic P mineralization was highly correlated with organic N and C in the incubation experiments. The average ratio of N/P mineralized was 7.6, and the ratio of C/P mineralized was 80. Based on this relationship and estimates of C turnover, annual P mineralization rates of 3.0 and 6.4 g P/m² could be predicted at the Pawnee and Matador Sites, respectively.

Soil pH was a modifying influence in the above studies. Stability of soil organic P to mineralization decreased as soil pH increased. This effect is consistent with the finding of lower organic phosphorus content in calcareous soils than in acid soils (Greb and Olsen, 1967).

Several methods are used for the estimation of soil organic P, ranging from direct determination by acid and alkaline extraction to measurement of the increase in inorganic P after ignition. In a comparative study Hance and Anderson (1962) found that methods involving extraction of the soil with alkali before and after acid treatment caused less hydrolysis and gave the highest values with acid soils. Ignition values by the procedure of Saunders and Williams (1955) were usually greater than the highest extraction values, but in more than half the soils the differences were small and unimportant. They considered the ignition procedure the most useful because it is rapid and easy to perform, with no direct evidence that ignition gives erroneously high results.

Much of the total soil P is organic P in surface horizons. In Chernozemic soils, McKercher (1966) found 25–55% of the total P in this form. Of the organic P in these soils, 10–30% was inositol phosphate, 1–2% was phospholipids, and less than 1% was nucleic acids. The remaining 70–90% was unknown (McKercher and Anderson, 1968).

Organic P levels in Ascalon fine sandy loam at the Pawnee Site increase from 88 ppm at the surface to 110 ppm at 20 cm depth, accounting for 29% and 47%, respectively, of the total P at these levels (Table 8.1). The Sceptre clay at the Matador Site attains an organic P level of 332 ppm in the surface 15 cm (56% of total P), which decreases to 108 ppm in the 15–30-cm layer (25% of total P).

Soil organic P at the Matador Site was differentiated into stable organic P (compounds resistant to weak acid hydrolysis) and labile organic compounds [compounds easily hydrolyzed by weak acids and mostly soluble in 0.5 M $NaHCO_3$; Stewart et al. (1973)]. Faunal P, microbial P, and plant P were divided into these two categories. Large seasonal decreases in the soluble organic P corresponded to increases in aluminum-bound and calcium-bound inorganic P. In these soils soluble organic P was 4–20 times the amount of inorganic P extracted by the same solution. The increase in organic P coincided with the period of rapid growth, high microbial activity, and low amounts of litter P. Dormaar (1972) found a large buildup of organic P during the winter months. These results confirmed the dynamic nature of a portion of the soil organic P.

Stewart et al. (1973) hypothesized that rapid turnover of organic P was the result of higher phosphatase activities associated with increased root activity and microbial populations and very low soil solution P levels. Bacterial numbers and phosphatase levels were highly correlated in a field experiment. Phosphatase enzymes are adsorbed and stabilized in soils, and their activity is an indication of past biological activity and potential for P mineralization. A reliable and widely used phosphatase assay procedure uses nitrophenyl phosphate as the substrate (Tabatabai and Bremner, 1969). This procedure avoids complications due to soil adsorption of the freshly hydrolyzed P by measuring the release of nitrophenol, which is less subject to adsorption losses in the soil.

Plant roots provide one source of phosphatases and seem to adapt to low levels of P by the extracellular production of increased amounts of these enzymes. The influence of 16 grasses and forbs on soil phosphatase activity was investigated by Neal (1973). The growth of dominant, codominant, and increaser species did not significantly alter the phosphatase activity already established in the soil. However, the presence of species of plants classed as invaders, because they often become the dominant species on overgrazed sites, significantly increased soil phosphatase activities. This suggests that the relative ability to mineralize organic P may be a factor to be considered in plant successional relationships.

8.8 Model Simulation Results

A simulation model incorporating the preceding concepts was used to study P cycling at two native grassland ecosystems: the Pawnee and Matador Sites. This facilitated comparison of model dynamics under different vegetation [warm-season grass (*Bouteloua gracilis*) vs. cool-season grass (*Agropyron smithii*)], edaphic (fine sandy loam soil vs. heavy clay-textured soil), and climatic conditions. Satisfactory simulation results under widely differing driving variables constitute validations of the hypotheses on which the model is based. Conversely, the simulation of P flow through decomposers and organic P forms had to be balanced by alteration of hypotheses and parameters to agree with observed flows and thereby indicated areas for further research. Both the Pawnee Site and Matador Site versions of the model are complete since driving variables representative of conditions at the respective sites were used.

8.8.1 Pawnee Site Results

Daily soil temperature and moisture values for the 1971 and 1972 seasons at the Pawnee Site were determined by interpolation from observed values. The 1972 temperature and moisture regimes were particularly suitable for demonstrating model dynamics because there were two periods of active growth separated by a drought. Above- and belowground biomass flows for *B. gracilis* were simulated in a simple producer submodel. Simulated P flows for the 1972 season at the Pawnee Site are itemized in Table 8.2. Simulated P concentrations in *B. gracilis* live tops and roots for 1972 are shown in Figure 8.3a. Graphs of soil water to 20 cm and live-top biomass are shown in Figure 8.3b. Phosphorus concentrations in live tops were initially high and declined during active growth. Phosphorus concentrations in roots declined as P was translocated to the tops but recovered when soil water became adequate for P uptake.

An unexpected result of these simulations was the sensitivity of top P concentration to periods of adequate soil water for P uptake. The close relationship between water status and P translocation from roots to tops is illustrated in Figure 8.4. After each major rainfall there is a corresponding peak in the rate of P translocation to the tops. The realism of this aspect of model operation is confirmed by field measurements in native grasslands of eastern Montana (Black and Wight, 1972). Their studies showed P concentrations in aboveground herbage were highly correlated with available water throughout several seasons. Within a week after major rainfall events top P concentrations increased.

Some key aspects of P cycling are illustrated by the dynamics of the labile inorganic pool. Figure 8.5 shows labile inorganic P in the top 20 cm of soil, cumulative plant uptake, decomposer uptake, and inputs from organic P mineralization. These simulation results indicate a complete turnover of labile inorganic P during one season. Decomposer uptake is nearly five times higher than root uptake. Root uptake and mineralization are balanced under unstressed conditions, as shown by the small changes in labile inorganic P.

Table 8.2.   Cumulative phosphorus flows (in g P/m²) for simulation runs of 210 days at the Pawnee Site and 204 days at the Matador Site (belowground flows accumulated for all soil layers)

Phosphorus flows	Pawnee	Matador
Plant P flows		
labile inorganic:live roots (902–932:904)	0.526	1.200
live roots:tops (904:906)	0.201	0.717
live roots:crowns (904:916)	0.107	0.314
live tops:dead tops (906:926)	0.148	0.658
dead tops:litter (926:936)	0.148	0.433
crowns:litter (916:936)	0.070	0.158
live roots:stable organic (904:908–938)	0.064	0.098
live roots:labile organic (904:909–939)	0.150	0.229
litter:labile inorganic (936:902)	0.047	0.149
litter:labile organic (936:909)	0.143	0.448
litter:stable organic (936:908)	0.047	0.149
Decomposer P flows		
labile inorganic to decomposers (902–932:907–937)	2.457	3.160
decomposer to labile organic (907–937:909–939)	2.059	3.206
decomposer to stable organic (907–937:908–938)	0.363	0.565
Organic P turnover		
stable organic to labile organic (908–938:909–939)	0.486	0.676
input to labile organic	2.838	4.559
input to stable organic	0.474	0.812
labile organic to labile inorganic (909–939:902–932) (mineralization)	3.020	4.283

Root uptake from the various soil layers is presented in Table 8.3. Nearly two-thirds of the root uptake comes from the top 7.5 cm. Values for annual P uptake were divided by the number of uptake days and live-root biomass to calculate peak uptake rates per gram of root. Comparable rates were obtained at both sites. Higher uptake from surface layers is a result of higher P concentrations in soil solution in these layers (see Table 8.1).

The response of the P cycle to increased supplies of labile inorganic P is indicative of the amount of P stress. To test this for the Pawnee Site, additions of 1.5 g P/m² to the surface layer and 1.0 g P/m² to the second layer were simulated at the beginning of the season. These additions resulted in a 1.9-fold increase in the plant and decomposer available P in the top 20 cm of soil. Simulated-plant root uptake in response to this treatment (Table 8.4) increased from 0.53 to 0.64 g for the season; top P concentrations showed a midseason peak of 0.30% compared with 0.24% P. These simulation results are consistent with field observations that under unstressed conditions P supplies are not seriously limiting.

8.8.2 Matador Site Results

Rooting depth by native species at the Matador Site (Sceptre clay) was somewhat greater than that at the Pawnee Site (Ascalon). The Matador simula-

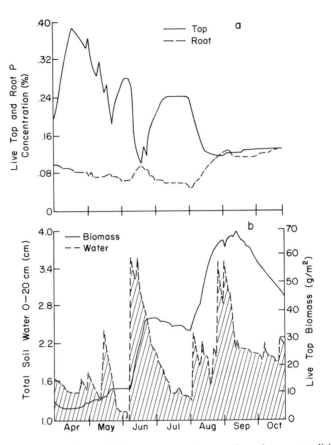

Fig. 8.3.   Simulation results with 1972 temperature and moisture conditions at the Pawnee Site: (a) P concentrations in live tops and roots of *Bouteloua gracilis;* (b) live-top biomass (dry wt) of *B. gracilis* and cm total soil water in the 0–20-cm layer

tion was, therefore, structured for six 5-cm belowground layers. Root-distribution studies with radioactive P indicated that over 90% of the root activity was in the upper 30 cm. Daily temperature values at these six soil layers for the 1969 season were estimated from observations at 5-, 20-, and 50-cm depths on 12 dates. Soil-water values were estimated from observations for 0–15-cm and 15–30-cm depths on 18 dates. Daily biomass flows at the Matador Site were generated from a matrix of values of live tops, dead tops, litter, crowns, and root biomass reported for eight dates.

The major phosphorus flows are itemized in Table 8.2 for comparison with the Pawnee Site simulation. When comparing the P uptake at the two sites, 0.526 g P/ m^2 at the Pawnee Site and 1.20 g P/m^2 at the Matador Site, the biomass turnover should be considered. The total annual turnover of tops, crowns, and roots was 536 and 1233 g/m^2 for the Pawnee and Matador Sites, respectively; thus the P requirements at the Pawnee and Matador Sites were almost identical (~1 mg P/g biomass).

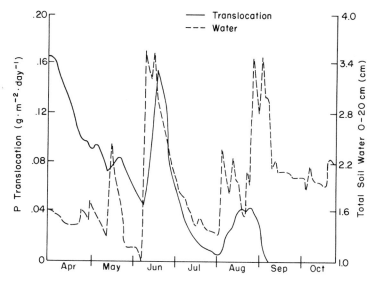

Fig. 8.4.   Simulation results of phosphorus translocation rates and cm total soil water (0–20 cm) for the 1972 season at the Pawnee Site

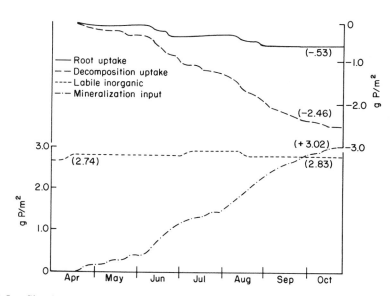

Fig. 8.5.   Simulated seasonal changes in labile inorganic P (0–20-cm layer) and cumulative loss to root uptake and decomposer uptake and gain from mineralization of organic P for 1972 at the Pawnee Site

Table 8.3.   A comparison of simulated values for root uptake and organic phosphorus mineralization at Pawnee and Matador Sites

Soil layer (m)	Live root biomass (g/m²)	Annual root P uptake (g/m²)	Labile inorganic P (g/m²)	Mineralization (g/m²)	Uptake days (days)	Uptake rate ($\mu g\ P \cdot g^{-1} \cdot day^{-1}$)
Pawnee Site						
0.000–0.025	111	0.283	0.813	0.651	28.0	91
0.025–0.075	80	0.155	0.958	0.831	25.5	76
0.075–0.125	23	0.045	0.605	0.784	31.2	63
0.125–0.200	12	0.043	0.365	0.753	64.9	55
Total	226	0.526	2.741	3.019		
Matador Site						
0.00–0.050	145	0.533	0.232	1.539	47.8	77
0.05–0.100	145	0.344	0.248	1.088	47.4	50
0.10–0.150	47	0.092	0.248	0.719	47.2	41
0.15–0.200	47	0.113	0.114	0.487	83.5	29
0.20–0.250	29	0.080	0.092	0.362	79.9	17
0.25–0.300	29	0.038	0.031	0.088	76.1	17
Total	442	1.200	1.010	4.283		

Phosphorus concentrations in live tops and roots are shown with live top biomass in Figure 8.6. For the Matador Site June and late August of 1969 were very dry. The P concentrations reflected limited uptake during these periods. The patterns of total plant P in live tops, dead tops, and live roots are shown in Figure 8.7. The peak in live root P in early August obtained from the favorable soil-water status then and accorded with field data (Stewart et al., 1973).

The balance between mineralization and plant and decomposer uptake is shown in Figure 8.8. The Sceptre soil had little labile inorganic P in the root zone (1.0 g P/m² in the 0–30-cm depth). Simulation results indicate a four-fold seasonal turnover of this pool. Net mineralization of 1.12 g/m² is close to the seasonal plant P requirement. Mineralization and uptake rates are so balanced that the

Table 8.4.   Effects of simulated increases in labile inorganic phophorus (g P/m²) on plant uptake and mineralization rates

P additions	Root uptake	Total seasonal aboveground plant P	Mineralization	
			Total	Net
Pawnee Site				
0.0	0.53	0.20	3.02	0.56
2.5	0.64	0.25	3.07	0.55
Matador Site				
0.0	1.20	0.72	4.28	1.12
0.5	1.51	0.90	4.31	1.01
2.5	2.79	1.54	4.35	0.77

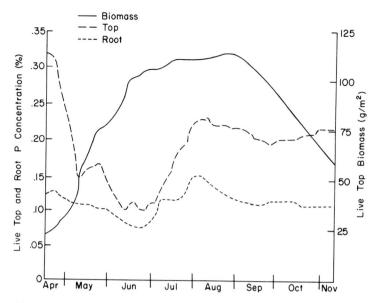

Fig. 8.6.   Simulation results of *Agropyron dasystachium* and *A. smithii* live-top and root
P concentrations (percent) and live-top biomass (dry wt) for 1969 at the Matador Site

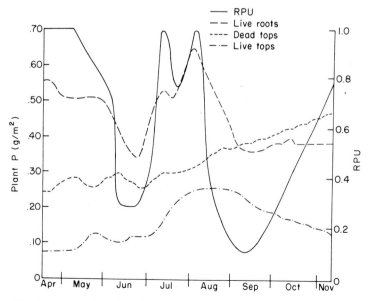

Fig. 8.7.   Simulation results of moisture status (indicated by values of RPU) and total
plant phosphorus in live tops, dead tops, and roots for 1969 at the Matador Site

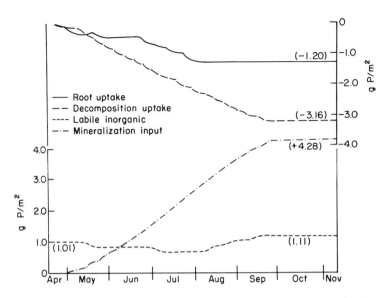

Fig. 8.8.   Simulated changes in labile inorganic P (0–30 cm) and cumulative loss to root uptake and decomposer uptake and gain from mineralization of organic P for 1969 at the Matador Site

labile inorganic P level drops to 0.68 g/m² at the end of July and recovers to 1.11 g/m² at end of season.

Simulation results for root uptake and mineralization rates for the six soil layers are shown in Table 8.3. Over two-thirds of the total root uptake is from the top 10 cm of soil. The significance of mineralization of organic P is illustrated by the dynamics of the top layer. The uptake of 0.53 g by plant roots and 0.91 g by decomposers from a pool of only 0.23 g in the top 5 cm of soil is balanced by 1.54 g of mineralized organic P.

Response of the Matador Site model was tested by additions of 0.5 and 2.5 g P/m² into the labile inorganic pool at the beginning of the season (Table 8.4). Simulated plant uptake increased 26% with the 0.5-g application and 230% with the 2.5-g application. Actual P additions required to increase labile inorganic P this much would be five times higher because of precipitation into mineral forms. Mineralization of organic P was not affected by these additions. Net mineralization, however, decreased 69%, reflecting increased decomposer uptake, which is consistent with observations of decreased phosphatase activity at increased solution P concentrations (Neal, 1973; P. Nannipierri, personal communication).

The effect of variable plant productivity on P cycling was tested by separately increasing top biomass turnover and root productivity. Three plant productivity regimes were compared with the control (Table 8.5). When both aboveground and root productivity were reduced by 66.6%, P uptake was reduced by 25% and translocation, by 35%. With reduced plant demand net mineralization also decreased, reflecting increased decomposer uptake with less plant competition. When aboveground productivity was doubled with no increase in root productivity, P uptake increased by only 9%, but translocation increased by 40%. When top biomass was doubled and root productivity increased by 133.3%, comparable to field responses to the addition of water and N (Stewart et al., 1973), P uptake increased by 57% and translocation nearly doubled.

8.8.3 Model Sensitivity

The operation of the P model should reflect the long-term stability in levels of labile inorganic and organic P evidenced in field observation in native grasslands while still allowing active dynamics. Steady-state conditions were achieved in both versions of the model by adjustment of key parameters. The most critical parameter for long-term stability was the maximum rate of mineralization of organic P. For the Pawnee Site maximum mineralization rates were varied from 0.05 day^{-1} in the top soil layer to 0.006 day^{-1} in the deepest soil layer. Apparently, factors unaccounted for are important in mineralization processes. This adjustment in mineralization rates with soil depth was not necessary for the Matador Site. Hunt (1977; also this volume, Chap. 6) found that the apparent maximum decomposition rate at the Pawnee Site decreased eightfold from upper to lower soil layers. These results indicate a promising avenue for further research.

The sensitivity of the seasonal dynamics to variation in five key parameters was systematically examined in the Pawnee Site study. The change in pool size of labile inorganic P at end of season was compared after tests where one parameter at a time was decreased to 90% and increased to 110% of the control value (Table 8.6). Cumulative values of root P uptake, translocation of P to live tops, P uptake by decomposers, and season-end decomposer P were similarly compared. Model results were more sensitive to soil parameters than to plant and decomposer parameters. Plant uptake was most affected by variation in c_k, the parameter that controls P diffusion in the soil.

Table 8.5. Simulation effects of variable plant productivity on phosphorus (g P/m²) uptake and mineralization at the Matador Site

Treatment	Plant uptake	Total seasonal aboveground plant P	Mineralization	
			Total	Net
Control	1.20	0.72	4.28	1.12
Aboveground and root productivity ⅔	0.90	0.47	4.26	0.98
Aboveground productivity × 2; root productivity unchanged	1.31	1.01	4.36	1.19
Aboveground productivity × 2; root productivity ⁴⁄₃	1.88	1.39	4.36	1.47

Table 8.6. Test of parameter sensitivity [values indicate percent increase (+) or decrease (−) resulting from a 20% increase in the respective parameter between 90% and 110% of control value]

Parameter	Phosphorus solubility constant C12	Soil P diffusion parameter CAPK	Root uptake parameter $U_{max\ b}$	Translocation rate constant C46	Microbial death rate C79
Labile inorganic P, 0–20 cm	+3.7	−4.4	−1.6	+0.1	−4.8
Root P uptake	−7.3	+16.5	+5.6	+0.1	−0.9
Live-top P	−9.6	+21.7	+8.0	+4.0	−1.0
Decomposer P uptake	−2.9	−0.2	−0.2	0.0	+11.3
Decomposer P	−2.2	−0.7	0.0	−0.5	+7.4

8.9 Conclusions

The objective of this study—to develop a P-simulation model that would operate under a wide range of abiotic and biotic inputs—was accomplished. The model allowed us to integrate effects of temperature, moisture, soil properties, plant phenology, and decomposition of organic matter to a degree not hitherto possible. The application of this model to two grassland sites and comparison of P flows with field measurements pinpointed gaps in knowledge of processes such as mineralization. For instance, in the field investigation of P cycling at the Matador Site, Halm et al. (1972) recognized the importance of organic P flows in the supply and cycling of plant P, but were unable to quantify the flows. The P-simulation model predicts these values and emphasizes their importance for balancing P cycling through the ecosystem. Likewise, the simulation of P cycling at the Pawnee Site provides a quantification of organic P flows.

New research has been initiated in organic P mineralization, relationship of microbial P turnover to microbial biomass, influence of phosphatases on P mineralization, and other variables in response to the informational gaps revealed in model development. However, the model in its present form provides a means of comparison between sites. The model has already been successfully applied to simulate P cycling in a tallgrass prairie ecosystem (W. J. Parton and C. V. Cole, unpublished manuscript). It should be possible to use the conceptual framework provided by the model to compare the salient aspects of P cycling between grasslands and other ecosystems.

The model in its present form is a submodel in ELM that provided a wider range of driving variables from abiotic, producer, decomposer, and consumer submodels. Discussions of the P model in the context of ELM are found in Chapters 1, 9, and 10.

Acknowledgments. The authors wish to express their thanks to John Denton, Phosphorus Laboratory, USDA, for his assistance in manuscript preparation, to Drs. George Cole and Rudolph Bowman of Natural Resource Ecology

Laboratory, for assistance in computer programming, and to the many others at the Natural Resource Ecology Laboratory who assisted in the development of this chapter.

This chapter reports on work supported in part by National Science Foundation Grants GB-31862X, GB-31862X2, GB-41233X, and BMS73-02027 A02 to the Grassland Biome, U.S. International Biological Program, for "Analysis of Structure, Function, and Utilization of Grassland Ecosystems" and by the Canadian National Research Council, International Biological Programme.

References

Alexander, M.: Introduction to Soil Microbiology. New York: John Wiley and Sons, 1961, 172 p.

Amer, F., Bouldin, D. R., Black, C. A., Duke, F. R.: Characterization of soil phosphorus by anion exchange resin adsorption and P^{32} equilibration. Plant Soil **6**, 391–408 (1955)

Anway, J. C., Brittain, E. G., Hunt, H. W., Innis, G. S., Parton, W. J., Rodell, C. F., Sauer, R. H.: ELM: Version 1.0. US/IBP Grassland Biome Tech. Rep. No. 156. Fort Collins: Colorado State Univ., 1972, 285 pp.

Ares, J.: Dynamics of the root system of blue grama (*Bouteloua gracilis* (H.B.K.) Lag.) J. Range Mgmt. **29**, 208–213 (1976)

Babiuk, L. A., Paul, E. A.: The use of fluorescein isothiocyanate in the determination of bacterial biomass of grassland soil. Can. J. Microbiol. **16**, 57–62 (1970)

Bar-Yosef, B., Kafkafi, U., Bresler, E.: Uptake of phosphorus by plants growing under field conditions: I. Theoretical model and experimental determination of its parameters. Soil Sci. Soc. Am., Proc. **36**, 783–788 (1972a)

Bar-Yosef, B., Kafkafi, U., Bresler, E.: Uptake of phosphorus by plants growing under field conditions: II. Computed and experimental results for corn plants. Soil Sci. Soc. Am., Proc. **36**, 789–793 (1972b)

Biddulph, O., Biddulph, S., Cory, R., Koontz, H.: Circulation patterns for phosphorus, sulfur and calcium in the bean plant. Plant Physiol. **33**, 293–300 (1958)

Biederbeck, V. O., Campbell, C. A.: Influence of simulated fall and spring conditions on soil system. I. Effect on soil microflora. Soil Sci. Soc. Am., Proc. **35**, 474–479 (1971)

Black, A. L., Wight, J. R.: Nitrogen and phosphorus availability in a fertilized rangeland ecosystem of the Northern Great Plains. J. Range Mgmt. **25**, 456–460 (1972)

Bowen, G. D., Rovira, A. D.: Are modelling approaches useful in rhizosphere biology? In: Modern Methods in the Study of Microbial Ecology. Rosewall, T. (ed.). Swedish IBP, 1973, pp. 443–450

Bowen, G. D., Bevege, D. I., Mosse, B.: Phosphate physiology of vesicular-arbuscular mycorrhizas. In: Endotrophic Mycorrhizae. Mosse, B., Tinker, P. B. (eds.). New York: Academic Press, in press, 1977

Carter, O. G., Lathwell, D. J.: Effects of temperature on orthophosphate absorption by excised corn roots. Plant Physiol. **42**, 1407–1412 (1967)

Chang, S. C.: Assimilation of phosphorus by a mixed soil population and by pure cultures of soil fungi. Soil Sci. **49**, 197–210 (1940)

Chapin, F. S., III: Phosphate absorption capacity and acclimation potential in plants along a latitudinal gradient. Science **183**, 521–523 (1974)

Clark, F. E.: Partitioning of added isotopic nitrogen in a blue grama grassland. In: The Belowground Ecosystem: A Synthesis of Plant-Associated Processes. Range Sci. Dep. Sci. Ser. No. 26. Marshall, J. K. (ed.). Fort Collins: Colorado State Univ., in press, 1977

Clark, F. E., Paul, E. A.: The microflora of grassland. Adv. Agron. **22**, 375–435 (1970)

Cole, C. V., Olsen, S. R.: Phosphorus solubility in calcareous soils. I. Dicalcium phosphate activities in equilibrium solutions. Soil Sci. Soc. Am., Proc. **23,** 116–118 (1959a)

Cole, C. V., Olsen, S. R.: Phosphorus solubility in calcareous soils. II. Effects of exchange phosphorus and soil texture on phosphorus solubility. Soil Sci. Soc. Am., Proc. **23,** 119–121 (1959b)

Cole, C. V., Grunes, D. L., Porter, L. K., Olsen, S. R.: The effects of nitrogen on short-term phosphorus absorption and translocation in corn *(Zea mays)*. Soil Sci. Soc. Am., Proc. **27,** 671–674 (1969)

Dahlman, R. C., Kucera, C. O.: Root productivity and turnover in native prairie. Ecology **46,** 84–89 (1965)

Dormaar, J. F.: Seasonal pattern of soil organic phosphorus. Can. J. Soil Sci. **52,** 107–112 (1972)

Doxtader, K. G.: Microbial biomass measurements at the Pawnee Site: Preliminary methodology and results. US/IBP Grassland Biome Tech. Rep. No. 21. Fort Collins: Colorado State Univ., 1969, 16 pp.

Greb, B. W., Olsen, S. R.: Organic phosphorus in calcareous Colorado soils. Soil Sci. Soc. Am., Proc. **31,** 85–89 (1967)

Greenwood, D. J., Wood, J. T., Cleaver, T. J., Hunt, J.: A theory for fertilizer response. J. Agr. Sci. **77,** 511–523 (1971)

Hagen, C. E., Hopkins, H. T.: Ionic species in orthophosphate absorption by barley roots. Plant Physiol. **30,** 193–199 (1955)

Hagen, C. E., Leggett, J. E., Jackson, D. C.: The sites of orthophosphate uptake by barley roots. Nat. Acad. Sci., Proc. **43,** 496–506 (1957)

Halm, B. J.: The phosphorus cycle in a grassland ecosystem. Ph.D. thesis, Dep. Soil Sci., Univ. Saskatchewan, Saskatoon, Canada, 1972, 170 pp.

Halm, B. J., Stewart, J. W. B., Halstead, R. H.: The phosphorus cycle in a native grassland ecosystem. In: Isotopes and Radiation in Soil–Plant Relationships Including Forestry, Proc. Symp., Vienna, December 13–17, 1972, STI/PUB/292, Vienna, IAEA 1972, pp. 571–589

Halstead, R. L., McKercher, R. B.: Biochemistry and cycling of phosphorus. In: Soil Biochemistry, Vol. 3. Paul, E. A., McLaren, A. D. (eds.). New York: Marcel Dekker, 1975, pp. 31–63

Hannapel, R. J., Fuller, W. H., Bosma, S., Bullock, J. S.: Phosphorus movement in a calcareous soil: I. Predominance of organic forms of phosphorus in phosphorus movement. Soil Sci. **97,** 350–357 (1964a)

Hannapel, R. J., Fuller, W. H., Fox, R. H.: Phosphorus movement in a calcareous soil: II. Soil microbial activity and organic phosphorus movement. Soil Sci. **97,** 421–427 (1964b)

Hance, R. J., Anderson, G.: A comparative study of methods of estimating soil organic phosphorus. J. Soil Sci. **13,** 225–230 (1962)

Hunt, H. W.: A simulation model for decomposition in grasslands. Ecology **58,** in press, 1977

Koontz, H., Biddulph, O.: Factors affecting absorption and translocation of foliar applied phosphorus. Plant Physiol. **32,** 463–470 (1957)

Larsen, S.: Soil phosphorus. Adv. Agron. **19,** 151–210 (1967)

Loneragan, J. F.: Nutrient concentration, nutrient flux, and plant growth. 9th Internat. Congr. Soil Sci., Trans. **2,** 173–182 (1968)

Loneragan, J. F., Asher, C. J.: Response of plants to phosphate concentration in solution culture: 2. Rate of phosphate absorption and its relation to growth. Soil Sci. **103,** 311–318 (1967)

McGill, W. B., Paul, E. A., Shields, J. A., Lowe, W. E.: Turnover of microbial populations and their metabolites in soil. Bull. Ecol. Res. Committee (Stockholm) **17,** 293–301 (1973)

McKercher, R. B.: The distribution and significance of various categories of organic and inorganic phosphorus in soils. Ph.D. thesis, Univ. Aberdeen, Scotland, 1966

McKercher, R. B., Anderson, G.: Content of inositol penta- and hexaphosphates in some Canadian soils. J. Soil Sci. **19**, 47–55 (1968)

Mosse, B.: Advances in the study of vesicular-arbuscular mycorrhiza. Annu. Rev. Phytopathol. **11**, 170–196 (1968)

Neal, J. L., Jr.: Influence of selected grasses and forbs on soil phosphatase activity. Can. J. Soil Sci. **53**, 119–121 (1973)

Olsen, S. R., Flowerday, A. D.: Fertilizer phosphorus interactions in alkaline soils. In: Fertilizer Technology and Use, 2nd ed. Olson, R. A., Army, T. J., Hanway, J. J., Kilmer, V. J. (eds.). Madison, Wisc.: Soil Sci. Soc. Am., Inc., 1971, pp. 153–185

Olsen, S. R., Kemper, W. D.: Movement of nutrients to plant roots. Adv. Agron. **20**, 91–151 (1968)

Olsen, S. R., Watanabe, F. S.: Diffusion of phosphorus as related to soil texture and plant uptake. Soil Sci. Soc. Am., Proc. **27**, 648–653 (1963)

Olsen, S. R., Watanabe, F. S.: Effective volume of soil around plant roots determined from phosphorus diffusion. Soil Sci. Soc. Am., Proc. **30**, 598–602 (1966)

Olsen, S. R., Watanabe, F. S., Danielson, R. E.: Phosphorus adsorption of corn roots as affected by moisture and phosphorus concentration. Soil Sci. Soc. Am., Proc. **25**, 289–294 (1961)

Olsen, S. R., Kemper, W. D., Van Schaik, J. C.: Self-diffusion coefficients of phosphorus in soil measured by transient and steady-state methods. Soil Sci. Soc. Am., Proc. **29**, 154–158 (1965)

Pomeroy, L. R.: The strategy of mineral cycling. Annu. Rev. Ecol. Syst. **1**, 171–190 (1970)

Porter, R. J.: Bacterial Chemistry and Physiology. New York: John Wiley and Sons, 1947, 1073 pp.

Power, J. F.: Mineralization of nitrogen in grass roots. Soil Sci. Soc. Am., Proc. **32**, 673–674 (1968)

Quirk, J. P.: Aspects of nutrient absorption by plants from soils. Aust. Soc. Soil Sci. Pub. No. 4, 1967

Reuss, J. O., Innis, G. S.: A grassland nitrogen flow simulation model. Ecology **58**, 379–388 (1977)

Reuss, J. O., Cole, C. V., Innis, G. S.: Nutrient subsystem: Phosphorus and nitrogen models. In: Process Studies Workshop Report. US/IBP Grassland Biome Tech. Rep. No. 220. Jameson, D. A., Dyer, M. I. (eds.). Fort Collins: Colordado State Univ., 1973, pp. 123–148

Sadler, J. M.: Influence of applied phosphorus on the nature and availability of inorganic phosphorus in a catenary sequence of Saskatchewan soils. Ph.D. thesis, Dep. Soil Sci., Univ. Saskatchewan, Saskatoon, Canada, 1973

Sanders, F. E., Tinker, P. B.: Mechanism of absorption of phosphate from soil by endogene mycorrhizas. Nature (London) **233**, 278–279 (1971)

Saunders, W. M., Williams, E. G.: Observations on the determinations of total organic phosphorus in soils. J. Soil Sci. **6**, 254–267 (1955)

Sayre, J. D.: Mineral accumulation in corn. Plant Physiol. **23**, 267–281 (1948)

Scaife, M. A., Smith, R.: The phosphorus requirement of lettuce. II. A dynamic model of phosphorus uptake and growth. J. Agr. Sci. **80**, 353–361 (1973)

Sheedy, J. D., Johnson, F. L., Risser, P. G.: A model for phosphorus and potassium flux in a tall-grass prairie. Southwest. Naturalist **18**, 135–149 (1973)

Shields, J. A., Paul, E. A.: Decomposition of ^{14}C-labelled plant material under field conditions. Can. J. Soil Sci. **53**, 297–306 (1973)

Singh, J. S., Coleman, D. C.: A technique for evaluating functional root biomass in grassland ecosystems. Can. J. Bot. **51**, 1867–1870 (1973)

Singh, J. S., Coleman, D. C.: Evaluation of functional root biomass and translocation of photoassimiliated ^{14}C in a shortgrass prairie ecosystem. In: The Belowground Ecosystem: A Synthesis of Plant-Associated Processes. Range Sci. Dep. Sci. Ser. No. 26. Marshall, J. K. (ed.). Fort Collins: Colorado State Univ., in press, 1977

Sosebee, R. E., Wiebe, H. H.: Effect of phenological development on radiophosphorus translocation from leaves in crested wheatgrass. Oecologia **13**, 103–112 (1973)

Sparrow, E. B.: Adenosine triphosphate (ATP) in grassland soil: Its relationship to microbial biomass and activity. Ph.D. thesis, Colorado State Univ., Fort Collins, 1973, 161 pp.

Stewart, J. W. B., Halm, B. J., Cole, C. V.: Nutrient cycling: I. Phosphorus. Canadian Comm. Int. Biol. Prog. (Matador project) Tech. Rep. No. 40. Saskatoon: Univ. Saskatchewan, 1973

Tabatabai, M. A., Bremner, J. M.: Use of *p*-nitrophenyl phosphate for assay of soil phosphatase activity. Soil Biol. Biochem. **1,** 301–307 (1969)

Tanaka, A.: Studies on the nutrio-physiology of leaves of rice plants. J. Faculty Agr. Hokkaido Univ. **51,** 449–550 (1961)

Thompson, L. M., Black, C. A., Zoellner, J. A.: Occurrence and mineralization of organic phosphorus in soils, with particular reference to associations with nitrogen, carbon, and pH. Soil Sci. **77,** 185–196 (1954)

9. Sensitivity Analyses of the ELM Model

R. Kirk Steinhorst, H. William Hunt, George S. Innis,
and K. Paul Haydock

Abstract

The sensitivity of an ecosystem simulation model, ELM, to variation in parameter values is studied by perturbing parameters in groups defined by a fractional factorial design followed by perturbing parameters singly and by calculating partial derivatives according to the method of Tomović. Four output variables are used as sensitivity indicators. Total actual evapotranspiration is not very sensitive to any of the 68 parameters studied. The consumer models exhibit a few sensitivities, while net primary productivity and soil respiration are sensitive to a variety of parameters. In some cases particular parameters to which the model is sensitive are identified. The model shows several sensitivities expected on the basis of knowledge of the grassland ecosystem.

9.1 Introduction

The many parameters of ELM are known with varying degrees of precision. Some have long been established in the literature; others are estimates by grassland researchers; still others have been measured empirically in Grassland Biome experiments. A complete sensitivity analysis involving the sensitivity of each output variable to changes in each parameter is impractical because of the size of the model. However, knowledge of the important parameters to which the model is sensitive directs future research since these parameters must be estimated precisely. In addition, location of parameters to which the model is either sensitive or nonsensitive does one of two things: (a) suggests that a particular part of the grassland system might be sensitive or nonsensitive or (b) if the corresponding ecosystem parameter is already known to be one to which the system is sensitive or nonsensitive, provides a kind of validation of the model (e.g., Miller, 1974).

Patten (1969) used sensitivity analysis of a relatively simple (six-compartment) model of a fisheries food web to illustrate direct and indirect effects of perturbations on the system and drew conclusions about the consequences of various hypothetical management strategies. Kaye and Ball (1969) illustrated that sensitivity analysis can be used to investigate the role of various processes in determining the dynamics of radionuclides in a tropical ecosystem. In response to criticisms about poor parameter specification in the Urban Dynamics model (Forrester, 1961), Britting and Trump (1973) adapted the model to Lowell, Massachusetts, to illustrate its functioning on a smaller scale. It was shown that the dynamics were similar to expectations. The authors felt that this provided a sensitivity test of the model. In fact, this type of exercise is a kind of validation or verification rather than sensitivity analysis.

Classical methods (Bellman, 1954; Leondes, 1973) of locating critical parameters for differential equation systems do not work for a large nonlinear difference equation model such as ELM. For small nonlinear systems, there are several examples of sensitivity analysis. Blaquiére (1966) describes a sensitivity analysis for a nonlinear system (a vacuum-tube oscillator), but it has only a few parameters. The sensitivity of a 21-compartment mosquito population model involving 10 parameters was studied by individually varying each parameter by ± 10 percent and calculating the relative change in the number of eggs expected from an emerging adult and in the ratio of sterile:fertile males present (Miller et al., 1973). Smith (1970) used the same approach to test the six parameters in a water, aquatic plant, and herbivore system model. Wiens and Innis (1974) used the same approach with an avian bioenergetics model.

For a general definition of sensitivity, consider any dynamic model

$$\dot{\mathbf{X}} = \mathbf{f}(\mathbf{X}, \mathbf{P}, \mathbf{Z}, t), \qquad \mathbf{X}(0) = \mathbf{X}_0, \tag{9.1}$$

where $\mathbf{X}$ is a vector of state variables, with the dot representing differentiation with respect to time; $\mathbf{f}$ is a vector of functions, $\mathbf{P}$ is a vector of parameters, $\mathbf{Z}$ is a vector of driving variables, and $\mathbf{X}_0$ is a vector of initial conditions. Following Kowal (1972), we could consider the sensitivity of the solution $\mathbf{X}$ to variations in $\mathbf{f}, \mathbf{P}, \mathbf{Z}$, and $\mathbf{X}_0$ and to technical matters such as numerical solution schemes. The discussion in this chapter is limited to analyses of sensitivity with respect to elements of $\mathbf{P}$. Sensitivities with respect to $\mathbf{f}$, especially, are quite difficult to deal with. This problem stems from the fact that changes in $\mathbf{f}$ may generate revisions of the hypotheses that are basic to the model. Sensitivity of a model to changes in the hypotheses borders on the comparison of distinct models, and although such a comparison might be interesting, it is not within the scope of this chapter.

We discuss two techniques for studying the sensitivity of the model: (a) the approach given by Tomović (1963) and Tomović and Vukobratović (1970) that allows a time trace of sensitivity defined in terms of partial derivatives and (b) perturbation of parameter groups using a fractional factorial design (Shannon, 1975), followed by perturbations to investigate single parameters and simple interactions between parameters. Examples of the use of the Tomović approach appear in the works of Vermeulen and de Jongh (1977) and Burns (1975). The Tomović technique defines sensitivity of output variable x (an element of $\mathbf{X}$) to parameter p as $\partial x / \partial p$. For the parameter-perturbation technique, sensitivity (s) is defined as:

$$s = \frac{\Delta x}{\Delta p} \qquad \text{or} \qquad s = \frac{\Delta x / x}{\Delta p / p}.$$

To apply any technique one must first identify those parameters that are of interest as well as the model output variables that will be used as indicators of sensitivity. Since ELM has well over 1000 parameters, and since each 1-yr simulation run costs about \$11.00, it was necessary to select a subset of the full parameter set for sensitivity analysis. This set was chosen by asking each of the seven principal contributors to choose approximately 10 parameters from his submodel. These contributors were asked to choose five parameters that they thought, using their knowledge of the model, were ones to which the system should be sensitive and to choose five others "at random." Table 9.1 lists these parameters and presents a brief description of their effect and their "normal" and "perturbed" values (as needed for the second technique of studying sensitivity).

Indicators of sensitivity are also needed to apply these techniques; those chosen were: (a) net primary productivity (n_p), the sum of gross photosynthesis minus shoot, crown, and root respiration for all producers, (b) secondary productivity (s_p), the sum of assimilation minus respiration for grasshoppers and seven mammals (cow, coyote, jackrabbit, grasshopper mouse, deer mouse, 13-lined ground squirrel, and kangaroo rat), (c) actual evapotranspiration (a_e), water lost by evaporation from the standing crop, litter, and soil and by transpiration from live shoots, and (d) CO_2 evolution from the soil (e_c), the sum of respiration from live roots, crowns, and decomposers. These indicators were chosen to span the major modeling segments and to cover in each segment an item of principal interest.

9.2 Tomović's Sensitivity Analysis

In 1963 and again in 1970 Tomović described an approach to sensitivity analysis that is fundamentally different from the incremental perturbations described below. Given a differential equation system such as Eq. (9.1), he defines the sensitivity of the ith state variable to the jth parameter as

$$s_{ij}(t) = \frac{\partial x_i}{\partial p_j}, \tag{9.2}$$

where x_i and p_j are components of the **X** and **P** vectors, respectively. Using Eq. (9.1) and assuming that the order of differentiation may be changed without affecting the results, he derives the equation

$$s_{ij}(t) = \sum_k \frac{\partial f_i}{\partial x_k} s_{kj}(t) + \frac{\partial f_i}{\partial p_j} \tag{9.3}$$

where the sum on k is over all components of **X** appearing in f_i.

Since the functions, f_i, are given by the model there is no theoretical difficulty in computing the factors and terms on the right-hand side of Eq. (9.3). However, there are a number of technical considerations associated with the application of

Table 9.1. Parameters used in sensitivity analysis, their definitions, nominal values, and perturbed values (see text).

Parameter name	Definition and immediate effect (in the direction perturbed)	Nominal value	Perturbed value	Macro-parameter
P_1	Thermal conductivity of dry soil. Decreases temporal variability in soil temperatures.	0.00070	0.00063	H
P_2	Constant assumed for g dry wt of biomass per m^2 of leaf area. Increases evaporation and decreases transpiration.	100.0	110.0	I
P_3	Depth to which soil loses water by bare soil evaporation. Increases bare soil evaporation.	4.0	6.0	A
P_4	Parameter affecting the ratio of actual to potential evaporation. Increases bare soil evaporation.	15.0	16.5	E
P_5	Maximum rate of drainage when soil water is below field capacity. Decreases drainage.	0.020	0.018	J
P_6	Parameter affecting the fraction of water loss coming as bare soil evaporation. Decreases transpiration and increases evaporation.	0.80	0.72	C
P_7	Reflectivity of the plant canopy. Increases potential evapotranspiration. Increases photosynthesis.	0.18	0.16	F
P_8	Parameter affecting the ratio of actual to potential transpiration. Decreases transpiration.	24.0	21.6	D
P_9	Fraction of shortwave radiation transmitted by the atmosphere on a clear day. Increases photosynthesis.	0.84	0.92	G
$P_{10}^{(I)}$	Water-absorptive capacity of roots in the Ith soil water layer. Reduces transpiration water loss from the middle soil water layers relative to that from the upper layers.	0.13 0.14 0.37 0.18 0.07 0.06 0.05	0.16 0.17 0.34 0.16 0.06 0.06 0.05	B
P_{11}	Nominal mortality rate of adult grasshoppers. Increases mortality.	0.00354	0.00389	H

Parameter	Description	Value 1	Value 2	Code
$P_{12}(J)$	Proportion of live shoots of producer J[a] available as food for grasshoppers. Reduces food availability.	0.850 0.850 0.750 0.200 0.050	0.765 0.765 0.675 0.180 0.045	B
P_{13}	Julian date after which grasshoppers may oviposit nondiapause eggs. Postpones egg laying.	225.0	248.0	D
P_{14}	Accumulated degree-days required for grasshopper eggs to reach diapause. Postpones diapause.	75.0	82.5	C
P_{15}	Temperature below which grasshopper eggs cease development. Hastens embryonic development.	8.0	7.2	G
P_{16}	Parameter affecting fecundity of grasshoppers. Decreases fecundity.	0.125	0.112	I
$P_{17}(I,J)$	Ratio of shoot material clipped to that consumed by grasshopper age class I for producer category J. Reduces damage to plants.			J

Age class	Producer category		
immatures	warm season grass	0.50	0.45
adults	warm season grass	1.50	1.35
immatures	cool season grass	0.50	0.45
adults	cool season grass	1.50	1.35
immatures	forbs	0.50	0.45
adults	forbs	1.00	0.90
immatures	shrubs	0.50	0.45
adults	shrubs	0.50	0.45
immatures	cactus	0.10	0.09
adults	cactus	0.10	0.09

Parameter	Description	Value 1	Value 2	Code
P_{18}	Minimum temperature for hatching of grasshopper eggs. Postpones hatching.	24.0	26.4	A
P_{19}	Proportion of assimilated food respired by grasshoppers. Increases consumption.	0.63	0.69	F
P_{20}	Ratio of forage density to grasshopper density below which sexual maturation is retarded. Increases the likelihood that sexual maturity will be prevented by a lack of food.	10.0	11.0	E

Table 9.1. Continued.

Parameter name	Definition and immediate effect (in the direction perturbed)	Nominal value	Perturbed value	Macro-parameter
$P_{21}(I)$	Ratio of active metabolic rate to the basal rate for mammal I[b/]. Increases consumption.	2.0 1.4 2.0 1.4 1.4 2.0 2.0	2.4 1.7 2.4 1.7 1.7 2.4 2.4	J
$P_{22}(I)$	Ratio of food wasted or trampled to that consumed by mammal I[b/]. Reduces waste.	0.750 0.010 0.010 0.07 0.07 0.07 0.20	0.675 0.009 0.009 0.06 0.06 0.06 0.18	B
P_{23}	Minimum value for disgestibility of cattle forage. Decreases the effect of digestibility on intake.			G
$P_{24}(I)$	Nominal home range size for consumer I[b/]. Increases home range (m^2).	10×10^4 4.5×10^4 8.0×10^4 2400.0 4000.0 7500.0 8000.0	9×10^4 5×10^7 8.8×10^4 2640.0 4400.0 8250.0 8800.0	D
$P_{25}(I)$	Expected number of offspring per birth per parent for mammal I[b/]. Increases litter size.	1.02 4.0 5.0 4.0 4.0 8.0 3.0	1.22 4.8 6.0 4.8 4.8 8.0 3.0	I
$P_{26}(I)$	Nominal death rate for mammal I[b/]. Increases death rate.	0.00080 0.00290 0.0035 0.0070 0.0038 0.0038	0.00088 0.00319 0.0038 0.0077 0.0042 0.0042	C

Parameter	Description			Class
$P_{27}(I)$	Food density below which mammal $I^{b/}$ starves. Decreases death rate.	10.0	8.0	F
		0.0130	0.0104	
		3.5	2.8	
		3.5	2.8	
		3.5	2.8	
		3.5	2.8	
$P_{28}(I)$	Twenty-day moving average of the product of insolation and maximum air temperature above which reproduction may occur in mammal $I^{b/}$. Lengthens reproductive period.	5000.0	4000.0	H
		5000.0	4000.0	
		6000.0	4800.0	
		5000.0	4000.0	
		4800.0	3840.0	
		4800.0	3840.0	
		4000.0	3200.0	
$P_{29}(I)$	Fraction of digestible intake lost as urine in mammal $I^{b/}$. Increases urine excretion.	0.060	0.066	E
		0.030	0.033	
		0.020	0.022	
		0.050	0.055	
		0.050	0.055	
		0.010	0.011	
$P_{30}(I)$	Parameter determining weight below which mammal $I^{b/}$ dies from starvation. Decreases resistance to starvation.	0.250	0.225	A
		0.110	0.099	
		0.110	0.099	
		0.030	0.027	
		0.0270	0.0243	
		0.0350	0.0315	
		0.0300	0.0270	
P_{31}	Maximum attainable ecological growth efficiency of decomposers. Increases growth of decomposers.	0.60	0.66	E
P_{32}	Parameter determining the effect of temperature on the decomposition rate. Slows decomposition.	-5.66	-5.77	B
$P_{33}(J)$	Proportion of labile constituents in the standing dead of producer $J^{a/}$. Increases the rate of decomposition of surface litter.	0.21	0.23	A
		0.21	0.23	
		0.25	0.28	
		0.21	0.23	
		0.17	0.19	

Table 9.1. Continued.

Parameter name	Definition and immediate effect (in the direction perturbed)	Nominal value	Perturbed value	Macro-parameter
P$_{34}$	Parameter determining the effect of temperature on the maintenance energy requirement of active decomposers. Increases respiration rate.	2.51	2.76	F
P$_{35}$	Rate of incorporation of surface litter into the soil. Increases the level of litter.	0.00014	0.00013	G
P$_{36}$	Parameter determining the effect of water tension on the decomposition rate. Increases decomposition between -0.5 and -14.8 bars.	0.68	0.75	J
P$_{37}$	Maximum rate of weight loss for starving decomposers. Decreases respiration rate.	0.010	0.009	D
P$_{38}$	Determines the effect of nitrogen content on the fraction of labile constituents in live shoots. Reduces the amount of easily decomposable material in flows from live shoots to litter.	1.348	1.213	H
P$_{39}$	Proportion of starving decomposers able to survive through becoming dormant. Decreases death from starvation.	0.150	0.165	C
P$_{40}$(I)	Array of parameters affecting the decomposition rate of the labile and resistant constituents of all substrates. Speeds decomposition.	0.30 8.94 0.15 4.47 0.32 9.54 0.18 5.36 0.040 1.19	0.33 9.83 0.16 4.92 0.35 10.49 0.20 5.90 0.044 1.31	I
P$_{41}$	Parameter affecting uptake of inorganic phosphorus by decomposers. Decreases uptake.	0.72	0.86	G
P$_{42}$	Parameter affecting uptake of inorganic phosphorus by roots. Decreases uptake.	1.43	1.72	F
P$_{43}$	Parameter affecting uptake of inorganic phosphorus by decomposers. Increases uptake.	0.042	0.034	H
P$_{44}$	Parameter affecting uptake of inorganic phosphorus by roots. Decreases uptake.	5.5	4.4	B

Parameter	Description			
P_{45} (I)	Parameter determining the rate of mineralization of organic phosphorus in nutrient stratum I. Increases mineralization.	0.024 0.024 0.024 0.0055	0.029 0.029 0.029 0.0066	E
P_{46}	Rate constant for the uptake of inorganic phosphorus by live roots. Increases uptake.	17.9	14.3	A
P_{47}	Parameter affecting rate of uptake of phosphorus by roots. Slows uptake.	0.327×10^{-5}	0.260×10^{-5}	C
P_{48}	Parameter affecting rate of uptake of phosphorus by decomposers. Slows uptake.	0.18×10^{-3}	0.14×10^{-4}	J
P_{49}	Parameter determining rate of translocation of phosphorus from live roots to live shoots. Decreases translocation.	0.0022	0.0018	D
P_{50}	Maximum rate of transfer from one phenophase to another. Speeds plant development.	0.20	0.22	G
P_{51}	Proportion of clear sky radiation forward scattered in the plant canopy. Decreases photosynthesis.	0.25	0.28	J
P_{52} (J)	Ratio of the removal of live shoots by grazing to net photosynthesis above which translocation and respiration of producer $J^{a/}$ are affected. Increases the effects of grazing.	0.110 0.160 0.140 0.120 0.100	0.099 0.144 0.126 0.108 0.090	E
P_{53} (J)	Minimum fraction of net photosynthate translocated from the shoots of producer $J^{a/}$. Decreases translocation to crowns and roots.	0.30 0.70 0.40 0.40 0.30	0.27 0.63 0.36 0.36 0.27	A
P_{54} (J)	Parameter controlling the timing of senescence in producer $J^{a/}$. Hastens senescence.	12000.0 12000.0 11000.0 11000.0 12000.0	10800.0 10800.0 9900.0 9900.0 10800.0	C

Table 9.1. Continued.

Parameter name	Definition and immediate effect (in the direction perturbed)	Nominal value	Perturbed value	Macro-parameter
$P_{55}(J)$	Parameter controlling the timing of vegetative regrowth of producer $J^{a/}$. Hastens regrowth in the fall.	2.0 6.0 5.0 3.0 10.0	2.2 6.6 5.5 3.3 11.0	B
$P_{56}(J)$	Quantum efficiency of producer $J^{a/}$. Increases photosynthesis.	0.070 0.040 0.060 0.050 0.0080	0.077 0.044 0.066 0.055 0.0088	H
P_{57}	Maximum fraction of crown that can flow to shoots. Decreases early spring growth.	0.0020	0.0018	D
$P_{58}(J)$	Maximum fraction of net photosynthate translocated from the shoots of producer $J^{a/}$. Reduces translocation to crowns and roots.	0.90 0.90 0.90 0.80 0.90	0.81 0.81 0.81 0.72 0.81	I
$P_{59}(J)$	Ratio of shoots to shoots plus crowns plus roots above which the flow from the crowns to shoots of producer $J^{a/}$ ceases. Decreases early spring growth.	0.040 0.020 0.060 0.060 0.040	0.036 0.018 0.054 0.054 0.036	F
P_{60}	Parameter affecting the absorption of ammonium by belowground litter. Increases absorption.	1.00	1.25	I
P_{61}	Parameter affecting the absorption of ammonium by live roots. Increases absorption.	0.10	0.25	J
$P_{62}-P_{65}$	Parameters determining the effect of inorganic nitrogen concentration on the rate of uptake by live roots. Increases uptake.	2000.0 400.0 84.0 4.8	2500.0 500.0 70.0 4.0	G

		0.00067	0.00055	A
P_{66}	Rate constant for the mineralization of soil organic nitrogen. Decreases the concentration of soil inorganic nitrogen.			
P_{67}	Parameter limiting the rate of conversion from ammonium to nitrate. Reduces nitrification.	1.25	0.80	D
P_{68}	Parameter controlling the effect of the concentration of nitrogen in belowground litter on the rate of conversion to soil ammonium. Slows mineralization.	0.45	0.55	B

a/ The five producers categories considered are warm season grass, cool season grass, forbs, shrubs, and cactus in that order.

b/ The seven consumers are the cow, coyote, jackrabbit, grasshopper mouse, deer mouse, thirteen-lined ground squirrel, and kangaroo rat in that order.

this technique to large models. First, as Eq. (9.3) indicates, s_{ij} is defined in terms of s_{kj}, $k = 1, 2, \ldots, n$. Even though only a few x_k values would usually appear explicitly in f_i, for each of those an equation such as Eq. (9.3) must be developed. Now in the equation for s_{kj}, one would expect other state variables to appear and thus require additional equations such as Eq. (9.3). Indeed, one can imagine that each state variable x_k effects each other state variable x_i via some path or paths; thus computation of s_{ij} would entail calculation of s_{kj} for each k, $k = 1, 2, \ldots, n$. For the current implementation of ELM this would amount to about 125 additional difference equations.

A second technical consideration, specifically, that of converting Tomović's approach from the differential to the difference form, is trivial. By starting with difference equations rather than differential equations in Eq. (9.1), one defines s by partial differentiation and derives the expected result—formally, an Euler approximation to Eq. (9.3).

Third, the partial differentiation required in Eq. (9.3) is occasionally tricky. For example, if p determines the value of the independent variable at which $f(x,p)$ changes from a to b, then

$$\frac{\partial f}{\partial p} = (b - a)\, \delta(v - p)$$

where δ is the Dirac delta that, while mathematically sound, is difficult to treat computationally. Each time v moves from one side of p to the other, $\partial f/\partial p$ jumps by an amount whose integral is $\pm(b - a)$.

An example of the problem of applying Tomović's approach to a general situation is illustrated by considering the effect of p_{56} (see Table 9.1) on n_p, where n_p is the accumulation of three flows for all five producer categories, that is

$$n_p = \sum_1^5 \text{gross photosynthesis}_i - \text{crown respiration}_i - \text{root respiration}_i$$
$$- \text{shoot respiration}_i.$$

The ratio $\partial n_p/\partial p_{56}$ contains the direct effects of p_{56} on photosynthesis and respiration using the equation above for n_p. Gross photosynthesis depends, among other things, on standing crop of photosynthetic material. Soil water is influenced by standing crop of photosynthetic material via transpiration, and soil temperature is influenced via the effects of shading. Soil water and soil temperature influence root dynamics and nutrient uptake, which in turn influence the respiration terms in the n_p equation. Thus even if we restrict attention to one quantum efficiency [say, $p_{56}(1)$] and if we assume $p_{56}(1)$ to have no influence on the productivity of the other producers (an erroneous assumption) and if we also assumed the above discussion of the p_{56} influences on the system to be complete (which is also erroneous), we would have to add 4 (for the terms in n_p), plus 7 (for soil-water levels), plus 13 (for soil-temperature levels), plus 8 (for N uptake), plus 4 (for P uptake) difference equations to the sensitivity calculation. As stated earlier, the extent to which the model segments are interconnected would assure that almost every state variable could conceivably be sensitive to any parameter. This problem is manifested in the Tomović approach by the appearance of a large number of state variable sensitivity indicators as unknowns in the equation for the sensitivity indicator of interest.

We conclude that problems with Tomović's approach to sensitivity analysis relegate it to a tool for analyzing model components, functional forms, and simple models.

9.3 Parameter Perturbations

In this approach a parameter is perturbed (changed from its nominal value), the model is run, and changes in the sensitivity indicators are noted. To determine the nature of interactions among parameters, more than one parameter at a time is perturbed. However, several problems arise. First, the problem of determining which of several perturbed parameters is responsible for an observed change is handled by making a number of model runs, each with a different set of parameters perturbed, and by using standard analysis of variance techniques to attribute the observed model response to particular parameters. Second, even operating with a reduced set of 68 parameters (Table 9.1), there are 2^{68} ways to perturb the set of 68, more ways than we can afford to test. Thus we arranged the parameters into 10 groups of macroparameters. The control condition for a macroparameter consists of each parameter at its control value, and the perturbed condition consists of each parameter at its perturbed value. For the macroparameter approach to lead to a meaningful analysis, the principal contributors to the model attempted to divide the parameters into groups such that parameters within a macroparameter would not affect similar or related processes. Thus if a sensitivity indicator were insensitive to a macroparameter, it should not be due to canceling effects of parameters within it. Also, if a sensitivity indicator were sensitive to a macroparameter or combination of macroparameters, the parameter causing the result could be more readily identified.

Even using macroparameters instead of individual parameters, there are 2^{10} (= 1024) combinations of control and perturbed values. Since this number of runs would be expensive, a $\frac{1}{2}^5$ fraction of these runs was made according to a fractional factorial design (Cochran and Cox, 1957; Kempthorne, 1973). The main effect of a macroparameter is defined as the difference between the average outcome of all model runs at the control level and at the treated level. First-order interaction effects, corresponding to $(\partial^2 x / \partial p_1 \partial p_2)$ terms in the Tomović definition of sensitivity, are defined as the difference between the effect of p_1 at the control level of p_2 and the effect of p_1 at the treated level of p_2. Similarly, interactions through the second, third, . . . , ninth orders are defined as differences of next lower order interactions. Including the mean (average of all model runs), the main effects, the two macroparameter interactions, and ending with the *ABCDEFGHIJ* (ninth-order or 10-way) interaction, there are

$$\binom{10}{0} + \binom{10}{1} + \cdots + \binom{10}{9} + \binom{10}{10} = 2^{10}$$

effects defined. All of these effects can be addressed with the full factorial design, but only selected ones can be addressed with a fractional factorial design. A fractional factorial design was selected to separate as far as possible main effects and first-order interactions. The design is Plan 32.10.4 in "Fractional factorial experiment designs for factors at two levels," National Bureau of Standards Applied Mathematics Series, No. 48 (April 15, 1957) with their treatment variables B, D, E, F, K equated to our macroparameters A, C, E, F, J.

The design is a full factorial in A, C, E, F, J (all treatment combinations appear). The analysis of variance, ANOVA, has one line for each combination of A, C, E, F, J. There are thus 2^{10} effects defined above and only $2^5 = 32$ lines in the ANOVA. The effects A, C, E, F, J, AC, AE, ..., $ACEFJ$ are aliased or confounded with other effects. In practice, higher-order interactions can often be assumed negligible, and the significance or nonsignificance of a particular F statistic can be interpreted as the result of a main effect or lower-order interaction. The interpretation of the results of the ANOVA is to some degree subjective. In attributing effects to particular macroparameters or interactions, one must consider the chain-like nature in the definition of effects. If the BFG interaction is significant, then the BF or FG or some related interaction or main effect is probably also significant. Strictly speaking, the F tests provide a ranking of the sensitivities on a mean square basis. Since the model simulations are nonstochastic, interpreting the F statistics as tests of significance is not warranted.

To apply the fractional factorial method, 1-yr model runs were made using the 1970 Pawnee Site initial conditions and driving variables. Light grazing was imposed. The $\frac{1}{2}^5$ fractional factorial requires running the 32 combinations of control and perturbed values for the 10 macroparameters listed in the left-hand column of Table 9.2. The macroparameters are labeled A–J, as in Table 9.1. The treatment combination indicated in Table 9.2 means that if the letter appears, that macroparameter is perturbed in that run. The treatment code labeled (1) is the control run. After the 32 model runs had been made, the output variables (n_p, s_p, a_e, and e_c) were calculated and analysis of variance was run on each (see Table 9.3). Since in many cases the higher-order effects were nonsignificant and the mean squares were not significantly different according to Bartlett's test, they were pooled into a single-error term.

9.3.1 Analysis of Results

The values of net primary production (Table 9.2) varied from 492 to 854 g dry weight $\cdot$ m^{-2} $\cdot$ yr^{-1}. Each model run led to greater production. This was not expected, but when one considers the direction in which the parameters were perturbed, the result is not surprising. Decreases in net primary production would occur if the parameters had been perturbed in the other direction.

Table 9.2.   Fractional factorial design responses

Treatment combination	n_p (g dry wt$\cdot$m$^{-2}\cdot$yr^{-1})	s_p (g C$\cdot$m$^{-2}\cdot$yr^{-1})	a_e (cm/yr)	e_c (g CO$_2\cdot$m$^{-2}\cdot$yr^{-1})
(1)	492	0.108	23.70	826
IJ	532	0.114	23.84	902
BDFG	582	0.074	23.56	768
BDFGIJ	642	0.079	23.64	835
BDEHI	658	0.114	23.64	770
BDEHJ	598	0.111	23.92	766
EFGHI	666	0.072	23.66	796
EFGHJ	612	0.070	23.92	788
CDGH	784	0.103	23.47	871
CDGHIJ	854	0.109	23.41	940
BCFH	664	0.097	23.59	753
BCFHIJ	721	0.104	23.60	816
BCEGI	630	0.085	23.82	768
BCEGJ	578	0.083	24.04	761
CDEFI	661	0.099	23.60	863
CDEFJ	597	0.099	23.86	857
ABGHI	745	0.104	23.74	815
ABGHJ	684	0.103	24.03	806
ADFHI	775	0.109	23.58	921
ADFHJ	710	0.110	23.94	913
ADEG	603	0.099	24.23	862
ADEGIJ	652	0.104	24.31	936
ABEF	552	0.103	24.63	795
ABEFIJ	596	0.107	24.73	870
ABCDI	674	0.111	23.81	895
ABCDJ	605	0.128	24.12	886
ACFGI	679	0.087	23.97	921
ACFGJ	631	0.087	24.22	910
ACEH	620	0.120	24.41	852
ACEHIJ	674	0.123	24.41	928
ABCDEFGH	747	0.083	23.93	785
ABCDEFGHIJ	823	0.088	23.85	853

The abiotic and producer parameters are likely to have the greatest effect on net primary production. From the analysis of variance (Table 9.3) one sees that second- and higher-order interactions are relatively small (e.g., see lines 235, 245, and 345 in Table 9.3 for all four sensitivity indicators). Thus significance arises mainly from first- and second-order effects. Most of the main effects are signifi-

Table 9.3. F-values for fractional factorial ANOVA

Line number	Low-order aliases	$n\frac{a/}{p}$ (g dry wt·m^{-2}·yr^{-1})
1	A, BCD, BEF, BGH, CEH, DFH, DEG, CFG, EIJ	1406.8
2	C, ABD, DEF, DGH, AEH, BFH, BEG, AFG, HIJ	4015.0
3	E, ABF, CDF, FGH, ACH, BDH, BCG, ADG, AIJ	1452.2
4	F, ABE, CDE, EGH, BCH, ADH, ACG, BDG, BIJ	427.3
5	J, AEI, BFI, CHI	3.0
12	AC, BD, EH, FG, DFIJ	1842.0
13	AE, BF, CH, DG, IJ	6.9
14	AF, BE, DH, CG, CDIJ, GHIJ	317.3
15	AJ, EI, BCDJ, BEFJ, BGHJ	1.6
23	CE, DF, AH, BG, ABDE	18.4
24	CF, DE, BH, AG, BEFG	25.4
25	CJ, HI, DGHJ	28.5
34	EF, AB, CD, GH, CGIJ	242.1
35	EJ, AI, BCDI, ABFJ	<1.0
45	FJ, BI, ACDI, CEGI	6.9
123	H, BEF, ADF, BDFGH, ABG, CDG, EFG, CIJ	14992.3
124	G, ACF, BDF, BCE, ADE, EFH, ABG, DIJ, CGHIJ	3463.8
125	ACJ, DFI, BGI, CEI	2.6[e/]
134	B, AEF, ACD, AGH, CFH, DEH, DFG, CEG, FIJ, CDEIJ	10.4
135	I, AEJ, BFJ, CHJ, DGJ	4813.2
145	AFJ, CDI, BEJ, DHJ, CGJ, ABI	6.2
234	D, CEF, CGH, BFG, AEG	4465.1
235	CEJ, CDI, DFJ, AHJ, EHI	0.006[e/]
245	CFJ, ADI, DEJ, BHJ, AGJ, FHI	0.006[e/]
345	EFJ, DHI, ABJ, CDJ, GHJ, AFI	1.0[e/]
1234	FH, BC, AD, ACEF, BDGH	30.1
1235	HJ, CI, ACEJ, DGHI, ABGJ	18.4
1245	DI, GJ, ACFJ, BDFJ	59.9
1345	FI, AEFJ, BCHI, ACDJ	1.0[e/]
2345	GI, DJ, CEFJ, CGHJ	1.7[e/]
12345	BHI, FHJ, CFI, ACEFJ	error

[a/] $F_{1,7}^{.05} = 5.59$ [d/] $F_{1,8}^{.05} = 5.32$

[b/] $F_{1,12}^{.05} = 4.75$ [e/] Pooled into error; d.f. = 1,1; $F_{1,1}^{.25} = 5.83$

[c/] $F_{1,9}^{.05}$ 5.12

s_p [b/] (g C·m⁻²·yr⁻¹)	a_e [c/] (cm/yr)	e_c [d/] (g CO_2·m⁻²·yr⁻¹)
46.2	9229.7	12809.5
1.3	182.5	1421.0
9.5	4692.6	4715.2
136.0	85.3	346.6
5.8	1342.0	4295.5
5.0	0.1	73.1
4.5	439.2	405.4
<1.0	8.9	228.1
<1.0	3.4	<1.0
1.4	343.5	243.2
<1.0	254.9	358.4
1.0	96.3	7.1
<1.0	68.1	77.3
2.2	15.7	12.7
<1.0	0.1	11.1
6.1	1857.5	1338.1
233.0	520.1	661.6
<1.0 [e/]	<1.0 [e/]	<1.0 [e/]
3.0 [e/]	1.2	22110.7
1.6 [e/]	<1.0	6653.9
<1.0 [e/]	1.0 [e/]	2.8 [e/]
6.1	2468.2	2933.2
<1.0 [e/]	<1.0 [e/]	2.5 [e/]
<1.0 [e/]	<1.0 [e/]	<1.0 [e/]
<1.0 [e/]	<1.0 [e/]	1.4 [e/]
6.8	354.3	191.2
<1.0 [e/]	35.7	4.7 [e/]
<1.0 [e/]	35.3	7.8
<1.0 [e/]	<1.0 [e/]	6.6
<1.0 [e/]	<1.0 [e/]	<1.0 [e/]
error	error	error

cant except macroparameter J, to which n_p is apparently not sensitive. However, n_p is not particularly sensitive to B, which is expected because neither the abiotic (p_{10}) nor the producer (p_{12}) parameter in B should greatly affect production. The large sensitivity seen on line 2 is likely due to one of the interaction effects or to the soil evaporation parameter in C rather than the plant phenology parameters (p_{54}). Also, n_p is sensitive to D (line 234), probably because of the abiotic parameter p_8, which, when decreased, makes more water available to the plants. From the result on line 4, perturbation of macroparameter F apparently has only a slight effect on production. However, from the single-perturbation runs of Table 9.4, one sees that F has a greater effect than D or A, implying that the three variable interactions dampen the effect of F in this ANOVA. The macroparameters A and E have a moderate effect on production. Variable G has some effect, but this effect cannot be separated from the interaction ADE or from EFH without considering Table 9.4, where G has a greater effect on n_p than do D and I. Hence it appears that these second-order interactions aliased with G are not important. In addition, n_p is very sensitive to macroparameter H, probably due to p_{56}. The model is similarly very sensitive to macroparameter I. The perturbations of p_2 and p_{58} reduce water loss and decrease shoot-to-root transfer, respectively, both of which lead to greater production.

Lines 12, 14, and 34 in Table 9.3 indicate sensitivities of varying degrees to one or more first order interactions. The contributions from particular interactions are difficult to separate because of the design and further analyses are required. General experience would lead us to hypothesize that EH is the main contributor in line 12 and either DH or CG in line 14 because the associated main effects are the largest. For EH the effect is probably due to a compensation between P_4 in E and P_{56} in H. For line 14, the last four runs in Table 9.4 complete the information necessary to calculate the effects for the interactions AF, BE, CG, and DH, which are calculated as n_p in the run in which both macroparameters were perturbed, plus n_p in the run in which neither was perturbed, minus n_p in each of the runs in which only one was perturbed. The highest absolute value for interaction is between C and G (8.7), followed by AF (6.5), DH (1.5), and BE (0.7). In a similar vein, the variation detected on line 34 of Table 9.3 is most difficult to explain. Further work has not been carried out, but we suspect either GH or CD.

In summary, macroparameters B and J are not ones to which the model is particularly sensitive, nor are the first-order interactions FH, BC, AD, CE, DF, AH, BG, CF, DE, BH, AG, and GI. On the other hand, the effects of macroparameters D, H, and I are quite important. With less assurance we can say that the model is sensitive to A, C, E, F, and G. Among the parameters included in the analysis, the ones requiring the most precise estimation are the: (a) depth of the soil's A horizon p_3, (b) the quantum efficiency of photosynthesis p_{56}, (c) the conversion factor for leaf-area index p_2, (d) the upper p_{58} and lower p_{53} limits of shoot-to-root translocation, and (e) the ratio of transpiration water loss to potential evapotranspiration p_8.

Secondary production is dominated by cattle. The large decomposer production component is treated in the soil-respiration analysis below. The values for

Table 9.4. Changes in sensitivity indicators[a] in response to perturbations of single parameters and macroparameter combinations

Perturbation[b]	Δn_p	Δs_p*10^4	Δa_e*10^2	Δe_c
A	27.4	112.9	58.7	63.5
P_{46}	0.0	0.0	0.0	0.0
P_3	27.4	33.5	59.8	68.3
P_{33}	-0.1	0.1	0.5	1.5
P_{18}	-0.1	88.7	-0.3	0.3
P_{30}	0.1	-10.0	-0.1	-0.1
P_{53}	1.2	-1.8	-2.2	-4.4
P_{66}	-2.1	-0.1	0.1	-3.6
B	6.3	2.6	13.0	-58.2
P_{44}	0.0	0.0	0.0	0.0
P_{10}	5.3	2.5	14.1	-12.3
P_{32}	0.0	0.2	-2.4	-45.0
P_{12}	0.0	0.0	0.0	0.0
P_{22}	0.7	-0.1	0.4	-1.0
P_{55}	0.4	0.2	0.9	-0.2
P_{68}	0.0	0.0	0.0	0.0
C	42.0	24.0	-6.8	12.8
D	38.4	11.5	-18.2	17.4
E	-14.0	-29.2	29.0	-27.2
F	42.0	-132.4	1.6	-2.2
G	60.2	-253.4	-9.8	-21.9
H	71.8	7.6	-7.5	-23.4
I	42.5	24.4	1.4	43.4
J	-4.6	30.3	13.9	31.0
AF	75.9	21.5	55.0	59.5
BE	-7.0	-26.8	39.5	-84.2
CG	93.5	-239.2	-17.8	-10.3
DH	111.6	18.6	-28.3	-4.5

[a] The dimensions of the sensitivity indicators and their values in the control run are as in Table 2.

[b] Indented entries are parameters. The other entries are macroparameters or combinations thereof. Perturbations are given in Table 1.

secondary production in the 32 runs ranged from 0.070 to 0.128 g C $\cdot$ m^{-2} $\cdot$ yr^{-1} (Table 9.2), the cattle component being about 75% of the total. The response range of 0.056 g C $\cdot$ m^{-2} $\cdot$ yr^{-1} (Table 9.2) represents about one-half the change that occurs between light and heavy grazing. The component s_p is largest under perturbation *ACEHIJ* because this treatment combination did not reduce cattle weight gain while small mammal production increased. Treatment combination *EFGHJ* yielded the lowest production because *F* and *H* are the only macroparameters that would increase production and affect small mammals only.

The macroparameters for which secondary production appears to be sensitive are *F* and *G*, that is, p_{27}, which in the direction perturbed increases intake relative to the minimum food level for starvation causing increased gain, and p_{23}, which decreases intake and thus gain. Macroparameters *A* and *E* have a slight effect; *E*, however, has two parameters p_{29} and p_{52}, which may cancel each other. The main effects of macroparameters *B*, *C*, and *I* are negligible. In macroparameter *B*, p_{22} adjusts the amount of food wasted. It has more effect on producers than consumers. There is a slight sensitivity to *D*, *H*, and *J*, but the effects of home range p_{24} in *D*, the timing of reproduction p_{28} in *H*, and the basal-metabolism factor p_{21} in *J* do not seem to be important. Likewise, the model is not very sensitive to first-order interactions except for possible sensitivities indicated on line 12 (probably due to the *AC* and *EH* interactions) and line 13 (probably due to *AE* and *DG* interactions). The only higher-order interaction of possible significance is on line 1234, that is, *ACEF*, which is confounded with *FH*, *BC*, and *AD*. Even though secondary productivity responds to parameter perturbations at least as much as do primary production and CO$_2$ evolution (Tables 9.2 and 9.4), the *F* values in Table 9.3 show the ANOVA to be less successful in identifying which parameters were responsible. Perhaps the mammal model is more likely than other models to respond to a variety of parameters in complicated ways; that is, high-order interactions among parameters may be more important.

The sensitivity of the abiotic portion of the model was studied by considering changes in actual evapotranspiration (a_e). The parameter changes in these model runs did not greatly affect total evaporation and transpiration. On a rank order basis, there are some sensitivities indicated with respect to the effects of some parameters and their interactions, but, practically speaking, the response to perturbation (23.41 cm to 24.73 cm, Table 9.2) is not sufficient to make much difference biologically. For example, the greatest sensitivity appears to be to macroparameter *A*, which includes a parameter for the depth to which the soil dries by evaporation p_3. The main effect means are 23.7 cm for the control and 24.1 cm for the perturbed run. This 0.4-cm difference is not of practical significance. In this water-limited system, the cumulative amount of water loss by the end of the year is about equal to total precipitation. If one were to view the two components, evaporation and transpiration, the proportions might change so that a sensitivity for evaporation or transpiration might be more pronounced. Also the timing of water loss is noticeably affected, as determined by inspection of graphs produced during the 32 runs, and this timing is of great significance to the producer and decomposer submodels.

The soil-respiration response is a combination of respiration from the decomposer section and smaller respiration values from the producer model. The decomposition contribution is almost an order of magnitude larger than the others. The maximum simulated was 940 g $CO_2 \cdot m^{-2} \cdot yr^{-1}$, and the minimum was 753 g $CO_2 \cdot m^{-2} \cdot yr^{-1}$ (Table 9.2); thus the range is about 20% of the magnitude of the response.

Carbon-dioxide evolution is at the end of the system in the sense that it is dependent on the outputs of all the rest of the submodels. Thus the macroparameters are not dominated by one or two components or microparameters, but in fact any parameter affecting decomposers, temperature, soil water, or productivity may have an important effect on e_c.

Even though this is the most difficult part of the sensitivity analysis, the higher-order interactions produce smaller response than the main effects and first-order interactions. Pooling several higher-order terms that were nonsignificant gives an error term with eight degrees of freedom. Macroparameters A and B appear to have major effects with D, E, I, and J showing smaller ones. From Table 9.4 we see that the responses to these macroparameters perturbed singly are substantially in the same order as indicated by the relative mean squares (F-values) in Table 9.3. Therefore, the effects of the aliased interactions may be discounted.

Perturbing macroparameter A increases primary production despite soil-water decrease and also increases the susceptibility of surface litter to decomposition. These are counteracting influences; hence it is not readily apparent why A should have such an important effect on e_c. On the other hand, macroparameter E decreases both a_e and n_p and increases the ecological growth efficiency of decomposers, so that the observed significant decrease in e_c is not surprising. Macroparameters J and B have little effect on either a_e or n_p, but affect the rate of decomposition so that in these cases we can attribute the macroparameter effect to parameters directly affecting the decomposer model. Macroparameters I and D both decrease transpiration (increase soil water) and I also increases the decomposition rate, so that the importance of these macroparameters is expected. There are a number of less important sensitivities involved in first-order interactions.

The response of the respiration variable illustrates that the parameters affecting decomposition may need to be well known. The analysis also suggests that the possible sensitivity of respiration to a number of parameter changes renders the macroparameter approach less suitable in this case.

The fractional factorial design involves confounding at two levels. Parameter effects are masked by grouping parameters into macroparameters, and the main effects and interactions are not completely separable. Strictly speaking, the analysis of macroparameters may give little information about individual parameters, and conclusions made about these parameters in the analysis above are based on knowledge of the parameter functions in the model and on experience gained in tuning the model. Similarly, the confounding of effects in the fractional factorial design requires the conclusions about the relative importance of various main effects and interactions be partly a matter of judgment.

The series of model runs presented in Table 9.4 can be used to test the assumption that parameters within macroparameters do not cancel each other's effects. Consider the model runs perturbing each of the parameters in macroparameters A and B. For the eight cases (two macroparameters × four sensitivity indicators), the effect of the macroparameter is largely attributable to a single parameter. The effect of the most important parameter differs from the effect of the macroparameter by an average of 10% and by no more than 23% (Table 9.4). For the three cases in which the effect of the macroparameter was less than the effect of the most important parameter, they differed by no more than 8%. These results for macroparameters A and B support the conclusion that the macroparameters found to be nonsignificant in the ANOVA are made up of parameters with little effect individually.

Since the parameters within a macroparameter were chosen from different submodels, they affect different processes and, depending on the direction of perturbation, the effects should be additive. In three cases (Δn_p with B, Δs_p with A, and Δe_c with B; Table 9.4) the microparameters are acting additively. Other evidence of additivity of effects is seen in the AF, BE, CG, and DH trials (Table 9.4). In each case, the combined effect is very close to the sum of the component effects. Given a different direction of perturbation, it is probable that the effects of parameters within a macroparameter cancel. In this light it would be better to set the direction of perturbation for important parameters whose effects can be clearly predicted so that they do not cancel and, then, to randomly perturb the others as was done here.

In several cases the modelers predicted which parameter was responsible for the effect of a macroparameter before the sensitivity analysis. Table 9.4 provides information to test several of these predictions; for example, p_{32} was correctly identified as the parameter responsible for the effect of B on e_c, and p_3 for the effect of A on n_p. The only failure was in the prediction that p_{53} is also important in the effect of A on n_p, whereas its effect is actually far less than p_3 and less even than p_{66}, a parameter from the phosphorus submodel. The modelers were asked to provide lists of "primary" and "random" parameters for the sensitivity analysis, and Table 9.4 provides evidence that parameters from these two lists differ significantly in their effects. For example, the effect of p_{56}, a producer submodel primary parameter, on n_p was 73.7, while the effects of two random parameters, p_{53}, and p_{55}, were 1.2 and 0.4, respectively. A grasshopper-model primary parameter, p_{18}, increased secondary productivity by 38%, while a random parameter, p_{12}, had no effect. Similarly, a mammalian consumer primary parameter, p_{30}, decreased secondary productivity by 1.2%, while the random parameter, p_{22}, had no effect. The fact that modelers have some knowledge of which parameters most affect the model is not surprising, but it is an important point to confirm because that knowledge is used in selecting suitable combinations of parameters in the macroparameter approach.

Table 9.4 contains an example of the kind of anomalous model behavior that sensitivity analysis is designed to discover. Perturbing parameter p_3 in macroparameter A increases a_e over the control value, which would be expected to cause a decrease in soil water and a decrease in n_p. Nevertheless, n_p shows a considerable increase over the control. Examination of additional model output revealed that soil water in the layers directly affected by the parameter change was lower than in the control, as expected, and it appeared that the model was in error. However, discussion with the author of the abiotic submodel revealed that another parameter affecting the relative loss of water by transpiration from the various soil layers logically should have been changed in conjunction with p_3, and our failure to change it led to an increase in soil water in the deeper layers, accounting for the increase in n_p.

The 14 runs in which single parameters were perturbed (Table 9.4) demonstrate the interactive nature of processes in the model. The sensitivity of all four indicators to abiotic model parameters affecting soil water is not surprising, since the modeled system is water-limited. A change in soil water affects productivity and soil respiration directly and secondary production indirectly, through the change in productivity. Parameter p_{18} illustrates how a change in one part of the model has ramifications throughout the model. Raising p_{18} postponed the hatching of grasshoppers in the spring, which might be expected to reduce the size of the grasshopper population. But by hatching later, the nymphs avoided a period of unfavorable weather which caused considerable mortality in the control run. Thus both peak population and secondary productivity were greater when hatching was later. The greater grasshopper populations consumed more herbage, decreasing n_p slightly (-0.02%), but increasing clipping and wasting of green herbage and thus litter production ($+0.06\%$). The increased litter levels intercepted more rainfall, which left less water for infiltration into the soil, thus reducing the amount of water lost from the soil by transpiration and bare-soil evaporation (-0.013%). The change in soil-water loss did not greatly affect soil-water levels because the reduction in infiltration was offset by the decreased water loss. The increase in e_c (0.03%) results from the increase in the input to litter of easily decomposable material. Perturbing p_{18} would have a different effect in a year in which conditions were good during the period hatching was delayed, which illustrates that the results of a sensitivity analysis may depend on the driving variables used.

Interpretations can be given to the other cases in which a single parameter affects several sensitivity indicators, but in some instances we are less confident of the chain of effects because several possible effects with opposing influences may exist. For example, it may be difficult to interpret the effect of changing a producer-model parameter that increases production, because the increase in production could come in different producer categories (warm- and cool-season grasses, etc.) and at different times of the year. The timing of changes in production determines which consumers will be affected and in which direction. It is possible that an increase in a prey species could tide a predator over a lean period, with deleterious effects for another prey species, and the net effect could be a decrease in total secondary productivity. For example, producer submodel parameters p_{53} and p_{55} both increase n_p, but have different effects on secondary productivity (-0.2% and $+0.02\%$, respectively). There are also interactions between producer categories; for example, cool-season grasses may use water stored in the soil, making it unavailable for the following warm-season grasses.

9.4 Discussion

The Tomović method, while theoretically appropriate, is not of practical use for ELM as a whole. Due to the interrelationships of state variables, each sensitivity coefficient depends on a number of others. Simultaneously solving for a set of sensitivity coefficients requires a number of additional different equations to be computed. For a large model, this increase in the number of equations to be solved drastically affects computer costs.

The ANOVA was moderately successful in identifying the important sensitivities and was very successful in identifying macroparameters with little effect. We conclude that a relatively safe way to locate all of the sensitive parameters is to check individually those parameters not excluded by the fractional factorial approach. If the system is insensitive to a macroparameter, then no further analysis is required for any parameter in it (since parameters were grouped to avoid canceling effects). If the objective is only to locate some sensitive parameters, the fractional factorial approach is efficient. The assumptions that higher-order interactions are nonsignificant was substantiated by the nonsignificance of those higher-order interactions that were addressed by the design and subsequently pooled into error. The statistical conclusions largely agree with intuitive evaluation of sensitivity given by the modelers, independent of the statistical analysis. The advantage of the statistical approach is objectivity and quantification of magnitude of effect. The ANOVA technique is effective in screening the sensitivity of outputs to a large number of parameters and parameter interactions. Since the method utilizes a fractional factorial design, there will always be some difficulty in attributing significance to one of the several possible confounded effects. In some cases it was possible to select the most probable source of sensitivity, while in others it was not. Single parameter perturbations are useful in refining judgments about certain confounded sensitivities. Obviously, there is a practical limit to the number of such runs that can be made. We suggest that the most powerful sensitivity analysis procedure for a model such as ELM is to use the fractional factorial analysis followed by subexperiments that may include other fractional factorial experiments, single-macroparameter tests, or single-parameter tests.

Of the parameters to which the model is sensitive, the abiotic ones seemed to be frequently important, but not uniformly. Actual evapotranspiration was statistically very sensitive, but no meaningful biological fluctuation occurred. Both primary production and e_c responded to parameter changes. The fractional factorial ANOVA led us to a number of conclusions about production, but not about e_c. Since soil respiration is affected by abiotic, producer, and consumer changes, the effects of macroparameters are less separable.

Overall, one can conclude that the model is not "fragile" in the sense of having great numbers of parameters that must be known exactly. There were, however, a number of parameters found that do require thorough understanding. Many of these have been carefully studied, but several, such as the shoot-to-root translocation-rate parameters, indicate areas where more research is needed.

Acknowledgments. The authors thank J. C. Anway, C. V. Cole, W. J. Parton, J. O. Reuss, C. F. Rodell, and R. H. Sauer for help in designing this research and interpreting the results.

This chapter reports on work supported in part by National Science Foundation Grants GB-41233X, and BMS73-02027 A02 to the Grassland Biome, U.S. International Biological Program, for "Analysis of Structure, Function, and Utilization of Grassland Ecosystems."

References

Bellman, R.: Stability Theory of Differential Equations. New York: McGraw-Hill, 1954, 166 pp.

Blaquiére, A.: Nonlinear System Analyses. New York: Academic Press, 1966, 392 pp.

Britting, K. R., Trump, J. G.: The parameter sensitivity issue in *Urban Dynamics*. In: Proc. 1973 Summer Computer Sim. Conf., Vol. 2. La Jolla, Calif.: Simulation Councils, Inc., 1973, pp. 1052–1059.

Burns, J. R.: Error analysis of nonlinear simulations: Applications to world dynamics. IEEE Transactions on Systems, Man, and Cybernetics SMC-5, 3, 331–340.

Cochran, W. G., Cox, G. M.: Experimental Designs. New York: John Wiley and Sons, 1957, 611 pp.

Forrester, J. W.: Industrial Dynamics. Cambridge, Mass.: M.I.T. Press, 1961, 464 pp.

Kaye, S. V., Ball, S. J.: Systems analysis of a coupled compartment model for radionuclide transfer in a tropical environment. In: Proc. 2nd Nat. Symp. Radioecology, Conf. 670503, TID-4500. Oak Ridge, Tenn. Nelson, D. J., Evans, F. C. (eds.), 1969, pp. 731–739.

Kempthorne, O.: The Design and Analysis of Experiments. Huntington, N.Y.: Robert E. Krieger Publ. Co., 1973, 631 pp.

Kowal, N. E.: A rationale for modeling dynamic ecological systems. In: Systems Analysis and Simulation in Ecology. Patten, B. C. (ed). New York: Academic Press, pp. 123–194.

Leondes, C. T. (ed.): Control and Dynamic Systems: Advances in Theory and Application, Vol. 9. New York: Academic Press, 1973, 514 pp.

Miller, D. R.: Model validation through sensitivity analysis. In: Proc. 1974 Summer Computer Sim. Conf., Vol. 2. La Jolla, Calif.: Simulation Councils, Inc., 1974, pp. 911–914.

Miller, D. R., Weidhaas, D. E., Hall, R. C.: Parameter sensitivity in insect population modeling. J. Theor. Biol. 42, 263–274 (1973)

Patten, B. C.: Ecological systems analysis and fisheries science. Trans. Amer. Fish. Soc. 98, 570–581 (1969)

Shannon, R. E.: Systems Simulation: The Art and Science. Englewood Cliffs, N.J.: Prentice-Hall, 1975, 387 pp.

Smith, F. E.: Analysis of ecosystems. In: Analysis of Temperate Forest Ecosystems. Reichle, D. E. (ed.) New York: Springer-Verlag, 1970, pp. 7–18

Tomović, R.: 1963, Sensitivity Analysis of Dynamic Systems. New York: McGraw-Hill, 142 p.

Tomović, R., Vukobratović, M.: General Sensitivity Theory. New York: American Elsevier, 1970, 258 pp.

Vermeulen, P. J., de Jongh, D. C. J.: Growth in a finite world—a comprehensive sensitivity analysis. Automatica 13, 77–84 (1977)

Wiens, J. A., Innis, G. S.: Estimation of energy flow in bird communities: A population bioenergetics model. Ecology 55, 730–746 (1974)

10. Critique and Analyses of the Grassland Ecosystem Model ELM

ROBERT G. WOODMANSEE

Abstract

The grassland ecosystem model ELM is discussed and criticized. The critique is based on the examination of the: (a) degree to which the objectives of the model were realized, (b) appropriateness of representation of selected biological, physical, or chemical processes modeled, (c) apparent adequacy of selected model responses, and (d) extent to which the goals of the modeling effort were realized.

The objective of the modeling activity was to develop a total system model of the biomass dynamics of a grassland that was representative of sites in the US/IBP Grassland Biome network and with which there could be relatively easy interaction. The goals of the modeling effort were to create a model that would serve as a communications device and organizer of information, be useful as a research instrument, and yield results helpful in elucidating biological phenomena in grassland ecosystems.

10.1 Introduction

A model is an abstraction of reality: an attempt to represent some of the important features of the real thing in a simplified way (Forrester, 1961). A simulation model is a dynamic mathematical model that mimics the functioning of a system by the process of step-by-step solution of the equations that describe the system. The Pawnee Site implementation of ELM is a simulation model of a shortgrass prairie ecosystem.

This discussion is an analysis and self-critique (Grassland Biome modeling staff) of the ELM modeling effort and includes criticisms of both the implementation and subsequent response of the model and the general modeling approach. The model is evaluated in reference to its objectives and goals. Objectives are defined as the specific guidelines that directed the development of the model. The objectives define the questions the model can address. Goals of the modeling effort refer to the desired improvements in understanding gained through development, analysis, and use of the model (Forrester, 1968).

The critique is based on the examination of the (a) degree to which the objectives of the model were realized, (b) appropriateness of representation of biological, physical, or chemical processes modeled, (c) apparent adequacy of the model response, and (d) extent to which the goals of the modeling effort were realized. No attempt is made here to evaluate the numerical techniques used to implement the model.

10.2 Model Objectives

The importance of establishing achievable, realistic objectives before a team modeling effort begins cannot be overstated (U.S. Forest Service, 1973). The importance is analogous to establishing a destination before commencing a journey. Their importance increases as the number of modeling team members or travelers increases. The consequence of establishing achievable, realistic objectives is the potential for order or for an integrated modeling effort with compatible submodels and reasonable or at least educational results. The consequence of unrealistic or unachievable objectives is turmoil with each team member or traveler defining his own direction. Precise objectives were especially important in the initial phase of the ELM modeling effort because members of the team were separated geographically. The implication here is not that once established, objectives should be adhered to automatically, but if they are changed, the reasons for alteration should be compelling and all participants must be aware of the changes. During development of ELM the original objectives were maintained.

10.2.1 Statement of Objectives

The global objective of this modeling activity was to develop a total system model of the biomass dynamics of a grassland representative of sites in the US/ IBP Grassland Biome network with which there can be relatively easy interaction.

A discussion of this objective is found in Chapter 1. This objective serves to specify the direction of the effort but is still too general to provide a basis for a number of decisions that have to be made in the development of the model. To provide a basis for these decisions, the following four specific questions were chosen as points that should be addressed by the first version of ELM:

1. What is the effect on net or gross primary production as the result of the perturbations: (a) variations in the level and type of herbivory, (b) variations in temperature and precipitation or applied water, and (c) the addition of nitrogen or phosphorus?

2. How is the carrying capacity (i.e., maximum sustainable domestic herbivore stocking density) in a grassland affected by these perturbations?

3. Are the results of an appropriately driven model run consistent with field data taken in the Grassland Biome Program, and if not, why?

4. What are the changes in the composition of the producers as the result of these perturbations?

The reader is certainly at liberty to judge the adequacy and appropriateness of the objectives themselves and to judge the model response in context of these objectives. Criticism of the model based on issues lying outside the objectives of the model is inappropriate. For example, an appropriate question would be what the effect is on net primary production of a 25% increase above mean annual precipitation. An inappropriate question would be what the effect is on net primary production of plowing in the spring of the year. This latter form of question is equivalent to expecting a topographic map (a model) to provide growing season information about the mapped area.

10.2.2 Critique of Objectives

The model represents a total system (ecosystem) by definition; it does include abiotic, producer, consumer, decomposer, and nutrient sections (Fig. 1.6, p. 8,9). The ELM model represents these functional components of the ecosystem in time, but not in horizontal space; in other words, the mean meter squared of the prairie under a given set of environmental conditions is considered to vary with time but not space. Some vertical flows and structure are represented. The nonrepresentation of horizontal spatial effects is not considered a severe limitation to modeling the dynamics of sessile organisms (relative to 1 m² in a relatively stable situation). However, representation of spatial effects is a limitation in modeling the dynamics of heterogeneously distributed (in time and space) consumers, horizontal water flow, and nutrient transport. A specific example from Chapter 9 indicates a significant increase in grasshopper biomass in an irrigated treatment compared to a control treatment. Field measurements indicated no such increase in the biomass was observed. One explanation for this disparity is that conditions were favorable for grasshopper hatching and survival in the irrigated treatment, but emigration may have occurred to the bordering prairie. Another example of this limitation is that plants are considered to compete for light as if they were uniformly mixed (Chap. 3). The model does not represent vegetational clumping.

The model does represent biomass dynamics by simulating carbon (g/m²) movement through the system. Carbon was chosen over biomass $[(CHO)_n]$ to assure that the system remained conservative. Biomass would be conserved only if carbon, hydrogen, and oxygen were each accounted for simultaneously. In addition to representation of biomass dynamics, the model simulates the dynamics of heat (temperature), water, nitrogen, and phosphorus.

The definition of "representative" in the objective is ambiguous in that "order of magnitude" could be interpreted to imply that the effect of a perturbation on a state variable could fall in the range of 10–1000 g C/m² and be acceptable, assuming that the correct value was 100 and the direction of response was correct for a given perturbation. Thus if irrigation increased peak herbage by 100 g C/m², the model would be within an "order of magnitude" with any increase between 10 and 1000 g C/m². Clearly, the "order of magnitude" criterion is too lax. A more appropriate but still rather ambiguous working definition of "representative" has been used in the modeling effort; that is, the model response should be in or near the 95% confidence intervals of the field data. In most cases variability of the data was wide and the confidence intervals broad. Saaty (1972) would certainly call these data sets "fuzzy," as did Innis (1972). However, in many cases data did not exist for comparison with model results, and the appropriate order of magnitude must be judged by intuition and experience.

A distinction should be made between two types of model results when judging whether or not the model is representative. One type represents whole-system response, with one of the best examples being CO_2 evolution from the decomposer compartment (Fig. 6.11 p. 175). Chapter 6 discusses the ramifications of the system components acting on decomposer compartments. The model results are an integration of interactions of most of the components represented in the system. The model, operating as the representation of a real ecosystem, is based on simulated biological, chemical, or physical mechanisms and processes. The conceptual development and formulation of simulated mechanisms may or may not satisfy all subject-matter experts (Forrester, 1968), but when all mechanisms of all submodels are combined into the whole represented ecosystem, the system-level responses are often reasonable and informative.

The other types of model results are those of a specific nature (somewhat less of an integration of all model components), for example, warm-season grass shoot growth (Chap. 3) (Fig. 3.11 p. 83) or soil-water content of different strata in soil (Chap. 2) (Fig. 2.7a–e, p. 46, 47). The example of shoot growth of warm-season grasses was deliberately chosen to illustrate that a particular response of a subsystem level component may in some cases appear spurious, but in other situations may be consistent with experience. For example, the validation test for warm-season grasses (experiment 1 in Table 10.1) shown in Figure 10.1 yields a seemingly more reasonable model response than the validation tests shown in Chapter 3 (Fig. 3.11, p. 83). However, the model responses are not directly comparable because Figure 3.11 represents live plus standing-dead aboveground material, which corresponds to field data in which the plant material was separated into these two categories. Figure 10.1 represents live and standing-dead aboveground material, which corresponds to field data on unseparated plant material.

Another reason for the discrepancy between two separate validation tests is that the model development was based on literature, experience, and data from 1970 and 1971 for the so-called ungrazed treatment (cattle excluded for a number of years prior to 1970). The procedures for validation of model results were to use the 1970 and 1971 data from a 3-yr data block (1970–1972) to adjust the parameters of the model to ensure good agreement between model results and field data. Data from 1972 were held in reserve to be used in comparison with results of the model that was run using 1972 meteorological data as driving variables. Unfortunately, intensive sampling caused excessive damage to the vegetation of the ungrazed treatment during 1970 and 1971. In 1972 the "ungrazed" treatment was moved to a location 3.2 km away to a site that was judged visually (aboveground appearance) to be similar to the original site. Subsequent analysis has shown that the 1972 site was on a different soil type and that the composition of the vegetation was dissimilar and the visual similarity used to choose the site was superficial. In addition, Sauer (Chap. 3) and Parton (Chap. 2) both suggest that summer convectional rainstorms can cause marked differences in soil water in short horizontal distances; and, indeed, soil-water measurements in 1972 show differences between the two ungrazed treatments suggesting important precipitation differences. Thus with soil water being monitored at one site and plant production at another, it is not surprising that some model responses (Fig. 3.11) appear erroneous. The model response shown in Figure 10.1 appears more reasonable and may well be more representative of the site where the driving variables were measured and parameters estimated (i.e., similar soil type).

The preceding discussion illustrates that the model evaluation is sensitive to field-sampling techniques, to the simulated environmental conditions (i.e., precipitation driving variables) and to the particular configuration of parameter values and initial conditions used to represent a site. The model is not readily rendered applicable to different sites without careful consideration of the environmental factors at that site. That is, to be representative of a site, the driving variables, initial conditions, and site-specific parameter values must represent the site in question.

Included in the discussion of "representative" was the statement that "the model is valid, provided it can predict the direction and order of magnitude of response of this system to certain perturbations, as well as predict the 'normal' dynamic." Nolan (1972) gives a detailed discussion of the concept of model validity. Perturbations were to be of a "reasonable" sort, and are discussed below in Section 10.2.3. The model was designed to predict both normal dynamics and variations that occur commonly in "tens of years." Definition of normal is ambiguous. Not included in the model formulation are mechanisms representing certain common environmental events, such as snowfall, runoff, outbreaks of pathogens or insects, animal migration, and wind erosion. The rationales for not including these mechanisms are as follows: (a) the infrequent occurrence of these events at the sites in the Grassland Biome study network (see Fig. 1.1) and (b) the insignificance of their effect in the time scale considered ($\leq \sim 5$ yr). The reader must judge the appropriateness of the rationale, however; the assessment should be based on the environmental conditions represented at the Pawnee Site.

Relatively easy interactions with the model have been achieved, assuming that one of the skilled ELM modelers has performed the interaction. Changes requiring alterations of parameters or groups of parameters can be made in minutes. Simple changes of coding in the structure of the model may be made in minutes or more complicated changes in not more than a few hours. Indeed, the limiting factor is "turnaround time" at the computer center.

10.2.3 Critique of Model Response to Specific Questions

The first question that the model was designed to address is what the effect is on net or gross primary production as the result of the perturbations: (a) variations in the level and type of herbivory, (b) variations in temperature and

Table 10.1. Results of a series of model runs, with comparisons to data available (from Chap. 1)

Experiment number	Treatment heading	Treatment description	Water total (cm)	Net primary production (g dry wt/m²)	Gross primary production (g dry wt/m²)	Light[a] interception (cal/cm²)	Live[b,c] peak (g dry wt/m²)	Date	WSC[b,c] peak (g dry wt/m²)	Deviation (%)[a]	Date
1	Treatment D; control	1972 weather	27.1	707.0	1086.0	24,950.0	170.0	256	65.25	38%	258
							187.0		47.0	25%	
2	Treatment E; water added	Water added to keep tension >−0.8 bars	88.0	3494.0	5151.0	50,643.0	620.0	286	382.0	62%	288
							388.0		175.7	45%	
3	Treatment F; N added	N fertilizer applied in June	27.1	612.0	1009.0	23,978.0	161.0	258	43.0	27%	260
							290.0		44.8	15%	
4	Treatment G; water and N added	E and F combined—different initial conditions	85.9	3190.0	4673.0	46,902.0	604.0	288	360.0	60%	288
							863.0		465.2	54%	
5	Treatment 2; light grazing	1972 weather 0.082 cows/ha for 6 mo	27.1	715.0	1087.0	21,166.0	139.0	258	94.5	68%	258
							182.0		101.5	56%	
6	Treatment 4; heavy grazing	0.2 cows/ha for 6 mo	27.1	823.0	1294.0	25,440.0	169.0	258	117.2	69%	258
							170.0		73.3	43%	
7	Treatment D + no consumers	All consumers removed 1972 weather	27.1	711.0	1092.0	25,147.0	172.0	256	66.0	38%	258
8	Treatment D + light grazing all year	0.04 cows/ha for 365 days	27.1	689.0	1071.0	24,302.0	165.0	256	62.8	38%	258
9	Treatment D + short heavy grazing	0.3 cows/ha for 140 days	27.1	552.2	901.0	19,000.0	119.0	188	37.0	31%	188
10	Treatment D + heat	Temperature raised 2°C	27.1	649.0	1060.0	24,920.0	163.0	258	62.0	38%	260
11	Treatment D + cool	Temperature lowered 2°C	27.1	706.0	1031.0	24,452.0	170.0	254	59.5	35%	256
12	Treatment D + drought	Rainfall reduced to 25% of 1972 level	6.86	49.0	134.0	7,507.0	63.2	174	12.2	19%	174
13	Treatment D DT = 1	1972 weather	27.1	713.0	1090.0	25,084.0	170.0	256	66.5	39%	258

Table 10.1. (continued)

Experiment number	CSG[b,c] peak (g dry wt/m²)	Deviation (%)[a]	Date	Forb[b,c] peak (g dry wt/m²)	Deviation (%)[a]	Date	Shrub[b,c] peak (g dry wt/m²)	Deviation (%)[a]	Date	Cactus[b,c] peak (g dry wt/m²)	Deviation (%)[a]	Date	CO₂[e] evolved (g)	Mammal producer	Grasshopper producer	Secondary producer
1	4.3	3%	254	12.9	8%	256	61.5	36%	254	26.8	16%	188	898.9	0.0245	0.0199	0.0444
	14.5	8%		20.0	11%		70.3[f]	38%		35.4[f]	19%					
2	20.7	3%	286	42.0	7%	288	145.8	24%	280	31.9	5%	272	1,591.0	0.025	0.029	0.054
	21.9	6%		41.3	11%		101.9[f]	26%		41.9[f]	11%					
3	10.3	6%	254	33.0	20%	256	38.2	24%	256	37.2	23%	262	1,034.0	0.025	0.021	0.046
	44.5	15%		69.0	24%		60.6[f]	21%		70.9[f]	24%					
4	20.4	3%	288	26.2	4%	288	150.2	25%	280	50.6	8%	181	2,130.0	0.025	0.025	0.050
	52.6	6%		68.0	8%		246.6[f]	29%		30.6[f]	4%					
5	2.5	2%	254	6.43	5%	256	23.6	17%	254	12.6	9%	150	953.0	0.093	0.019	0.112
	10.4	6%		14.7	8%		38.6[f]	21%		16.7[f]	9%					
6	4.8	3%	174	9.6	6%	256	2.1	1%	256	35.8	21%	198	902.0	0.201	0.018	0.219
	16.6	10%		13.6	8%		1.02[f]	0%		65.4[f]	38%					
7	4.9	3%	254	13.2	8%	256	61.2	36%	254	26.8	16%	188	891.0	0.000	0.000	0.000
8	3.5	2%	252	11.6	7%	188	61.5	37%	254	26.8	16%	188	906.0	−0.0337	0.0197	−0.014
9	3.2	3%	174	9.0	8%	144	58.8	49%	252	26.8	22%	188	953.0	0.260	0.020	0.280
10	7.8	2%	188	9.0	6%	146	67.0	41%	258	26.8	16%		1,016.0	0.025	0.000	0.025
11	6.4	4%	252	15.6	9%	254	62.0	36%	252	27.0	16%	262	774.0	0.0212	0.0010	0.0222
12	1.41	2%	172	6.4	10%	174	16.8	27%	174	26.5	42%	134	543.0	0.0254	0.0256	0.051
13	4.2	2%	253	12.4	7%	187	60.2	35%	256	26.8	16%	146	862.0	0.020	0.029	0.049

[a] Percentage of live peak (see text).
[b] Where multiple entries occur, model values are given above and field-determined means are given below.
[c] Model computes g C (grams of carbon), whereas g dry weight are quoted; g dry wt = g C · 2.5.
[d] Total available for all runs is 112,552.0 cal/cm².
[e] Model computes gC, whereas g CO₂ are quoted; g CO₂ = g C · 3.67.
[f] Averages of samples—through the growing season.

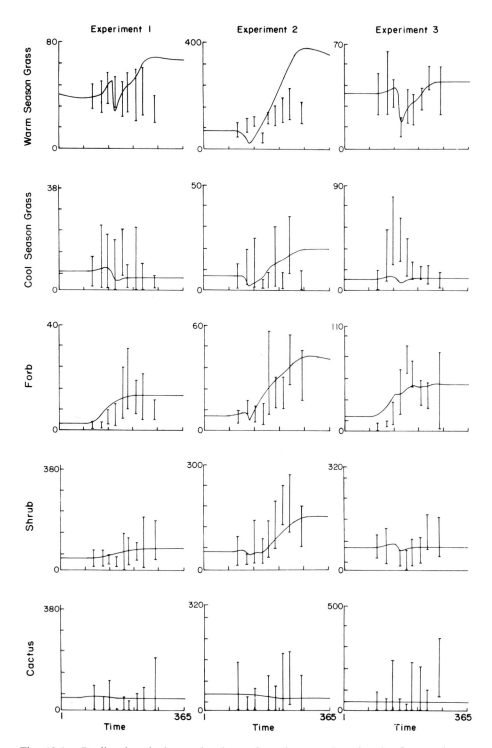

Fig. 10.1. Predicted and observed values of producer carbon for the five producer categories and six experimental areas (from Chap. 1)

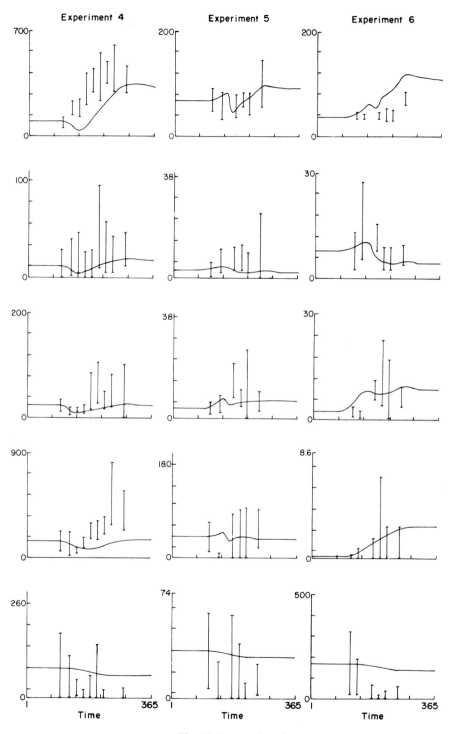

Fig. 10.1.   continued

precipitation or applied water, and (c) the addition of nitrogen or phosphorus. Before the model response to the simulated perturbations can be examined, the response to the simulated control treatment (treatment D, Table 10.1) must be evaluated.

Total water represents both precipitation and evapotranspiration for 1972 because no water was lost from the system through percolation to ground water or by runoff. Precipitation following the growing season is assumed lost to evaporation with none being stored in the soil profile. The 27.1-cm value is within the 95% confidence limits of field data. Net primary production has not been calculated from field data of 1972 for the control treatment, but the 707.0 g/m² value is indistinguishable from field data, depending on which method of calculation of net primary production is used (Singh et al., 1975). Gross primary production is calculated for a supposed maximum photosynthetic rate which is reduced by certain limiting factors (Sauer, Chap. 3). Net primary production is simply gross primary production minus respiration of the plant. The magnitude of the respiration value is dependent on the assumptions stated by Sauer (Chap. 3).

No explicit comparative data are available for light intercepted, CO_2 evolved, mammal production, grasshopper production, or secondary production. Model response for the maximum living plus dead plant material represented at one time (peak standing crop) in 1972 is within the 95% confidence limits of the field data. Comparison of model response with field data for the components of peak vegetation shows inconsistent results. Response for warm-season grasses was 18 g (39%) higher than field estimates. Field data for warm-season grasses are considered reliable because this functional group is dominated by *Bouteloua gracilis,* and measurements of that species have been relatively precise. Among the possible explanations for the lack of agreement might be the underestimate of respiration due to nonconsideration of nighttime respiration and consequent higher net primary production because the amount of gross primary production not respired goes to net primary production for a given species or functional group. Also as Sauer (Chap. 3) explains, the fall of standing-dead material (dead stem and leaf material that is attached to the root crown) to litter is inadequately represented. Another possible explanation is the effect on model response of inappropriate use of Liebig's law of the minimum (Odum, 1971), that is, selecting at a given time the most limiting of the environmental factors, soil water, temperature, nitrogen availability, phosphorus availability, and the effect of phenological stage of the plant group considered. Subsequent calculations in that time period in the simulation are based on the one most limiting factor, and the interactions of these environmental factors are not considered. Consequently, the collective effects of suboptimal environmental factors on plant growth are likely underestimated. These two explanations for model behavior are not presented as the known definitive reasons for lack of agreement of model response with field data, but rather as examples of how concepts used to develop the model affect model response.

Cool-season grasses (dominated by *Agropyron smithii*) response is low compared to the field data presented by more than a factor of 3. In general, when considering model runs under different simulated treatments, cool-season grasses are underestimated by a factor of 3–5, except in Table 10.1, experiment 2. The reason for the underestimate is unknown.

Representation of the forb group in the model as a single functional category is naive, with the results shown in Table 10.1 being misleading. The model was formulated using, as a basis of consideration, responses of plants to environmental factors such as temperature, soil water, radiation, response to available nitrogen and phosphorus, and phenology. But the assumption was made that all forbs responded similarly to the environmental factors. Often in the "normal" dynamics of the shortgrass prairie a group of plants collectively referred to as "cool-season forbs" (dominated by winter annuals) can significantly contribute to biomass (Hyder et al., 1975). The cool-season forbs are ignored in the model, and I contend that the information regarding the behavior of the group and their demonstrated importance warrant their inclusion. Detailed comparison of the model response and validation data is inappropriate due to their incompatability. However, the model suggests the appropriate magnitude of response of the group that can be considered warm-season forbs (forbs that grow vegetatively during the summer months).

Comparison of the model response to field data for shrubs and cacti indicates close agreement (Table 10.1). However, the model response cannot be considered to adequately represent the dynamics of these two categories of plants because the variability of the field data for these two groups is so great due to spatial heterogeneity (Dodd and Lauenroth, 1974).

The evolution of CO_2 (Table 10.1) represents both root and microbial respiration resulting from decomposition processes. The value is reasonable, but data are unavailable for precise comparison. The formulation in the decomposer submodel seems to be adequate for representation of decomposition processes in ELM. Hunt (Chap. 6) discusses further validation tests for the decomposer submodel.

Response of the mammals (exclusive of cattle) and grasshoppers in the control simulation is reasonable, but evaluation and criticism of the dynamics and exact responses are difficult due to the great variability in field data (Figs. 10.1 and 10.2). However, the lack of response of the simulated consumers (excepting the grazing treatments where cattle were simulated) to any perturbation shown in Table 10.1, especially experiment 12, indicate unrealistic insensitivity of the mammal submodel. The model must be used with caution in studying the dynamics of small mammal biomass or the interactions of small mammals with primary producers under varying perturbations. However, Anway (Chap. 4) presents evidence that other aspects of the performance of the mammal model are more satisfactory. The grasshopper submodel response appears to be more sensitive to perturbation, but the evaluation of the magnitude or direction of response is impossible due to the lack of data. Rodell (Chap. 5), however, presents evidence that the grasshopper model can predict year-to-year variations in the general level of grasshopper populations.

Question (1a) asks how net or gross primary production is affected by variations in the level and type of herbivory. With the exception of experiment 7 (Table 10.1) this question is addressed by showing results of grazing by cattle.

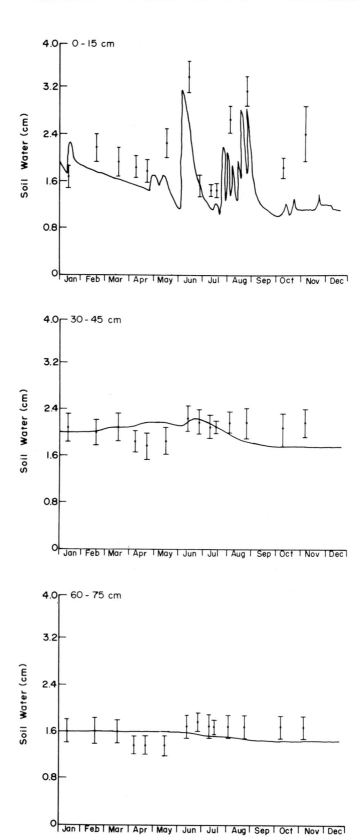

Fig. 10.2. Predicted and observed values of soil water for a situation similar to the control plot 1 (experiment 1) (from Chap. 1)

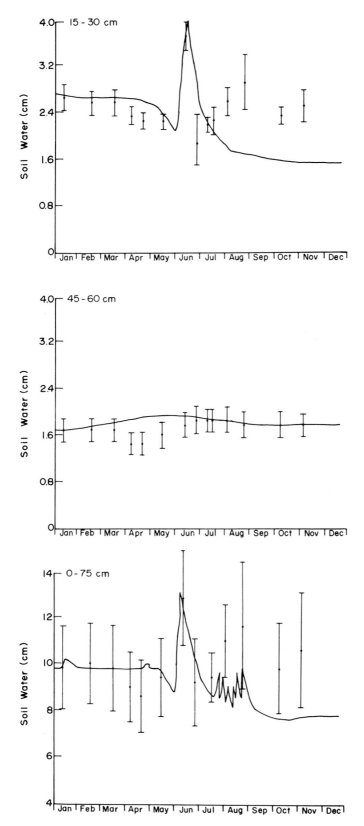

Fig. 10.2. continued

Experiment 7 indicates "no consumers," when actually what was removed or not included in the model was the small mammal component, cattle, and simulated grasshoppers. The model formulation does not include birds, reptiles, amphibians, any belowground herbivores (nematodes, macroarthropods, etc.), or above- or belowground macroarthropods other than grasshoppers. Birds and reptiles are considered trivial in terms of energy flow, but the other organisms may be more important, collectively, than mammals and grasshoppers (Coleman et al., 1976). These animals are implicitly included in the model in that the model response for primary production accounts for the smaller herbivores. For example, the values of various parameters in the model were adjusted so that model response would correspond to the 1970 and 1971 field data. The model response with the particular combination of the parameter values then accounted for the effects of herbivory by the nonmodeled animals. Thus the model is limited in its capacity to simulate responses to consumers other than cattle, small mammals, and grasshoppers.

The response of net primary production and gross primary production to experiments 6–9 of Table 10.1 are reasonable in terms of trends, but data are unavailable to test accuracy. The increased production in experiment 6 results from a hypothesized stimulatory effect on aboveground production of grazing [assuming the grazing effect is not too severe; Jameson (1963)]. Thus the model response does appear to represent effects of cattle grazing.

Question (1b) asks how net or gross primary production is affected by variations in temperature and precipitation or applied water. In Table 10.1, experiment 2, values presented for net primary production and gross primary production are probably high, judging from the data for live peak, although data for comparison are not available. Two of the reasons for simulated overproduction were discussed above, that is, nonconsideration of nighttime respiration and excessive use of Liebig's law. Another reason which is possibly important, especially with increased precipitation, is that in formulating the producer model the assumption is made that roots decrease exponentially with increased soil depth and that the water-absorption capacity is exactly proportional to the biomass of roots in a given stratum of soil (see Chap. 3 for details). This formulation may not adequately account for the effect of water stored deep in the soil profile. Experiments 10 and 11 in Table 10.1 are very difficult to evaluate due to a lack of data and complex interactions of submodels in response to temperature. The model response to severe drought is likewise hard to criticize since no data are available for comparison.

Question (1c) asks how net gross or primary production is affected by the addition of nitrogen or phosphorus. Model response of the primary producers (experiments 3 and 4 in Table 10.1) was a failure when nitrogen application was simulated. However, the decomposer submodel responded favorably, indicating that it and the nitrogen submodel operate satisfactorily. No simulation of phosphorus fertilization was attempted because information and experience acquired by Grassland Biome experimental scientists and modelers suggest there should be little or no response to application of this element.

Question (2) asks how the carrying capacity in a grassland is affected by the perturbations in question (1). No results of this calculation are presented here, but such calculations or inferences could be made easily.

Question (3) asks whether the results of an appropriately driven model run are consistent with field data taken in the Grassland Biome Program, and if not, why? This question is asked of each model run; therefore, the number of examples is large. Examples are given above in the discussion of question (1), and other examples are discussed in the earlier chapters of this volume.

Question (4) asks what changes occur in the composition of the producers as a result of these perturbations. We can trace the relative changes in the biomass of the represented functional groups of primary producers. Runs simulating up to 5 yr of system response have been made, and the results have been satisfactory and informative (W. J. Parton, personal communication). Long-term runs (10 yr) would be necessary to determine whether these changes would be permanent. Thus far we have not made such long-term runs.

10.3 Analysis and Criticism of ELM in Relation to the Desired Goals of Modeling in the Grassland Biome Program

Why build a simulation model of a grassland ecosystem? To examine this question we address each of several goals with one or two of many possible examples. This discussion represents ideas and concepts that have developed in the Grassland Biome Program. These ideas and concepts are now considered to be our working systems approach to the study of the structure and function of grassland ecosystems.

10.3.1 Models Can Be Built

One goal of modeling is to show that the tools of systems analysis (in this case, simulation modeling) can be applied to ecosystem science and can yield a reasonable model of the dynamics of a grassland system. This is an unsatisfactory goal in itself, but in the developmental stages of a modeling effort some emphasis must be placed on building the model. Some authors hold that it is naive to attempt to model systems as complex as ecosystems (de Wit, 1970). With successful initial development, the more satisfying and more justifiable goals of model development can be addressed.

We have shown through the ELM modeling effort that the model of a grassland ecosystem can be built. In addition to ELM two other total-system models of grassland ecosystems have been implemented through efforts of the Grassland Biome: (a) PWNEE (Bledsoe et al., 1971) and (b) a linearized version of PWNEE (Patten, 1972). Innis (1972) and Van Dyne (1975) have discussed some of the general problems involved in large, interdisciplinary modeling efforts, but they give few specific examples. It is attempted here, briefly, to point out a few of the specific problems of model building we have encountered, so that: (a) others who are attempting or planning to build a model may avoid or at least recognize some of the problems and (b) the modeling effort of the scale of ELM will be recognized as a very complex process that is not trivial to accomplish.

Modeling Team.   The primary modeling team included the leader, a mathematician, two soil scientists, and a group of postdoctoral fellows whose scientific training included meteorology, zoology, plant ecology, plant taxonomy, genetics, and soils, plus two excellent programmers. The systems analysis and computer programming ability of the members ranged, at the beginning of the modeling effort, from none to expert. The first task of the team was to educate all members in the fundamentals of simulation modeling, systems ecology, and programming techniques. Ideally, it would have been useful and possibly more efficient if all members of the modeling team had been accomplished modeler–ecologists to begin with. In 1971 and 1972 when the effort was launched, very few people with those combinations of qualifications existed. Also when too many experts are placed in a team situation, the product can be chaos. Further discussion of this point is beyond this paper and involves the domain of psychoanalysis and the study of group dynamics (Berne, 1963). Suffice it to say that an environment can be created to allow people to talk to each other about common problems, but no guarantee can be made that they will listen to each other. The ELM modeling team was moderately successful in accomplishing interaction among members, and the product was a model of an ecosystem.

Successful interactions of modelers with other scientists were also essential for completion of the modeling effort. It is imperative that scientists who are not modelers themselves but are contributors to the overall effort respect the scientific competence and judgment of the modelers. It is also necessary for the modelers to respect the confidence and judgment of the subject-matter specialists. As a result, at least partially, of the training and ability of the modelers the team was moderately successful in gaining this respect.

Model Development.   The development of the model progressed through the following sequence. The postdoctoral fellows, with one exception, arrived in the program over a period from late 1971 to early 1972. Indoctrination and initial model development occurred during early and mid-1972. The modeling team then disbursed with each member going to a different network site headquarters. The purpose for the disbursement was to allow each team member to become familiar with one of the grassland ecosystems within the Grassland Biome and establish a working rapport with network site personnel. Following a 1-yr residence at the network sites, the modeling team reassembled in Fort Collins and completed ELM.

The anticipated Grassland Biome Program goals of the concept of site residence by the modeling team members was highly successful, that is, the members became familiar with the different ecosystems in the biome and working rapport was established with site personnel. However, actual model development was hindered by the separation of the modeling team members.

Data.   Once a modeling team is assembled, objectives of the model developed, and the model structure agreed on, the problem of the availability of data arises. Routine data-collection procedures can be found in Swift and French (1972). Most of these data were collected with the objective of supplying state-variable data for validation of models. In the initial phases of the Grassland Biome Program no model existed to serve as a guide for which data were needed. Thus some data were collected that were inappropriate or not used for model development or validation. In many cases data that would have been useful were not collected. Furthermore, data concerning processes operating in the ecosystem were sparse. Often when data were collected that were appropriate for model development and validation, they were highly variable and had to be used with caution. In my opinion the Grassland Biome Program placed too much effort on the collection of vast quantities of different state-variable data with too little consideration of how the data were to be used, and my conclusion is based on hindsight. When the Grassland Biome began, experienced ecosystem modelers were scarce, and the tools of modeling were crude. Much time was spent perfecting the tools rather than building models useful for guiding research. Based on what has been learned about modeling in the past few years, there is no reason for subsequent ecological programs not building models, at least strong conceptual models to be implemented later into mathematical models before data collection is undertaken.

A more specific criticism of the ELM modelers regards their use of information of an admittedly dubious nature—"expert opinion." All modelers know that it is often necessary to estimate the value of some variable or constant because no data or other information are available or the information is unknown to the modeler. Often modelers assume the responsibility of making the estimate themselves rather than asking a subject-matter specialist to make a more educated guess. Progress in several sections of ELM model development may have been substantially hindered because the advice of knowledgeable experts was not used.

Representation of Interactions.   The mathematical representation of the interactions of controlling factors of biological processes is one of the most perplexing problems in the modeling of ecological systems. In ELM two basic methods of representing interactions were used almost exclusively. One method is to assume Liebig's law of the minimum (Odum, 1971) to be adequate for representing controls of many processes. Using this basic premise processes are assumed limited by the single environmental factor chosen by the model program from a set of factors that at a given simulation time step expresses the greatest limitation. The rationale or justification for this representation is that during the simulation, control will shift to different members of the set of environmental factors as conditions change in the modeled system. The other method frequently used to represent the controls of processes in ELM is to determine the effect of each of several controlling factors and then multiply those effects by each other to yield a total, combined control (Forrester, 1961). In both methods the set of possible controlling factors was generally expressed as a normalized (a dimensionless quantity ranging from 0 to 1) effect.

Each of these methods of expressing the interactions of controls has merit and each has limitations. The method using the single, dominant controlling factor is simple and convenient and in some situations may be adequate. However, experience suggests that in many ecological processes the interaction of several variables must be considered explicitly (Odum, 1971). The method of multiplying factors together to yield a single controlling value assumes interactions exist, but it does not account for nonlinear or synergistic responses. Unfortunately, there are few biological field data, especially in noncultivated systems, to help guide choosing the representation of interactions of components of ecosystems.

Circularity in Modeler–Model Response.   To show that a model can be built, it is necessary at some stage to document progress for the scientific public. This volume is the first formal ELM documentation. However, documenting a model is much less exciting than making conceptual improvements or gaming with the model to learn from it. Figure 10.3 is a simple representation of a group of desirable modeling processes. The cycle of model development seems to end only when funds or the modeler expires or moves (Dillon, 1971). An example of the difficulty caused by not breaking the cycle is our own effort. The modeling team was scheduled to disperse on June 30, 1974. We agreed to have manuscripts prepared by that date. The task of documentation proved much more difficult than expected due to the lack of precedent in preparing papers of this kind. Furthermore, members of the team did not cease changing the structure of the model. As a result, the final validation and sensitivity analysis runs were delayed, and the modeling team dispersed before all manuscripts were completed. The task of final preparation of the volume was much more difficult and publication was delayed.

10.3.2 The Model as an Organizer of Information

The structure of the model expresses a collation and organization of information about a grassland ecosystem in which the various components of the model operate at compatible levels of resolution. Level of resolution refers to the degree of conceptual detail of biological mechanisms represented in the model. De Wit (1970) discusses the importance of recognizing the various levels of resolution that might be considered in a model. The information needed to run ELM is diverse, including determination of variables such as Michaelis–Menten kenetics of ion uptake in plants, litter size of the grasshopper mouse, photosynthesis rates of *B. gracilis,* heat transfer through soils as related to bulk density, and the biomass microbes in the 15–30 cm soil layer. This information is used to represent the processes and controls of processes operating in the ecosystem that directly influence the flow of biomass (carbon) between trophic levels. Of concern in the model is that CO_2 fixed as plant biomass (g/m^2 dry wt) as a function of leaf biomass, radiation, temperature, soil water, and so on (Chap. 3).

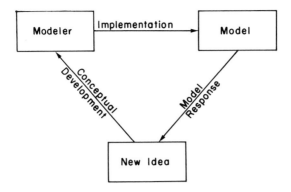

Fig. 10.3.   Circularity of modeler–model interactions

To predict the fixation, it is not necessary to ascertain stomatal resistance or CO_2 diffusion equilibria across water films in the mesophyll layers of leaves. If we were modeling solely the process of photosynthesis, we could afford to be concerned with these factors, but what is desired from the producer section of ELM is for plant biomass and productivity to be appropriately distributed through the season, and for plant biomass to be properly distributed among the biophagic and saprophagic consumers. Similarly, in the consumer section of ELM we wish to represent how animals use fixed carbon in relation to weight gains or losses. We feel that these dynamics can be adequately represented without partitioning the carbon to livers, fat, eyes, or other body parts. The point to be made is that the model organizes information, incorporates judgments about the hierarchical ordering of reality (de Wit, 1970), and adequately simulates numerous processes so that the carbon flow through the ecosystem can be studied.

The intent of modeling is to adequately represent the attributes of the system of concern under a given set of objectives. The intent is not to include all known information that might in some way influence the system and thus satisfy all experts that all known mechanisms are represented.

The submodels of ELM were designed to be compatible at the level of resolution that allowed flow of carbon through the grassland ecosystem to be traced. A desirable feature of ELM or any model of an ecosystem would be to have the model incorporate the best available information at an appropriate level of resolution; in other words, state-variable values and representations of system processes represent the best conceptual basis possible. I do not imply that the model must include only the best known information to respond appropriately or meet its objectives and be of use. But to increase the conceptual utility of a model as organizer of information useful to biologists, it is highly desirable for the model formulation and interpretation of results to include the most reliable or appropriate information available. Unfortunately, several constraints must be considered. The modeling techniques available must be sufficiently advanced to allow representation of the available information in a relatively simple manner. Furthermore, computer memory capacity must be large. An example of a situation in which these constraints are not met would be in the area of population dynamics where spatial relationships may be important. Currently representation of the movement of organisms from one location to another would require running essentially two more models representing different locations in space. The ELM model cannot represent this type of movement because time and cost are prohibitive and the memory capacity of the computer is too small to run even two models the size of ELM simultaneously.

Contained within ELM are formulations that incorporate our best information about the grassland ecosystem or biological processes therein. There are numerous formulations that exceed our best information, that is, new hypotheses, especially in areas where little information is known about fundamental processes. The ELM model also contains formulations that have not included our best information. Among the reasons for this are the lack of modeling techniques allowing utilization of the information, for example, spatial effects in heterogeneously distributed communities or horizontal movement of matter. Unfortunately, insufficient effort was exerted in some instances to make full use of available information, that is, interactive effects of process controls or inadequate utilization of literature or subject-matter experts.

10.3.3 The Model as a Communications Device

The model is an inter- and intradisciplinary forum for discussion using a common language. The conceptual model on which ELM is based serves as the basic, stripped-down skeleton of the format of the annual grassland research report being written within the Grassland Biome with collaboration from scientists of California, Nevada, South Dakota, and Colorado. The document entitled "The Structure, Function, and Utilization of the Annual Grassland Ecosystem" is intended as a general ecological reference or "user's manual" of that ecosystem in the United States. The synthesis effort includes the labor of more than two dozen scientists with backgrounds ranging from theoretical meteorology to plant genetics. The model is serving effectively as a primary communications device, enabling scientists with these diverse backgrounds to discuss in a congruent manner phenomena such as movement of nitrogen from the atmosphere into the soil through the plants to the animals and hence back to the soil, to be recycled or lost to ground water. Without such a formal communications device, scientists often become mired down in discussions of jargon, different points of view, and different levels of resolution, never getting to the subject at hand.

Unfortunately, the implemented mathematical model representing a whole grassland ecosystem serves as a communications device only for those who are intimately involved with the modeling effort. It had been hoped that by using the SIMCOMP simulation language (Gustafson and Innis, 1973), the coding and implementation would be sufficiently simple that people with only minimal training in FORTRAN would be able to understand the logic and formulation of the represented mechanisms. Progress toward achieving this goal has been made, but complete success has been elusive. The model size, well over 100 state variables and 300 flows, and the complexity of interactions between system components precludes easy assessment of the functioning of the model. Detail of a model has to be balanced against the relative and meaningful levels of resolution. Also, the coding of nine different individuals is represented in the model; thus nine different levels of skill and logic are to be found. However, we feel we have made a significant advance beyond a FORTRAN-coded model.

Another serious limitation of the implemented mathematical form of the model as a communications device with researchers outside of Colorado State University is that it is coded in the SIMCOMP 3.0 simulation language (Gustafson and Innis, 1973). Currently the version of SIMCOMP in which ELM is coded is specifically designed for Control Data Corporation (CDC) 6000 or 7000 series computer systems. However, there is considerable variation in CDC 6000 and 7000 systems, causing the transfer of SIMCOMP 3.0 to these systems to be difficult.

The model has proved very effective in serving as a focal point for communication about precisely stated information (isolating individual process formulations). In early model development, the effect of cattle on plants was expressed only through removal of herbage by the animals. Exercise of the model under heavy grazing conditions yielded results that suggested little or no damage suffered by the plants. This was clearly unreasonable. Through consultation with field scientists and examination of the mechanisms, modelers were able to focus attention on the animal–plant interaction. This examination led to more realistic representation of the interactions. As a result, the processes of trampling, wastage, and physiological damage to plants were implemented in the model. Another example is the importance of microbes in phosphorus cycling of grasslands. Soil scientists have long emphasized inorganic P transformation in the soil, but recently more emphasis has been placed on the importance of organic P turnover in native grasslands, as regulated by microbial activity (Halm 1972; Stewart et al., 1973; Chap. 8). The recognition of the importance of these factors was aided by the experimenter–model interaction.

10.3.4 The Model as a Research Tool

The Model as a Field Study, Literature, and Experience Integrator. The model is a research tool on par with field, laboratory, and literature studies. An example of this situation is the following. Through exercising the model and attempting to follow the dynamics of aboveground plant material, attention has been focused on the importance of the process of the fall of standing dead material to litter. Our ignorance of that process was clearly indicated. This flow has been isolated as a key process in grassland ecosystems. At the suggestion of biome modelers, field experiments are being designed to describe and quantify the process and its controls. Results of the experiments will refine and better quantify the current representation of the process in the model. From the experiment we hope to identify mechanisms not currently included in the model but that should be incorporated. Preliminary indications suggest that snowfall, physical properties of the snow itself, short-term wind speed (gusts), and high-intensity rainfall events may be important controls of the process and should be implemented into the model. Several such studies of processes operating within the ecosystem have been suggested by Grassland Biome modelers as a result of experience with the model. In some cases these experiments have been performed, but unfortunately, turnaround time (i.e., model → field experiment → model) has been on the order of 1–2 yr for simple experiments and longer for more difficult experiments. Results and hypotheses suggested by these studies are only now being incorporated in later versions of the model.

The Model Used to Study Poorly Understood Mechanisms of the Ecosystem.   The ELM model is a vehicle that can be used to study the interactions of difficult-to-study parts of the ecosystem, such as the belowground system. To study the interactions within the system, definition of the belowground system was essential. Using versions of the model as guides or often as "straw-men," the formalization of the description of the system has made important advances. In a study like that of the Grassland Biome, it is often difficult to know which advance was suggested by which effort, but working collectively using field experience and experiments, laboratory studies, and modeling, we have suggested some rather precise definitions of the functional structure of the belowground ecosystem.

Following definition of the system, meaningful investigations of the interactions of components have been accomplished. We have used the model to gain insight into such processes as root death, microbial activity and inactivity, the role of microorganisms in nutrient cycling, and root respiration, to mention only a few.

The Model as a Research Tool to Test Hypotheses.   Currently, George M. Van Dyne and Freeman M. Smith (personal communication) are leading an effort to test systems level hypotheses about the effects of grazing, fire, and pesticides, and other perturbations recognized in range management. The model is proving useful in guiding thinking related to system responses and interactions. Through this type of model use we hope to examine many systems-level responses and thus advance understanding of grassland ecosystems and help establish ecology as an empirical science.

Empirical science progresses on the falsification of hypotheses. A model, as a hypothesis, can contribute to empirical science in one sense only if it (the hypothesis) is falsifiable (G. S. Innis, personal communication). A common complaint that is leveled at models as large and complex as ELM is that the assumptions are so great and the parameters so many that they are not falsifiable. Proponents of this view point out that there are innumerable assumptions in the model that are almost certainly not met in any given application (simultaneity of initial conditions, all initial conditions from the "same" place, homogeneous system, snow effects can be ignored, etc.). Failure to meet these assumptions provides the model with the legitimate excuse that it was not designed to operate under the given conditions.

If the modeler insists on applying the model to a given situation, then he has at his disposal enough parameters of the kinds discussed in the previous chapters to force the model to produce almost any output. This argument seems to be based on an inappropriate analogy with polynomial approximation; given n distinct points in 2-space, each with a different first coordinate, there exists a polynomial of degree at most $n-1$ which passes through these n points. This argument, while worthy of consideration, must be treated carefully. In the first place, models are in general not polynomials, and while a similar argument holds for a large class of sets of functions, it does not apply with complete generality. Moreover, the coefficients in ELM generally have biological, ecological, or physical meanings. As such, they can be evaluated (at least approximately) via observations independent of the model. The modeler is not at liberty to assign values to those parameters that force the model to agree with observations.

The ELM model is eminently falsifiable. Each of the authors in this series has indicated situations in which the model, despite our best efforts to tune it, fails to illustrate the dynamics observed in the field. These failures are occasionally a result of our not satisfying the assumptions on which the model is based, but far more often, failure is traced to a specific hypothesis describing a biological mechanism that our data falsify. Where these mechanisms are representations of the state of the ecological art, the empirical science is advanced; where these mechanisms are representations of the state of the modeling art, the modeling science is advanced.

The Model as a Predictive Device.   The model is a "predictive" device that can be used as a management aid (Forrester, 1961). In many of the tests of the model that have been run thus far, response has been favorable. The model is often predictive in the sense that the perturbations that have been simulated have yielded reasonable results when compared to data or experience. Examples of these types of perturbation are grazing treatments of various intensities and seasons of grazing, irrigation or precipitation modification, and simulation of series of wet or dry years. Other simulated perturbations (fertilization) have yielded unsatisfactory results that have caused reexamination of model coding and conceptualization. Thus far when response has been unreasonable, we have been able to find errors in coding, misconceptions, or in some cases we have discovered previously unrecognized responses in the biological system. For example, Cole et al. (Chap. 8) and Hunt (Chap. 6) independently recognized that the relative rates of decomposition of organic matter decrease with soil depth. Through the process of model tuning these investigators were able to quantify the decomposition rates at various soil depths, a task that would be extremely difficult to accomplish in the field. Other examples, such as response of plants to grazing and the phosphorus cycle, illustrating this value of the use of models are given in Section 10.3.3.

When establishing whether the model is a predictive device that can be used as a management aid, we suggest that the decision can only be made by the person or groups of persons who are utilizing the model. At best, ELM or any model can be used as an aid to, rather than the basis of, decisionmaking. Users must recognize that predictions of a model may not be meaningful outside the context of the modeling goals and objectives. The goals of the modeling effort were to create a model that would serve as a communications device and organizer of information, be useful as a research tool, and yield results that could help in elucidating biological phenomena in grassland ecosystems. The objectives of ELM were established to help guide the development of the model and ensure that the task could be realistically accomplished. The objectives obviously placed constraints on the applicability of the model, but if those constraints are satisfied, then the model can attain its stated objectives, aid biological understanding, and suggest and evaluate management practices.

Acknowledgments.   The author wishes to thank Drs. H. W. Hunt and G. S. Innis for their careful review of this manuscript. This chapter reports on work supported in part by National Science Foundation Grants GB-31862X, GB-31862X2, GB-41233X and BMS73-02027 A02 to the Grassland Biome, U.S. International Biological Program, for "Analysis of Structure, Function, and Utilization of Grassland Ecosystems."

References

Berne, E. M. D.: The Structure and Dynamics of Organization and Groups. New York: Lippincott, 1963, 260 pp.

Bledsoe, L. J., Francis, R. C., Swartzman, G. L., Gustafson, J. D.: PWNEE: A grassland ecosystem model. US/IBP Grassland Biome Tech. Rep. No. 64. Fort Collins: Colorado State Univ., 1971, 179 pp.

Coleman, D. C., Andrews, R., Ellis, J. E., Singh, J. S.: Energy flow and partitioning in selected man-managed and natural ecosystems. Agro-Ecosystems **3,** 45–54 (1976)

de Wit, C. T.: Dynamic concepts in biology. In: Prediction and Measurement of Photosynthetic Productivity. I. Šetlík, (ed.). Wageningen, the Netherlands: Centre for Agr. Publ. Docu., 1970, pp. 17–23

Dillon, J. L.: Interpreting systems simulation output for managerial decision-making. In: Systems Analysis in Agricultural Management. Dent, J. B., Anderson, J. R. (eds.). New York: John Wiley and Sons, 1971, pp. 85–120

Dodd, J. L., Lauenroth, W. K.: Responses of *Opuntia polyacantha* to water and nitrogen perturbations in a shortgrass prairie. Proc. IVth Midwest Prairie Conf., Univ. North Dakota, Grand Forks, N.D., 1974

Forrester, J. W.: Industrial Dynamics. Cambridge, Mass: M.I.T. Press, 1961, 464 pp.

Forrester, J. W.: Principles of Systems. Cambridge, Mass.: Wright-Allen Press, 1968, 400 pp.

Gustafson, J. D., Innis, G.: SIMCOMP Version 3.0 user's manual. US/IBP Grassland Biome Tech. Rep. No. 218, 1973, 149 pp.

Halm, B. J.: The phosphorus cycle in a grassland ecosystem. Ph.D. thesis, Univ. Saskatchewan, Saskatoon, 1972

Hyder, D. N., Bement, R. E., Remmenga, E. E., Hervey, D. F.: Ecological responses of native plants and guidelines for management of shortgrass range. USDA Agr. Res. Serv. (ARS) Tech. Bull. No. 1503, 1975

Innis, G. S.: Simulation of biological systems: Some problems and progress. In: Proc. 1972 Summer Computer Sim. Conf., Vol. II. La Jolla, Calif.: Simulation Councils, Inc., 1972, pp. 1085–1089a

Jameson, D.: Responses of individual plants to harvesting. Bot. Rev. **29,** 532–594 (1963)

Nolan, Richard L.: Verification/validation of computer simulation models. In: Proc. 1972 Summer Computer Sim. Conf., Vol. II. La Jolla, Calif.: Simulation Councils, Inc., 1972, pp. 1254–1265

Odum, E. P.: Fundamentals of Ecology. Philadelphia, W. B. Saunders, 1971, 574 pp.

Patten, B. C.: A simulation of the shortgrass prairie ecosystem. Simulation **19,** 177–186 (1972)

Saaty, T. L.: Operation research: Some contributions to mathematics. Science **178,** 1061–1070 (1972)

Singh, J. S., Lauenroth, W. K., Steinhorst, R. K.: Review and assessment of various techniques for estimating net aerial primary production in grasslands from harvest data. Bot. Rev. **41,** 2, 181–232.

Stewart, J. W. B., Halm, B. J., Cole, C. V.: Nutrient cycling: I. Phosphorus. Canadian Comm. Int. Biol. Prog. (Matador Project) Tech. Rep. No. 40. Saskatoon: Univ. Saskatchewan, 1973

Swift, D. M., French, N. R. (coordinators): Basic field data collection procedures for the Grassland Biome 1972 season. US/IBP Grassland Biome Tech. Rep. No. 145. Fort Collins: Colorado State Univ., 1972, 86 pp.

U.S. Forest Service: Planning: A key to successful management. Mgmt. Notes **17,** 7–10 (1973) (Washington, D.C.: USDA, Forest Service, Division of Administrative Management)

Van Dyne, G. M.: Some procedures, problems, and potentials of systems-oriented, ecosystem-level research programs. In: Procedures and Examples of Integrated Ecosystem Research. Tech. Rep. No. 1. Uppsala, Sweden: Barrskogslandskapets Ecologi, Swedish Coniferous Forest Project, 1975, pp. 4–58.

Author Index

Numbers in *italics* refer to the reference sections.

Subject Index

New and Forthcoming Volumes in Ecological Studies

Vol. 20
Theories of Populations in Biological Communities
By F. B. Christiansen and T. M. Fenchel
1977. x, 144p. 68 illus. 5 tables. cloth

Vol. 21
Cells, Macromolecules and Temperature
Conformational Flexibility of Macromolecules and
Ecological Adaptation
By V. Y. Alexandrov
1977. xi, 330p. 74 illus. 30 tables. cloth

Vol. 22
Air Pollution
Phytotoxicity of Acidic Gases and Its Significance in
Air Pollution Control
By R. Guderian
1977. viii, 127p. 40 illus. (4 in color) 26 tables. cloth

Vol. 23
Lessepsian Migration
The Influx of Red Sea Biota into the Mediterranean by Way of the Suez Canal
By F. D. Por
1978. approx. 250p. 47 illus. 2 maps. 10 plates. cloth

Vol. 24
A Coastal Marine Ecosystem
Simulation and Analysis
By J. Kremer and S. W. Nixon
1978. approx. 250p. 80 illus. 13 tables. cloth

Vol. 25
Microbial Ecology of a Brackish Water Environment
Edited by G. Rheinheimer
1978. approx. 300p. 77 illus. 84 tables. cloth

Vol. 27
The Ecology of Some British Moors and Montane Grasslands
Edited by O. W. Heal and D. F. Perkins
1978. approx. 425p. 134 illus. 148 tables. cloth

Vol. 28
Pond Littoral Ecosystems
Methods and Results of Quantitative Ecosystem Research
in the Czechoslovakian IBP Project
By D. Dykyjova and J. Kvet
1978. approx. 450p. 18 illus. 87 tables. cloth

Springer-Verlag   New York   Heidelberg   Berlin

An Interdisciplinary Journal from Springer-Verlag

Environmental Management

Editor-in-Chief
Robert S. De Santo

Editorial Board
Karl E. Shaefer, Francesco di Castri, John MacKinnon, William Bennetta, John Yamamoto, Jerome P. Harkins, Percy Knauth, Ralph A. Fine, Joseph A. Miller, and Vytautas Klemas

Geared to real environmental problems and their solution, *Environmental Management* brings together ideas from such diverse areas as law, science, engineering, sociology, government, political science and psychology as they relate to applied environmental management and research, planning and analysis. The journal's objective is to foster an atmosphere of mutual strategy and understanding among decision-makers involved in these areas, and thus to help answer the urgent need for environmental protection we are faced with today.

Environmental Management is divided into four major areas. The *forum* section—presenting editorials, brief communications, essays and letters to the editor—is designed to promote communication between contributors and readers. The *profiles* sections offers descriptions of individuals and institutions active in environmental affairs. With articles by experts from every corner of the world, the *research* section provides insight into new materials and ideas, discusses present methods and case studies and sheds light on those facets of the field which create, control or complicate environmental management. Finally, the *literature* section gives information on current important world literature in the field.

International in flavor and interdisciplinary in scope, *Environmental Management* is truly a state-of-the-art publication. To read it routinely is to stay abreast of new discoveries in a vital and rapidly changing field.

For sample copies, descriptive and subscription information, write to:

Springer-Verlag New York Inc.
175 Fifth Ave.
New York, NY 10010